Our Sun: A Cold-Fusion Plasma Star
Illustrated Science Exploration by Rolf A. F. Witzsche

© Text Copyright Rolf A. F. Witzsche 2018
all rights reserved

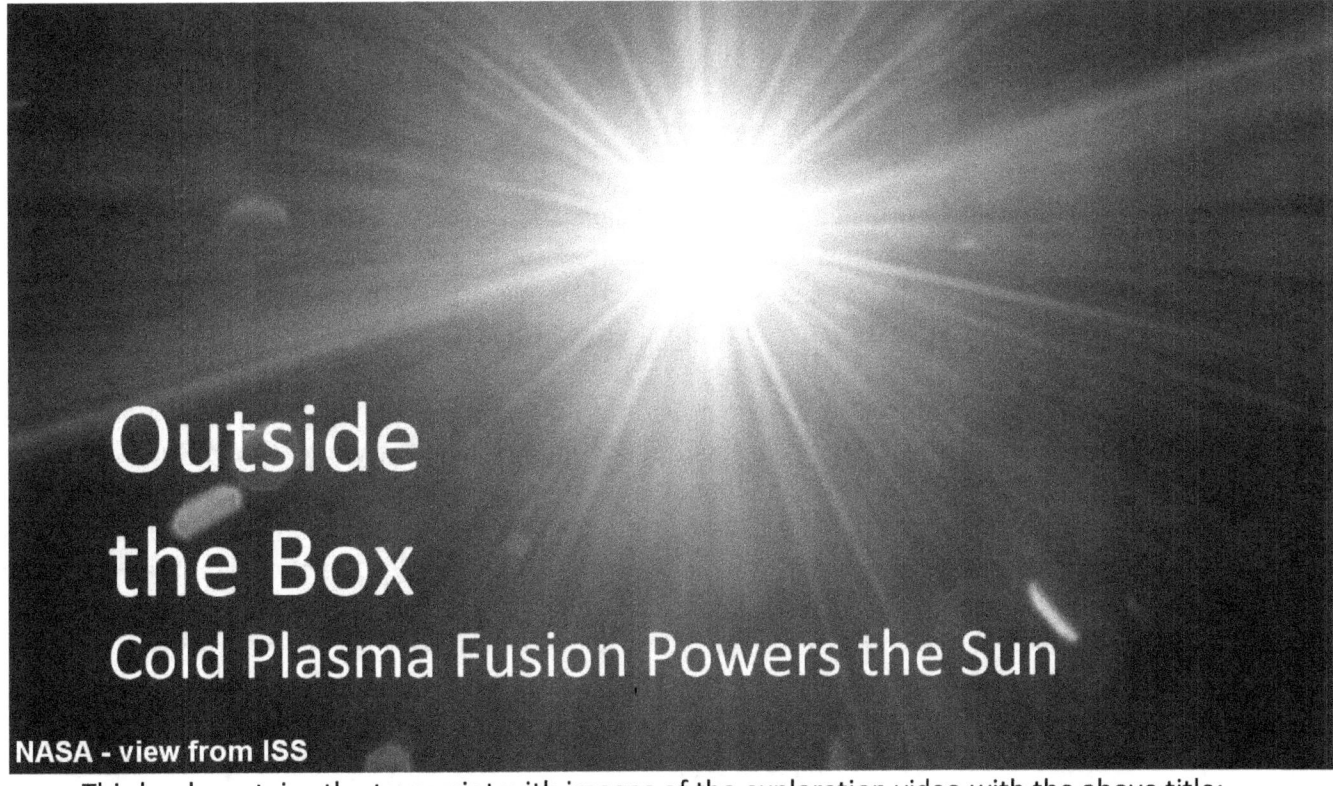

This book contains the transcript with images of the exploration video with the above title:
see: http://www.ice-age-ahead-iaa.ca/

Lead in:

If you want to discover how the Solar System and the Sun operate outside the 'box' of empiricist-constrained science, allow me to invite you to a tour of exploration.

The greatest opportunity of all times lays before us. but... Humanity has no future without an anti-entropic, highly efficient, high-density energy source.

Topics:
** Cold Nuclear Fusion Powers the Sun.
** It is impossible for the Sun to be a gas sphere.
** The Sun is a plasma star with electric nuclear fusion.
** The Sun we see, is not as we expect it to be.
** The universe is motivated by the plasma-sink process.
** Cold fusion drives the universe.
** Solar fusion IS cold fusion, we should replicate it.
** Alfven Wave oscillations.
** The coronal heating paradox.
** The accelerating solar winds paradox.
** Historic astrophysical evidence.
** Presently visible, large plasma structures,

** Solar wind, Ice Ages, and our future.
** When sunspots speak to us.
** How do we get out of the paradoxical trap?
** Solving the impossible paradox before us
** Deception of normality.

Ever since false 'science' has turned the nuclear-fusion-energy principle upside down decades ago and diverted relative research into dead-end pursuits that are enormously costly in capital and human resources, while this 'science' closes the door to the actual realization of nuclear fusion energy, it became imperative, for a crash program to be launched into the right direction.

The ITER experiment, in France, is a case in point. It is built on a principle that is a deception from the start and has never produced usable end-product power anywhere in the world. One of the primary purposes of the facility was to demonstrate that steady-state operation is physically possible. This too, is a deception. The fusion of tritium and deuterium atoms isn't the power-producing factor, but merely facilitates the unbinding of the excess neutron of the tritium. For steady state operation the tritium must be replaced. This is planned to be accomplished by utilizing the unbound neutron to split a Lithium-7 atom into helium4 and tritium with the release of another neutron that may not be energetic enough to cause a chain reaction. It is hoped that for each of the neutrons generated by the fusion process more than one replacement tritium atom can be harvested from the Lithium fission. This is what the fusion power process depends on, but this may not happen. We will know in 2035 what the answer will be. Likewise, it is presently unknown if the helium produced in the fusion reaction can be purged from the reaction chamber fast enough, before it dilutes the fuel and stops the fusion reaction. This typically happens after a second (the current world-record in fusion-burn duration).

Nuclear fusion power production is a dream that will likely never come to. It is a process that doesn't happen anywhere naturally, nor does it power the Sun.

If society aims for a real nuclear-fusion-energy production, the quest must begin with letting go of the empire-inspired false doctrines about the Sun, and to acknowledge the actually operating principles of the Sun, which render the Sun a plasma star powered by its interaction with interstellar plasma streams. For this plenty of evidence exists, reflecting numerous scientific principles, verified in space, on the ground, in labs, and by numerous other measurements.

The point is that when one aims for a crash program to succeed, it becomes imperative to begin with actual scientific discoveries that point in the correct direction, instead of rushing into an intentional dead-end project such as ITER is. Of course, it is hard to break with the doctrines of the masters of empire, and to accept advanced discoveries of demonstrated physical principles.

I have produced numerous videos on the subject. The challenge is great, but when it is met, real nuclear-fusion energy, or related efficient processes, are well with society' realization, and this in the short term - not multiple decades in the future.

Table of Contents

Introduction - Let me surprise you ... 20
    The Big Bang Creation ideology ... 21
    The electric cosmology ... 22
    The effects of cosmological theories ... 23
    The long suspected resumption of the Ice Age ... 24
    The transformation of our world ... 25
    Gigantic infrastructure development ... 26
    Science has been 'guided' with the payola ... 27
    The train of politically 'guided' science ... 28
    The exotic Big Bang concepts ... 29
    Deep in the Big Bang cosmology ... 30
    The Earth stands as a silent testimony ... 31
    Outside the topics of fairy tales of the Big Bang ... 32
    Disturbing to those who cherish the illusion ... 33
    Topics of truly gigantic plasma structures ... 34
    Will we built the 6,000 new cities that we need ... 35
    Will we create a new world for us ... 36
    The largest part of this spiritual challenge ... 37
    In the wonders of our humanity ... 38

Part 1: The Sun is NOT a gas sphere. The Sun is plasma ... 39
    The Sun is not a gas sphere ... 40
    Image: The Sun Topics ... 41
    Image: Plasma streams ... 42
    Image: galaxies as node points ... 43

Image: The Milky Way, a node 44

Image: The Milky Way is plasma powered 45

Image: Milky Way plasma domes 46

Image: The Milky Way in comparison 47

Image: Stars at node points 48

Image: Plasma focused onto the Sun 49

Image: Powered by interstellar plasma streams 50

Image: LaPoint and Peratt discoveries 51

Image: LaPoint, artificial Sun 52

Image: The Milky Way lookalike 53

Image: The Sun located in the Milky Way 54

Image: The x-ray Sun 55

Image: Star-size comparison 56

The Sun cannot exist 57

The solar-wind particles are plasma 58

Guides interstellar plasma onto our Sun 59

In the case of the Sun 60

Plasma is electricity 61

The Sun becomes surrounded 62

Fusion reactions on the surface of the Sun 63

Operational differences between the Sun and the Earth 64

Plasma streams that connect stars 65

An internally powered sun is not needed 66

Why is the Sun not a sphere of hydrogen gas? 67

The Sun is theorized to fuse hydrogen atoms together 68

The Sun is anything but a constant star 69

The heat generated at the core of the Sun 70

The Sun's energy cycle is oscillating ... 71

It shouldn't be possible for the solar wind ... 72

The corona around the Sun is hotter than the Sun ... 73

When we look at the sunspots on the Sun ... 74

Impossible for the Sun to be a gas sphere ... 75

The mass-density of the Sun ... 76

The premise that 99.999% of the universe does not exist ... 78

Plasma is the blood of the universe ... 79

The effects of the plasma streams in space ... 80

Secondary effects of flowing plasma ... 81

Plasma streams as solar wind ... 82

The Sun is a plasma star ... 83

The principle is efficient ... 84

Two main types of plasma particles ... 85

Electrons rebound ... 86

A density determined zone ... 87

A creative atom synthesizing 'engine' ... 88

When we look at the Sun ... 89

When the inflowing plasma stream becomes weak ... 90

Nuclear fusion happens on the surface of the Sun ... 91

The dynamic dance of electrons ... 92

When two protons are forced closely to each other ... 93

The neutron, essentially acts like a glue ... 94

By the process of protons becoming bundled ... 95

A hundred different elementary atoms ... 96

Even the hydrogen atom is, heavy ... 97

Hydrogen origin paradox ... 98

Part 2: The universe of plasma refuting the Big Bang ... 99

    The Big Bang creation is more a myth than a theory ... 100

    The entire universe exploded into being ... 101

    All mater, all energy ... 102

    The dust is deemed to have condensed ... 103

    The tale is full of holes ... 104

    The Earth itself, is a paradox under the Big Bang theory ... 105

    Light disproves the explosion theory of the Big Bang ... 106

    The red-shift of light from distant galaxies ... 107

    Red shift is the result of energy depletion ... 108

    A gigantic piece of evidence that disproves the Big Bang theory ... 109

    The Earth is proof that synthesizing atomic fusion ... 110

    Plasma is a sea of electrons and protons ... 111

    Plasma in space ... 112

    On contact, two protons snap together ... 113

    On contact, the electron is forced to rebound ... 114

    Atoms are formed by the dynamic 'dance' ... 115

    Atoms are electrically neutral plasma structures ... 116

    Atom-forming fusion increases mass density ... 117

    Electric nuclear fusion happens naturally ... 118

    Experiments at the Los Alamos National Laboratory ... 119

    Plasma compression may be a billion-fold ... 120

    Very large cosmic 'primer fields' ... 121

    Plasma compression may exceed trillions to 1 ... 122

    The Earth is our witness ... 123

    Large atomic elements decay over time ... 124

    The ratio of lead in uranium-containing rocks ... 125

The measured ratio disproves the Big Bang theory	126
The dating of the Earth, with the atomic clock	127
Gravitational accretion of cosmic dust from the Big Bang	128
The hydrogen-sun theory	129
Look at the volume of hydrogen that is needed	130
According to the false theory	131
Where did the huge volume of hydrogen come from	132
Out of the range of the cosmic abundance ratio	133
We see a Sun that is dark inside	134
A Sun that is a sphere of thinly dispersed plasma	135
From where does the Sun derive its energy	136
From plasma our world was formed!	137
The radiometric dating of the Earth	138
Plasma gets 'consumed' by the solar electric-fusion process	139
If plasma would merely flow into the Sun	140
For hydro-electric generating to work	141
The energy of the water	142
The synthesizing fusion on the surface of the Sun	143
Plasma-fusion maybe the sink that activates the source	144
When the plasma-flow into a sink activates the source	145
The American theoretical physicist David Bohm	146
The speed of light	147
In the sea of latent energy that is cosmic space	148
Quarks cannot be divided	149
David Bohm,	150
We see two very-long climate cycles expressed	151
The very long cycles can be seen as evidence	152

Plasma streams that have galaxies at their node points	153
The Andromeda galaxy	154
The other connecting stream from the Milky Way	155
The resonance waves become overlaid	156
The fusion-sun sink process	157
The galactic plasma density is at a low point	158
When the sink effect draws more than the supply line holds	159
The universe is motivated by the plasma-sink process	160
The need for the sink feature	161
In the process of creating the uranium atom	162
When plasma flows in interstellar space	163
Assume that the plasma sink is our Sun	164
In the Red Square Nebula	165
The plasma researcher David LaPoint	166
The flip ring is a magnetic ring	167
The features that we see in the Red Square Nebula	168
The flow of electricity is critical	169
By the resulting magnetic fields	170
Plasma surrounding our Sun becomes so dense	171
This is what causes our Sun to become a sink	172
The Sun is seen as a vast sea of granular cells	173
The graduals are cells of Primer Fields	174
The Sun is a vast sea of cellular primer field structures	175
Here again, the inflowing plasma	176
Nuclear fusion causes the Sun to act as a plasma sink	177
Synthesized atoms flow away with the solar wind	178
The synthesis extends far beyond the helium fusion stage	179

In the remarkably close agreement	180
The heavier elements, past the helium stage	181
The presence of 'heavy' elements in the solar atmosphere	182
Evident by the existence of noctilucent clouds	183
Cold fusion drives the universe	184
Electric fusion is cold fusion	185
Atomic elements are extremely rare in the universe	186
The cold-fusion process in the Sun is simple	187
When energy is invested	188
As the protons fuse	189
The resulting field reduction	190
The dynamics create an imploding effect	191
The resulting voids accelerates the remaining plasma	192
As more protons are being used up	193
As the fused proton and neutron clusters are joined	194
The entire electric scene becomes a great void	195
In comparison, the waterline carries the plasma streams	196
The same principle applies	197
Energy is carried away by electric transmission lines	198
The process heat is radiated as light	199
Solar fusion IS cold fusion	200
Cold fusion is efficient as a solar process	201
Artificial nuclear-fusion power processes fail	202
The easiest-to-fuse fusion fuel	203
Attempted power production processes all invariably fail	204
Natural solar processes are not yet utilized for power production	205
This does not mean that the real solar-energy processes	206

    Plenty of evidence exists 207

    A plasma-flow pattern is 'visible' in the ionosphere 208

    The sky is not the limit 209

Part 3: Historic evidence of large cosmic plasma structures 210

    Historic paradoxes 211

    The coronal heating paradox 212

    As if the corona had been heated 213

    The same effect cause the solar-winds to be visible 214

    In the standard internal-fusion theory 215

    On the electric-sun platform 216

    Interstellar plasma streams 217

    Verified in static experiments by David LaPoint 218

    The superheated corona can be seen as 219

    The accelerating solar-wind paradox 220

    Under the internal-fusion theory for the Sun 221

    In plasma solar physics accelerating solar wind not a paradox 222

    The solar wind originates in the confinement domes 223

    The plasma jet tunnels through the corona 224

    When plasma becomes highly concentrated 225

    When plasma is compressed 226

    Protons would push each other apart 227

    Plasma streams generate a magnetic field aground them 228

    Under the Big Bang model where only gravity is recognized 229

    Plato, the great science genius 230

    As the prisoner ventures past the exit 231

    Historic plasma evidence 232

    Some amazing aspects of evidence 233

Some extremely larger items of evidence	234
Some related evidence exists that suggests	235
Historic evidence found in the Giza pyramids	236
The baseline of the pyramids is perfectly aligned	237
The orientation of the sphinx	238
This distant timeframe coincides	239
The experiment-derived, magnetically-shaped dynamic flow geometry	240
The visible plasma geometry in the night sky	241
The idea for the building of the pyramids	242
Stone Henge	243
Replicated in the layout of the Stone Henge	244
The monument's features	245
Replicated in the form of a large monument	246
Aligned into 56 evenly-spaced filaments	247
The plasma-flow experiment	248
These plasma features were once seen in the night sky	249
We are presently at a deep low point	250
Familiar features that are visible today	251
Presently visible, large plasma structures	252
Two gigantic plasma structures from the heart of the Milky Way Galaxy	253
The plasma explorer Hannes Alfven had theorized	254
The primer fields geometry derived from high-energy experiments	255
The experiment-derived geometry illustrates	256
The magnetic bowl structure	257
The huge scale of the plasma structure	258
The bowl type magnetic field structure	259
Intergalactic plasma streams may remain forever invisible	260

String-bound groups, small and large 261

When dense plasma streams encounter atomic elements 262

The interaction energizes the dance 263

The same type of effect in the form of lightning 264

Part 4: The solar wind speaking of the coming Ice Age 265

Solar wind, Ice Ages, and our future 266

The solar synthesizing fusion 267

While they don't have a direct electric connection 268

The bound plasma particles that produce the atoms 269

Electrically neutral fusion products get blown along 270

The regulating feature 271

When the input streams diminish 272

When the fusion products clog up the cells 273

The longest duration of continuous fusion 274

The solar wind appears to fulfill the purifying function 275

In a chain-reaction negative feed-back loop 276

When the Primer Fields collapse 277

The solar wind also tells us something else 278

The loss of Ulysses 279

Ulysses gave us the most 'pristine' view 280

Ulysses saw the solar wind pressure diminished 281

The diminishing trend towards the solar cut-off 282

When the sunspots speak to us 283

We can observe the sunspot-principle in action 284

Backpressure limits the plasma-current flow rate 285

When active magnetic primer fields for the fusion cells diminish 286

Escaping plasma streams are seen as plasma loops 287

- When the escaping backpressure does not collapse the cells 288
- Giant eruptions are extremely rare 289
- The occurrences of prominences, or solar flairs 290
- Plasma is a near-perfect electric conductor 291
- When the solar flairs dramatically diminish 292
- The sunspots numbers are less dramatically affected 293
- From a science standpoint, the sunspots are valuable 294
- Sunspots also provide us a portal 295
- The time has come to get real 296
- Science has become a trap 297
- When doctrines require that one set reality aside 298
- The Big Bang theory was developed as a counter-theory 299
- Similar to more modern Global Warming theory 300
- Under the weight of countervailing doctrines 301
- Science still plays with epicycles in solar physics 302
- How do we get out of the paradoxical trap? 303
- We get out of the trap like Johannes Kepler did 304
- The mission named Ulysses 305
- The Ulysses satellite flew three orbits around the Sun 306
- What the Ulysses spacecraft has measured 307
- When sunlight from our Sun is expanded 308

Part 5: The plasma solar system evident in CO2 309
- Evidence that disputes the Big Bang theory 310
- The proof that the expansion is not happening 311
- The spectral uniformity tells us 312
- When a gas cloud expands for billions of years 313
- The atomic elements that cause the similarity 314

This means that plasma alone is the singularity ... 315

David Bohm named Einstein's successor ... 316

The similarity of the spectral lines ... 317

Cold Fusion Evidence in CO2 ... 318

If the Big Bang composition is universally similar ... 319

The heavy CO2 molecules ... 320

Evident by their distribution of the elements in space ... 321

CO2 concentration on the Earth is presently at the lowest level ... 322

CO2 never stays in the atmosphere for long ... 323

Because of the recycling loss ... 324

CO2 is extremely critical for all life on our planet ... 325

If the CO2 loss had continued past the Permian Period ... 326

The solar event occurred during a peak period ... 327

A type of solar lightning ... 328

The single event raised the global CO2 level 5-fold ... 329

A similar lightning strike occurred during the Jurassic Period ... 330

Massive Australian Precambrian/Cambrian Impact Structure ... 331

Lightning impact crater was carved 2000 kilometers wide ... 332

The big discharge-impact crater on Mars ... 333

The forming of the Deccan Traps in India ... 334

Evident in the carving of the Grand Canyon in Arizona ... 335

A similar canyon was cut on Mars in distant time ... 336

Life might have vanished on Earth ... 337

Saturn's rings are made up of water ice ... 338

The existence of Saturn's icy moons ... 339

Solar lightning events are more likely for Saturn ... 340

Titan, the largest of Saturn's moons ... 341

The great CO2 up-ramping that gave life a new chance 342

CO2 on Earth has diminished back to the CO2 starvation level 343

The normal solar induction of CO2 is insufficient to maintain life 344

We need to up-ramp the CO2 concentration ourselves, this time 345

Pumping some of the stored-up CO2 back out of the oceans 346

Part 6: The diminishing solar wind towards solar collapse 347

The weakening has dramatically increased 348

The measured 30% reduction of the solar wind pressure 349

The heliosphere attenuates galactic cosmic ray flux 350

A corresponding 20% increase in galactic radiation 351

The solar wind a feature of a regulating system 352

The KEY to the solar-wind paradox 353

The solar wind may be seen in terms of a tea kettle 354

The solar wind can be likened to steam being boiled off 355

The solar wind is the 'steam' that flows from the 'kettle' 356

Below the boiling point 357

Water can be heated many times hotter under high pressure 358

The Earth's climate is the coldest in 440 million years 359

The threshold line becomes important 360

When the Sun is inactive 361

Cosmic rays are fast moving protons or electrons 362

The dawn of humanity didn't begin until the Ice Ages began 363

The story of the human development 364

Another star-system named Cygnus X-3 365

The Cygnus X-3 pulses 366

The Cygnus X-3 source became active around 700,000 years ago 367

The 700,000-year timeframe 368

- The phase shift in human development 369
- The heliosphere is already becoming noticeably weaker 370
- The current electric weakening in the solar system 371
- Ulysses has measured 30% less 'steam' coming from the Sun 372
- We might not have much time left 373
- The magnetic dynamic-flow geometry of the Primer Fields 374
- The inactive Sun will in time become a dim star 375
- The energy output of the inactive Sun 376
- The world's agriculture will have to be relocated 377
- The observed rate of diminishment 378
- We are perfectly able to know with great certainty 379
- Harsh conditions under an inactive Sun 380
- The Ice Age choke-hold may have been a major contributing reason 381
- Event-driven reactions are dangerous traps 382

Part 7: Shaping the future is a spiritual issue 383
- The deception of normality 384
- Our solar system is collapsing in numerous ways 385
- Multi-level regulated astrophysical systems 386
- The first-level regulating stage 387
- The regulating system can't affect the very long trends 388
- The second-level of the regulating system 389
- The 22-year resonance cycle 390
- The third level of the regulating system 391
- The next event on the diminishing slope 392
- When the astrophysical principles become understood 393
- On the platform of understood principles 394
- The larger part is spiritual in nature 395

Will we become sufficiently human 396

Or will we remain stuck in the current smallness of thinking 397

In order to explore these types of questions 398

We need this exploration 399

I had thought at the time 400

As the challenge widened 401

The very notion of winning 402

As the kaleidoscope keeps turning, this time in Russia 403

Is the Earth really flat? 404

A miracle unfolds in India 405

Why should small games continue to dominate 406

The truly endless horizons are the horizons of our own creating 407

Still, the world grinds on 408

The cup bearer to the king is a dangerous pawn 409

When people must flee the land of the free 410

Hope is justified 411

We live in a universe of vast electric power 412

When we stop to keep our mental horizon blocked 413

You are invited to come and visit the great libraries 414

The final seven chapters of my series The Lodging for the Rose 415

More Illustrated Science Books by Rolf A. F. Witzsche 416

Introduction - Let me surprise you

Let me surprise you

A confusion of theories abound, about what our Sun is, and how it operates.

Primarily the confusion is divided into two groups of theories based on completely opposite concepts.

The Big Bang Creation ideology

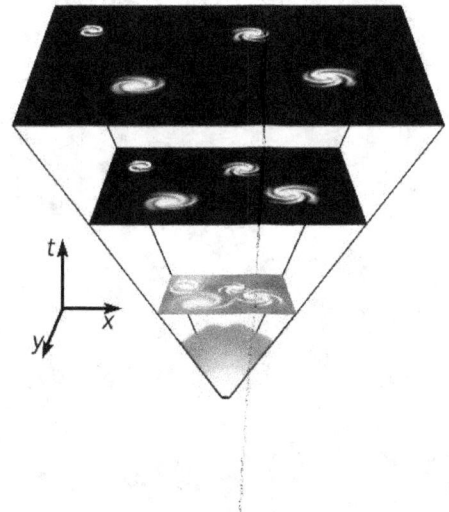

The Big Bang creation myth refuted by the electric solar fusion model

One concept is centered on the Big Bang Creation ideology. It speaks of a gigantic explosion 13.8 billion years ago, in which all the matter and energy of the entire universe was created at a single place in the first three minutes. The explosion is deemed to have furnished all the dust of the cosmos that was driven apart by the shockwave into all directions, where it condensed by gravity into planets, stars, and galaxies.

The electric cosmology

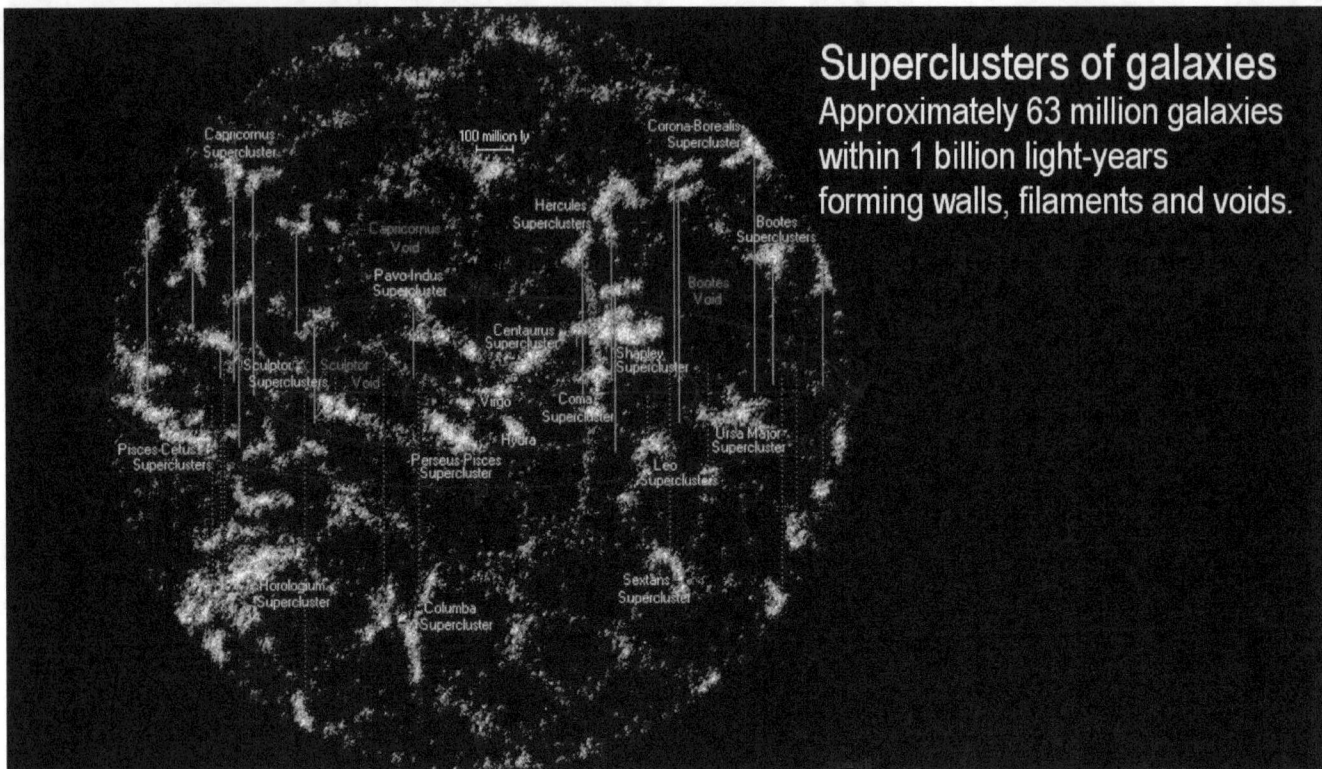

The opposite concept is the electric cosmology that speaks of a boundless universe that is pervaded by endless streams of plasma that interconnects all the galaxies and powers all the stars within them, including our sun. In the electric cosmology, plasma, which caries electric potentials, is deemed the lifeblood of the universe.

The two opposite concepts have each forged opposing trends of theories that affect how we respond to the dynamics of our Sun and its operating principles. They also affect how we develop our future.

It is here, in this context of our theories affecting the future, where things begin to get serious.

The effects of cosmological theories

The effects of cosmological theories have an enormous impact on the future of humanity, by the way we respond to the effects. The effect that we face in the near future promises to be so immense that it takes the duality of the scientific perceptions out of the realm of merely academic significance, and gives it an existential significance.

The long suspected resumption of the Ice Age

For example, the long-term exploration of the principles of plasma dynamics, applied to cosmic dynamics and solar dynamics, has yielded the astonishing recognition, with a high degree of certainty, that the long suspected resumption of the Ice Age that is deemed to be still thousands of years in the future, is much nearer and will likely start in the 2050s with the Sun going inactive. It will unfold extremely rapidly under a 70% dimmer and cooler Sun.

The transformation of our world

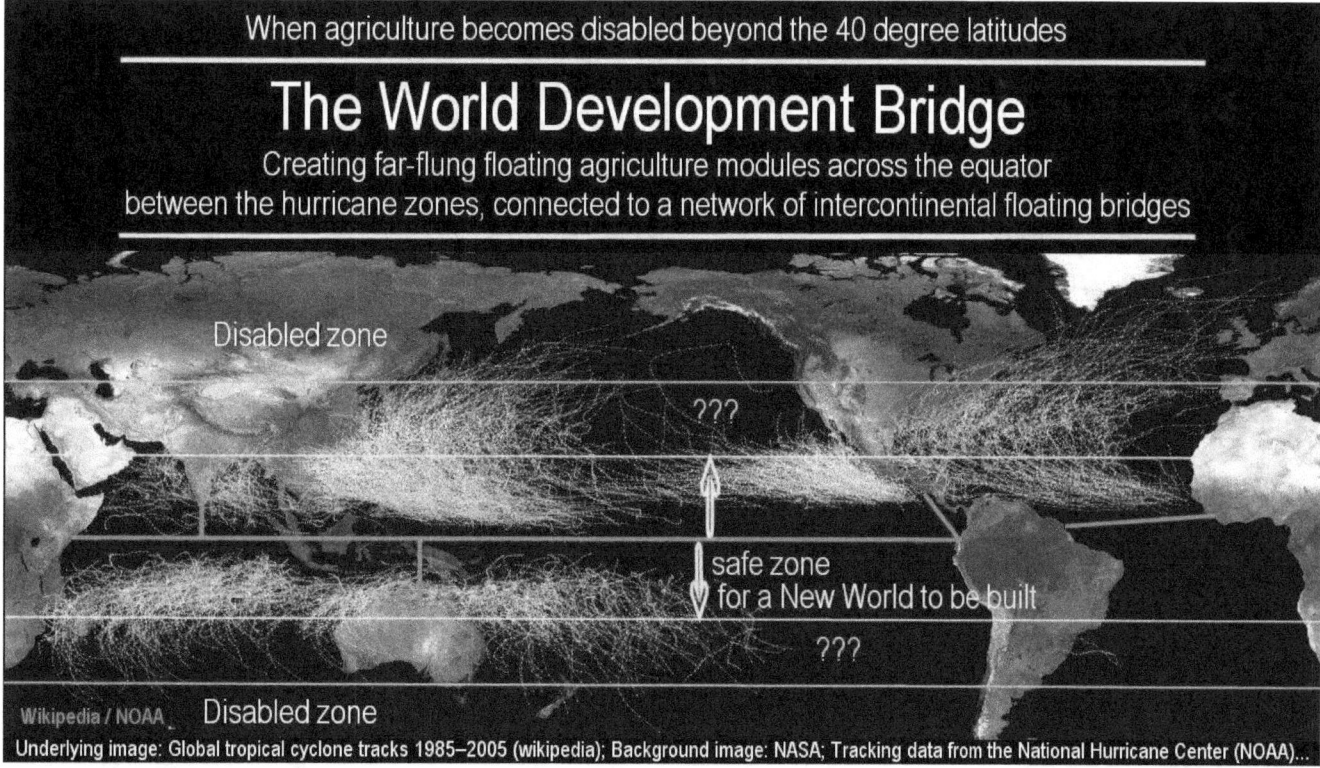

With the potential transformation of our world on this gigantic scale looming on the near horizon, the determination of what is truth is no longer a mere academic concern, but is a concern that will determine the future existence of humanity. The transformation of our world of the magnitude that arises from a 70% dimmer Sun, will render almost all countries above the 40 degree latitude unsuitable for agriculture if not completely uninhabitable. This means that much of the world's present agriculture will have to be relocated into indoor facilities or be placed afloat across the tropics, as little suitable land exists in the equatorial regions.

Gigantic infrastructure development

It is self-evident that the type of gigantic infrastructure development, as will be required - complete with floating cities to service the floating agriculture - won't be contemplated, much less be built, unless the science division is healed.

Obviously, of two opposite science concepts, only one can be real. But why do we have two opposite concepts? Isn't science a quest for understanding what is real? This is true, but only to the point where politics enter the scene.

Science has been 'guided' with the payola

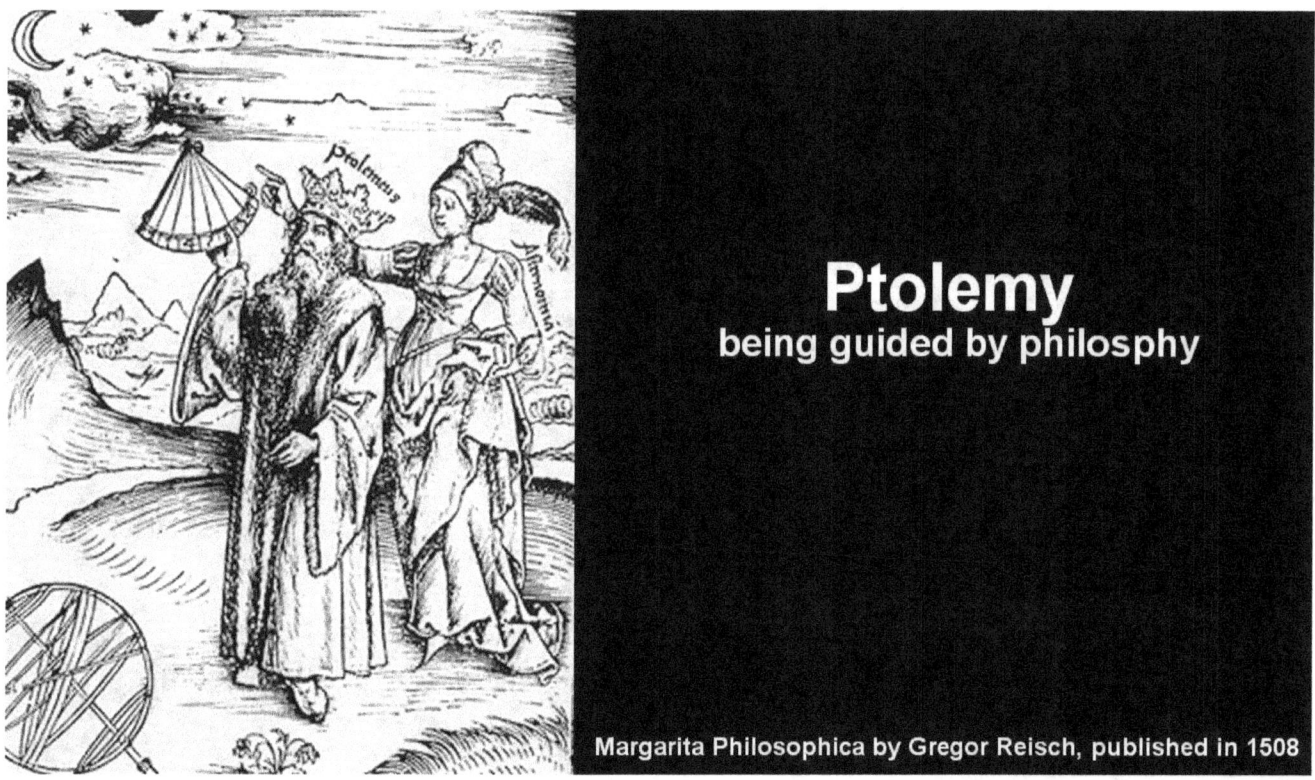

Throughout the pages of history, science has been 'guided' with the payola and other means to serve the doctrines of the various empires, which typically have the power to have their objectives met. On this train, truth falls by the wayside. It always has. This applies to cosmology too.

The train of politically 'guided' science

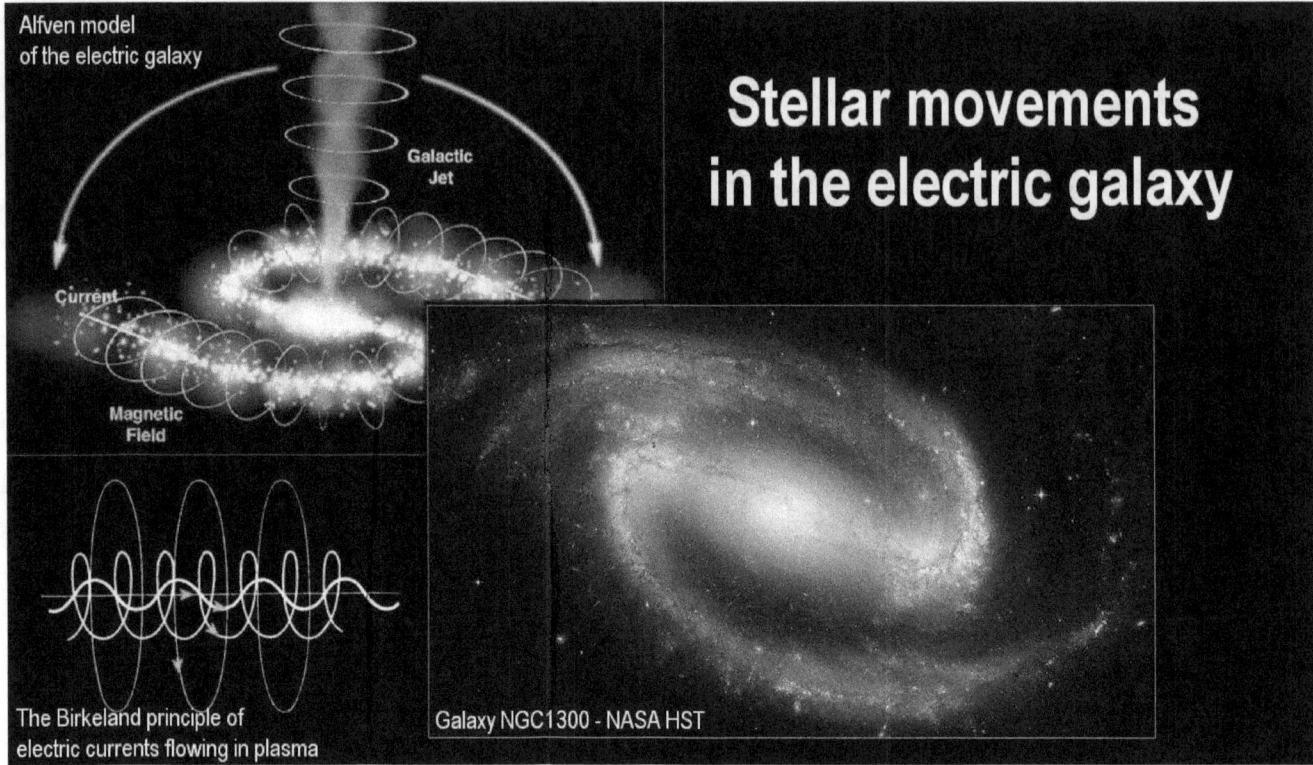

While the train of politically 'guided' science goes far back into history and has many tales attached, the counter science in cosmology appears to have been invented to counter the breakthrough discoveries in plasma physics by the Swedish, 1920 Nobel Price winner in Physics, Hannes Alfven.

The timing suggests that the Big Bang Creation theory was developed, and was massively promoted, as a counter-theory against the plasma universe.

The plasma universe offers an unlimited electric-energy future to humanity. The recognition of it would scrap the value of the private ownership of the world's energy resources that is one of the pillars of empire. In the Big Bang cosmology protects that pillar. Under the Big Bang doctrine, plasma is deemed not to exist. Of course it does exist. A vast body of physical evidence testifies that it does exist.

The exotic Big Bang concepts

Inversely, no real evidence exists for the exotic Big Bang concepts that are hugely played up and promoted as real, such as black holes, dark matter, and stellar explosions.

Deep in the Big Bang cosmology

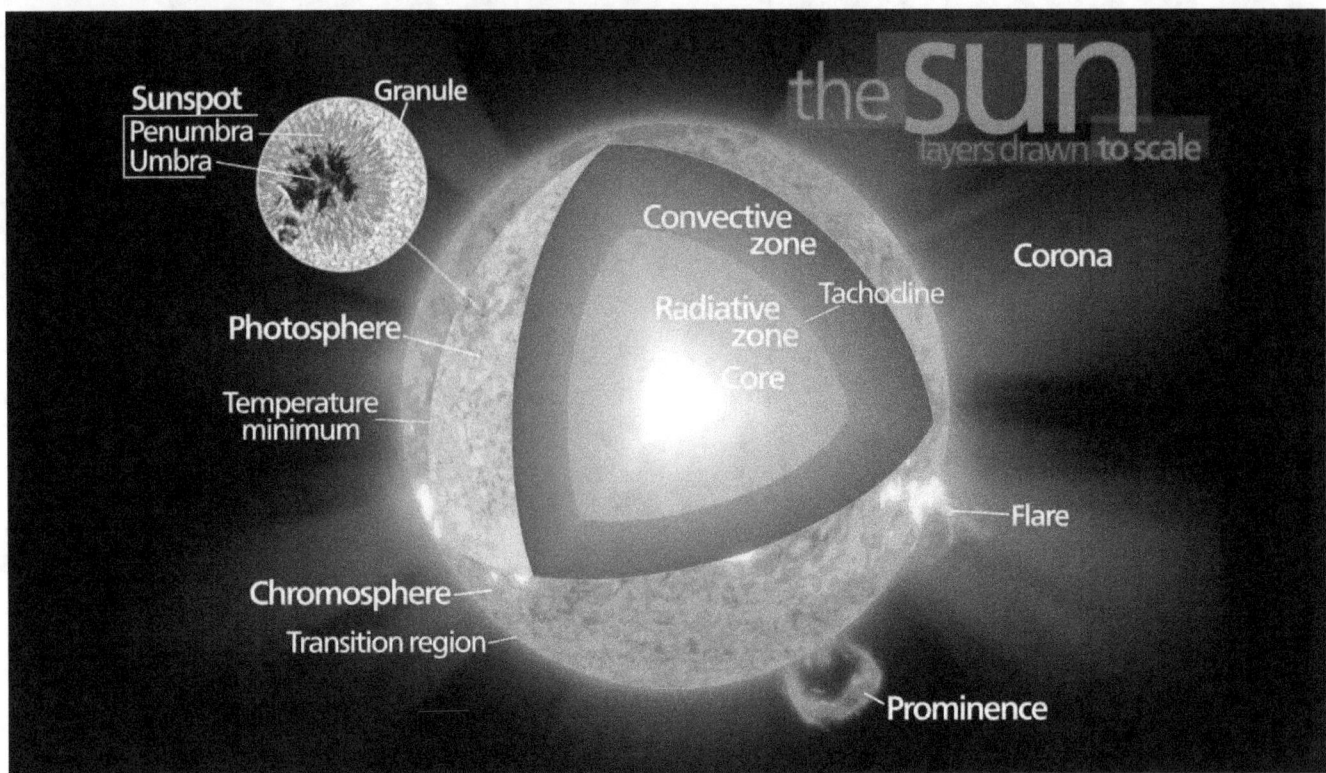

Deep in the Big Bang cosmology, where only gravity is allowed to be considered as a causative force, we find also the theory promoted that our Sun is a gas sphere, powered by nuclear reactions occurring in its core, which are deemed to fuse hydrogen atoms into helium atoms.

The theory of the internally powered Sun, renders our Sun a universal constant for all climate considerations, which literally closes the door to rational ice age concepts.

The trap that has been created with this prevents humanity from preparing its world for the next Ice Age to come, that may begin in 30 years. On this train of the Big Bang theory the depopulation objective is well served, which is a core policy within the oligarchic system of empire. The stated objective of the masters is, to reduce the population of humanity to less than 1 billion people, as required for the stability of a feudal system.

This is what stands behind the science duality. But how do we get out of this trap? Were do we go from here? The answer must be that we dare to explore what is real.

The Earth stands as a silent testimony

NASA - Earth from Apollo 16
wikipedia

Ironically, it is the Earth itself that weighs heavily against the Big Bang theory. The Earth stands as a silent testimony that its atoms were created brand new when the Earth itself was formed. This scientifically proven fact, backed with hard evidence from a wide range of sources, turns the Big Bang theory, and everything that is built on it, into a very imaginary fairy tale.

Since the fairy tale has spawned numerous theories and inspired many opinions, it becomes necessary, therefore, in exploring the truth, to separate what is demonstrably real, from the landscape of educated opinions and cultivated illusions, and that one does this in a comprehensive manner. The reason is that ultimately, the subject of truth, is a single package.

Outside the topics of fairy tales of the Big Bang

Topics

(1)
- ** It is impossible for the Sun to be a gas sphere, which the internal nuclear-fusion theory requires.
- ** The Sun is a plasma star with electric nuclear fusion occurring on its surface that operates at low temperatures.
- ** The Sun we see, is not as we expect it to be: It is a creative atom synthesising 'engine' that is only skin-deep.

(2)
- ** The universe is motivated by the plasma-sink process
- ** Cold fusion drives the universe
- ** Solar fusion IS cold fusion, we should replicate it
- ** The time has come to get real
- ** The coronal heating paradox
- ** The accelerating solar winds paradox

In the course of the exploration the video introduces a number of revolutionary concepts that may seem surprising, but which are critical for a rational understanding of the dynamics of the solar system outside the topics of fairy tales of the Big Bang cosmology. This means that the resulting presentation of the nature of our Sun as an energy source, won't be of a type that is taught in schools, institutions, and is presented in science documentaries for the television audiences.

Since the field of exploration that is presented here, has become largely unknown, but covers a number of related concepts, the video exploration is being presented as a series of 7 parts. The evidence with which the Earth refutes the Big Bang cosmology is presented in Part 2, which thereby becomes a part of the evidence for the plasma Sun, that has electric nuclear fusion occurring on its surface where its energy-radiation originates. But before we can get to this, Part 1 of the series is needed to establish what the Sun really is.

The internal-nuclear-fusion theory of the Sun, which is the generally accepted theory, has many flaws built into it, while no evidence actually exists that exclusively supports the theory. None whatsoever! That's shocking, isn't it? All historic and visible evidence supports instead the recognition of the Sun as a plasma star that is externally powered with cold-fusion nuclear synthesis occurring at its surface. This affects our climate, economics, politics, and how we relate to one-another as human beings.

When the truth becomes known, the world is changing. The old theories no longer apply. Recognized evidence discredits them.

Disturbing to those who cherish the illusion

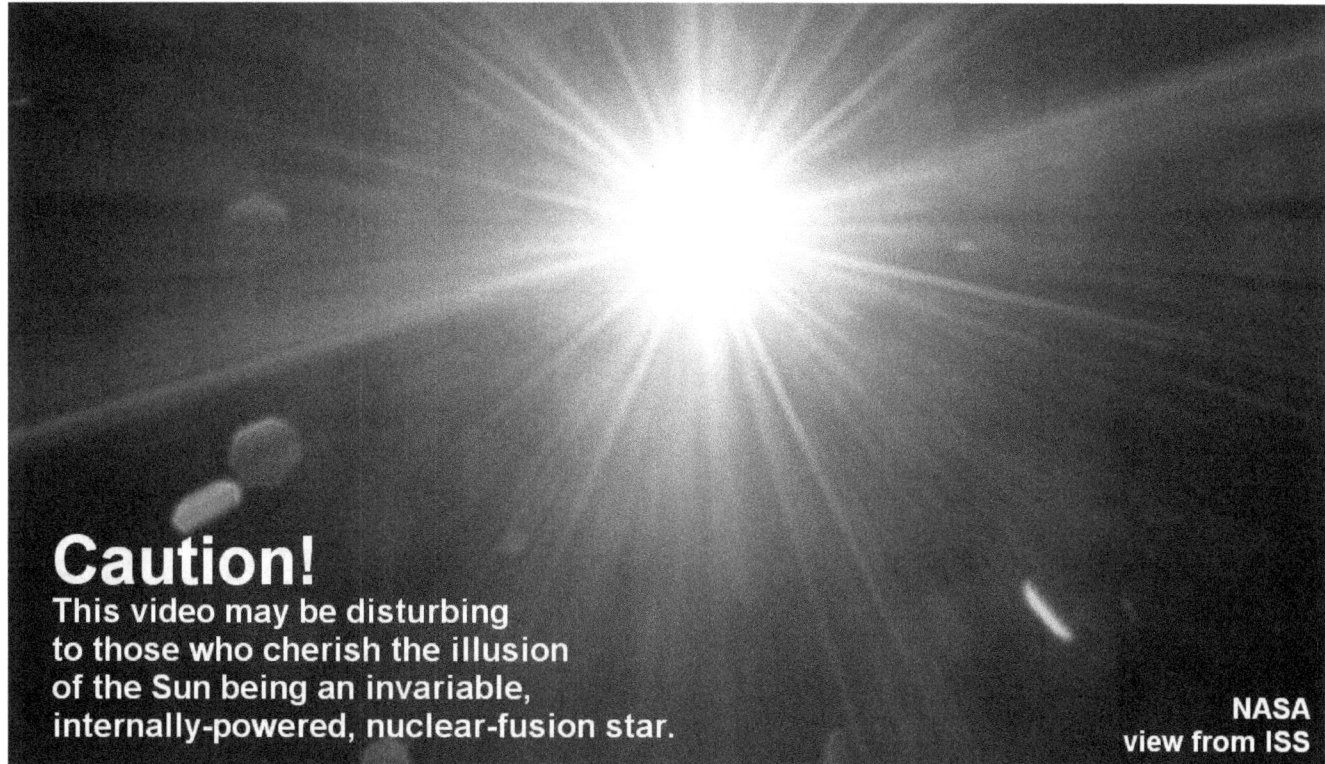

Caution! This video may be disturbing to those who cherish the illusion, of the Sun being an invariable, internally-powered, nuclear-fusion star.

Topics of truly gigantic plasma structures

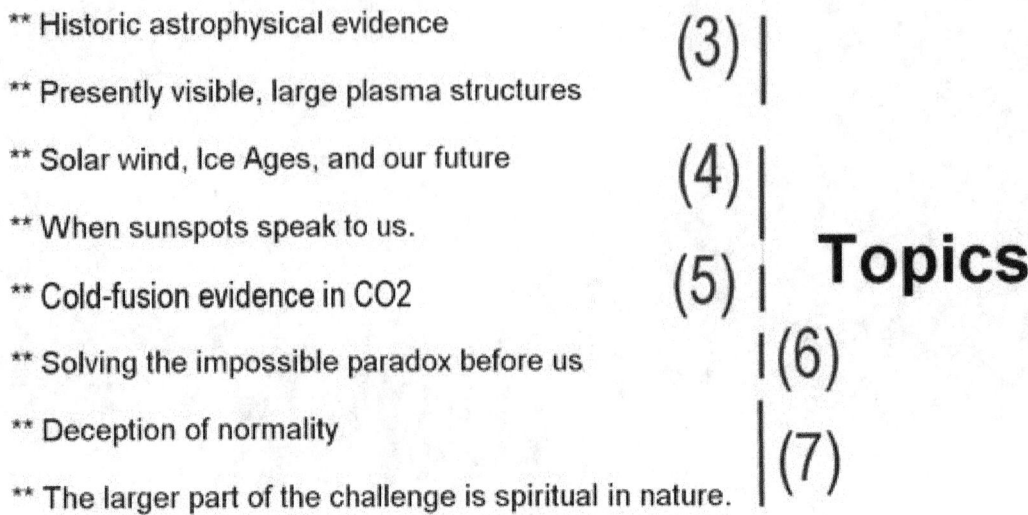

Part 3 presents evidence in historic events of large plasma structures in the sky, that were evidently visible in the past, but are no longer visible in our electrically weak times. Inversely, it also presents topics of truly gigantic plasma structures that were never visible, until recent times, when modern instrumentation made them visible to us. All of these prove that we live in the face of an electrically powered Sun.

Part 4 deals with the connection between the solar wind and ice ages in the electric universe.

Part 5 deals with the largest historic electric events in the solar system, their impact on life on Earth, and the imperative for us to save the future.

Part 6 deals with evidence for the near impending end of the current interglacial period, and the start of the next Ice Age in 30 years, and the certainty of the transition.

Part 7 deals with the biggest question that the entire series is leading up to. The question is whether we will respond to the scientific imperatives imposed on us by the future, to protect our existence.

Will we built the 6,000 new cities that we need

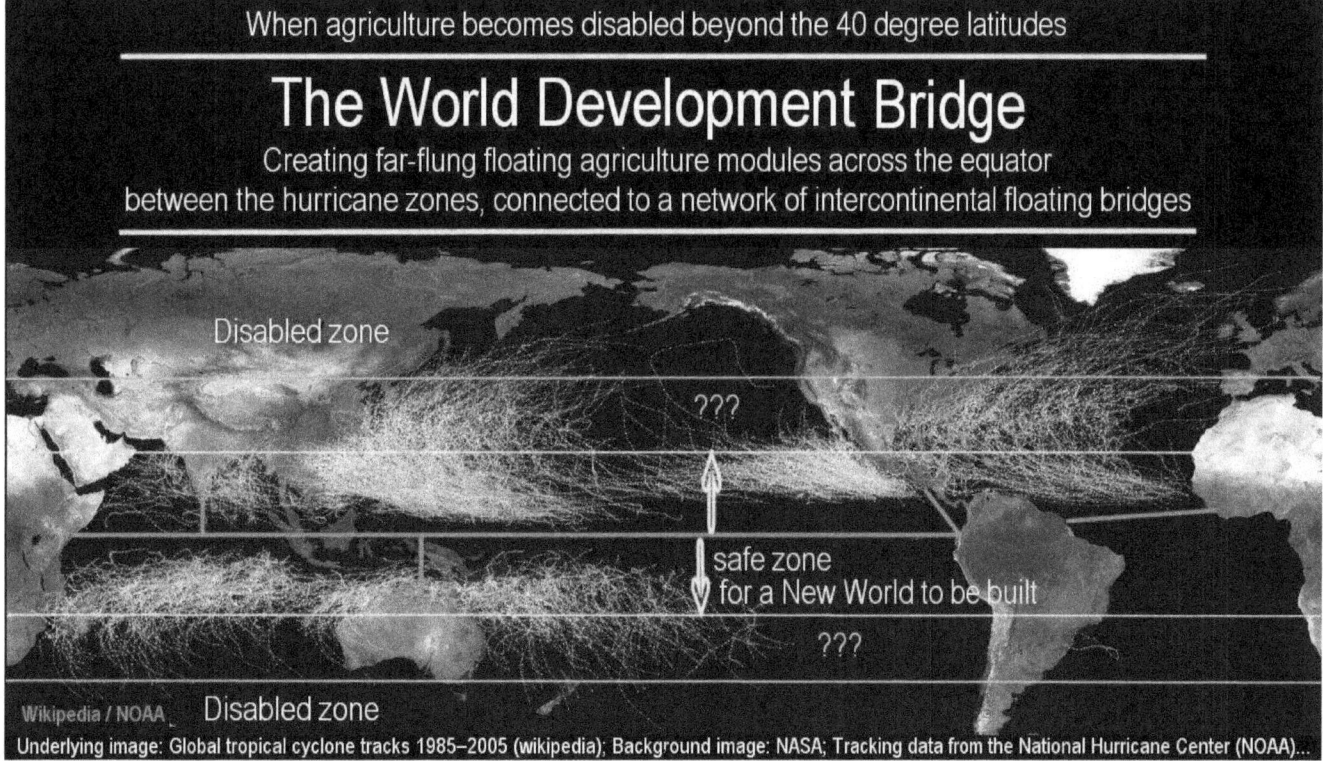

We have the materials, technologies, and energy resources on hand to build the worldwide infrastructures with which to secure the human landscape in the harsh time ahead when the Ice Age starts anew and large parts of the world become uninhabitable.

But will we use the resources we have? Will we built the 6,000 new cities that we need to enable the relocation of most of the great nations on the Earth? Will we do this and live? Will we built the 20,000 kilometers of intercontinental bridges along the equator, and the millions of acres of floating agriculture, that the bridges would connect to?

Will we create a new world for us

Will we do all of this and create a new world for us for a richer living under harsher conditions? Or will we do nothing and allow ourselves to be blown away with the winds of the cycles of the universe?

Part 7 deals with the difficulty in answering these questions.

The largest part of this spiritual challenge

In this context the entire scene becomes no longer merely a physical challenge, or a technological question, or even a science issue, but becomes a spiritual challenge. The largest part of this spiritual challenge will ultimately be, whether we will rouse ourselves to regard one another as human beings, with enough love for one another as the brightest diamond in the landscape of life, that the currently faint spark of love for one another becomes a fire of universal love.

In the wonders of our humanity

Then, when we get to this point, the path will be free for the brightest future imaginable. And for this imperative too, we have the resources already at hand in the wonders of our humanity.

Part 1: The Sun is NOT a gas sphere. The Sun is plasma

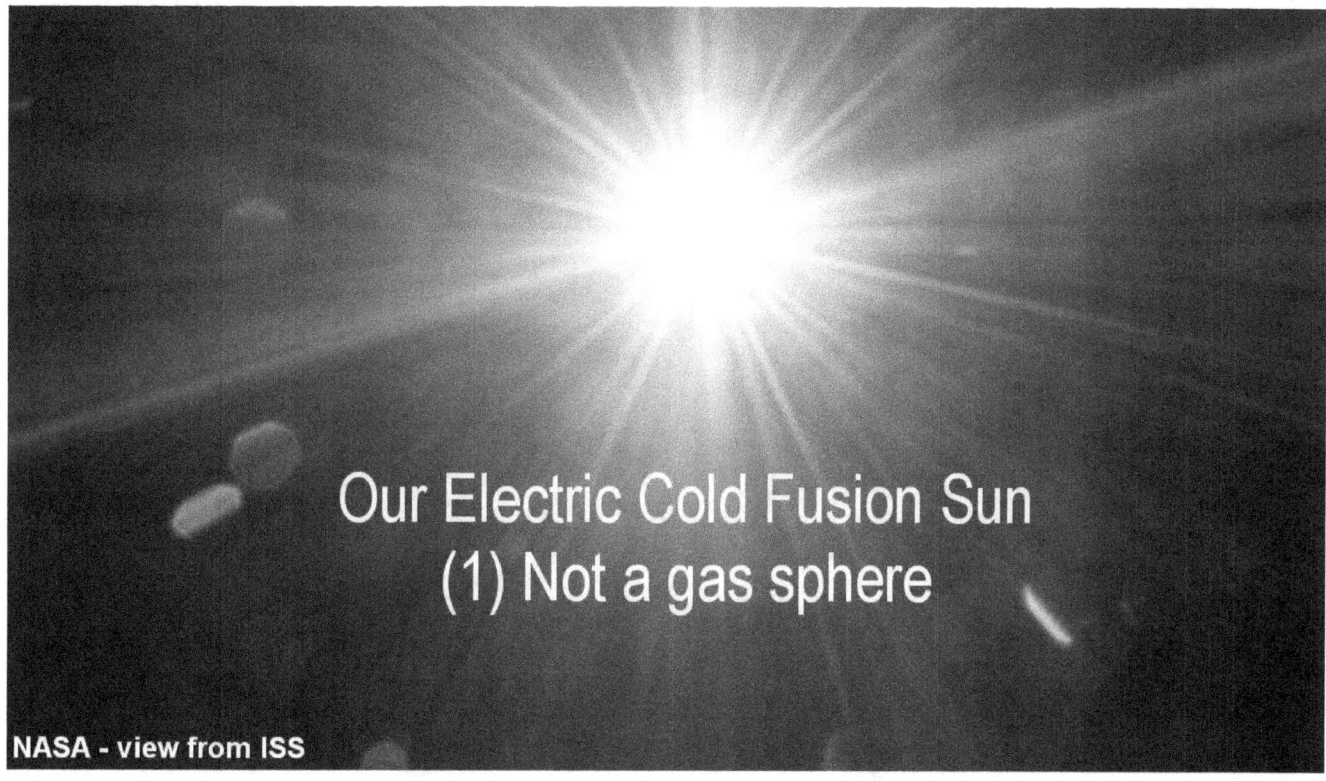

Our Electric Cold Fusion Sun (Part 1) A Plasma Star

The Sun is not a gas sphere

The Sun is not a gas sphere. It cannot be that.

It is impossible for the Sun to be a gas sphere, as the theory of the internally powered Sun requires it to be. But why is this impossible? The answer is simple. It couldn't operate in any other manner, because its operational principle is simply the most efficient one there is. Nor is it self-powered. It is powered by a principle that involves almost the entire universe. Physically, it is powered by plasma. Plasma is the life-blood of the universe. It is electrically charged. This makes it powerful. The electric force is 39 orders of magnitude stronger than gravity. The inclusion of plasma opens the empty box of conventional astrophysics where only gravity is deemed to rule and 99.999% of the universe is deemed not to exist.

Part 1: The Sun is NOT a gas sphere. The Sun is plasma.

Image: The Sun Topics

Image: The Sun Topics

Music: Beethoven, Courtesy of Orchestra and Choir of the LaRouche Youth Movement, with guests - July 2011 - Germany

Image: Plasma streams

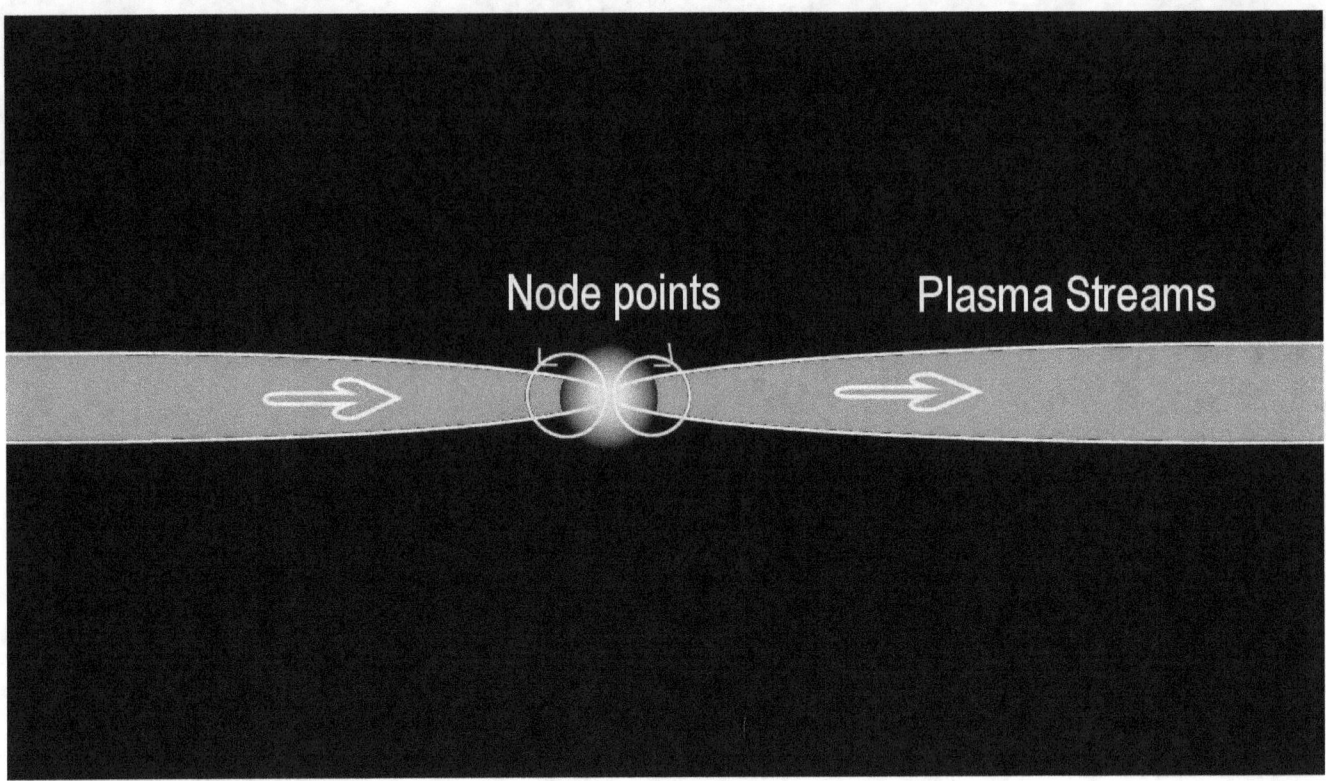

Image: Plasma streams

Image: galaxies as node points

Image: galaxies as node points

Image: The Milky Way, a node

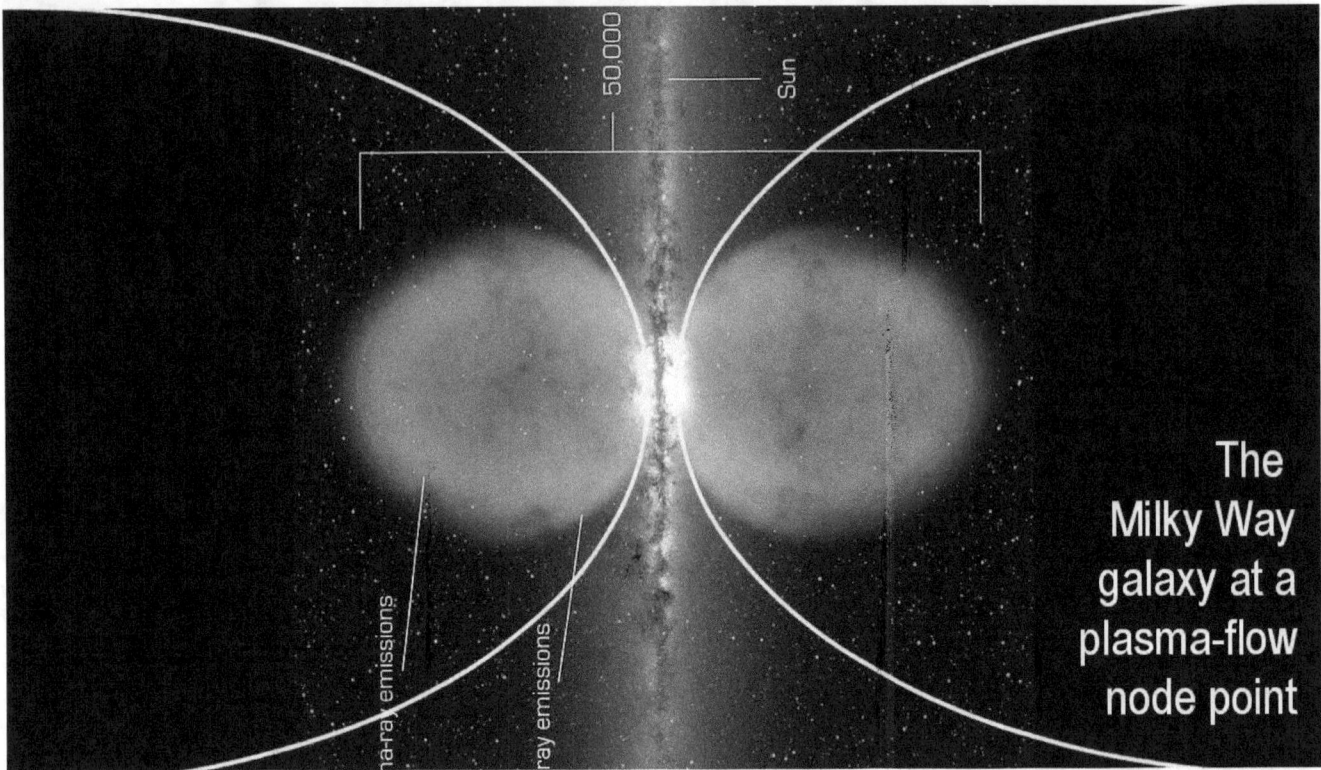

Image: The Milky Way, a node

Image: The Milky Way is plasma powered

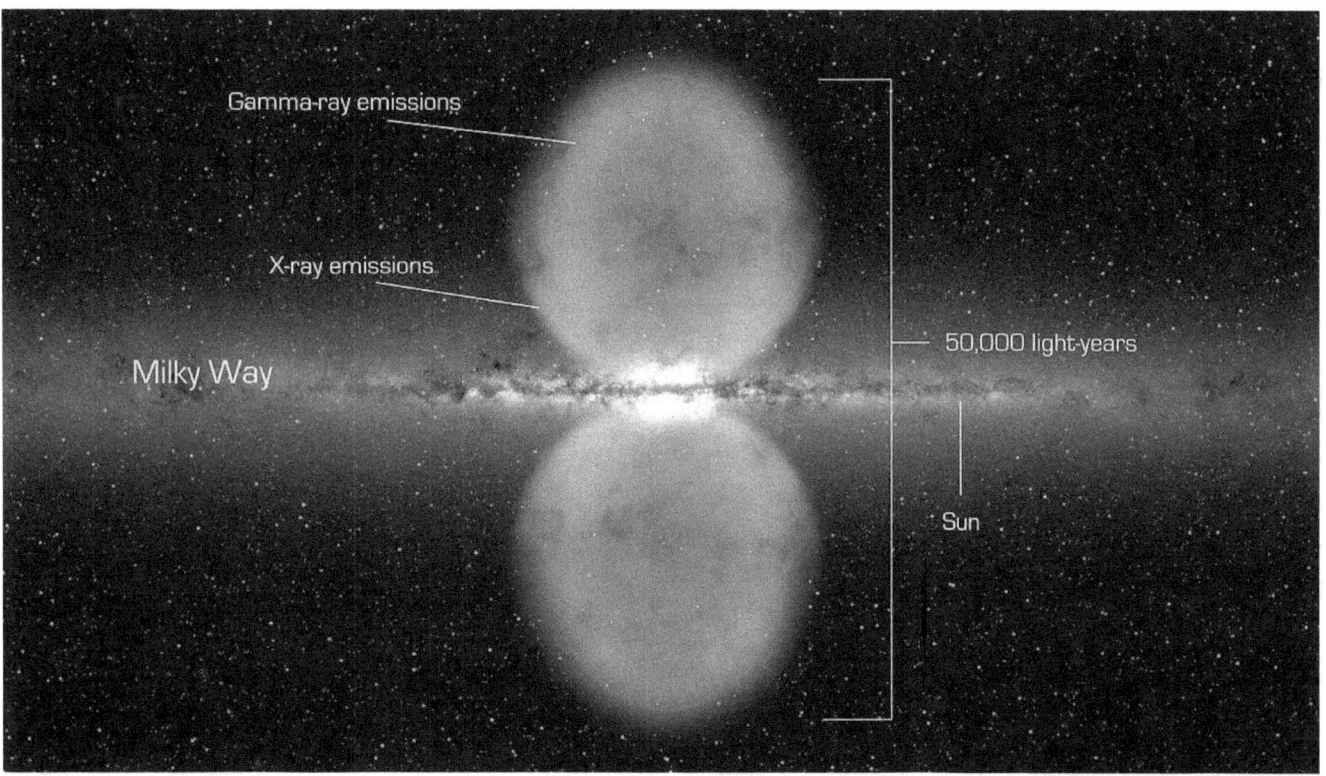

Image: The Milky Way is plasma powered

Part 1: The Sun is NOT a gas sphere. The Sun is plasma.

Image: Milky Way plasma domes

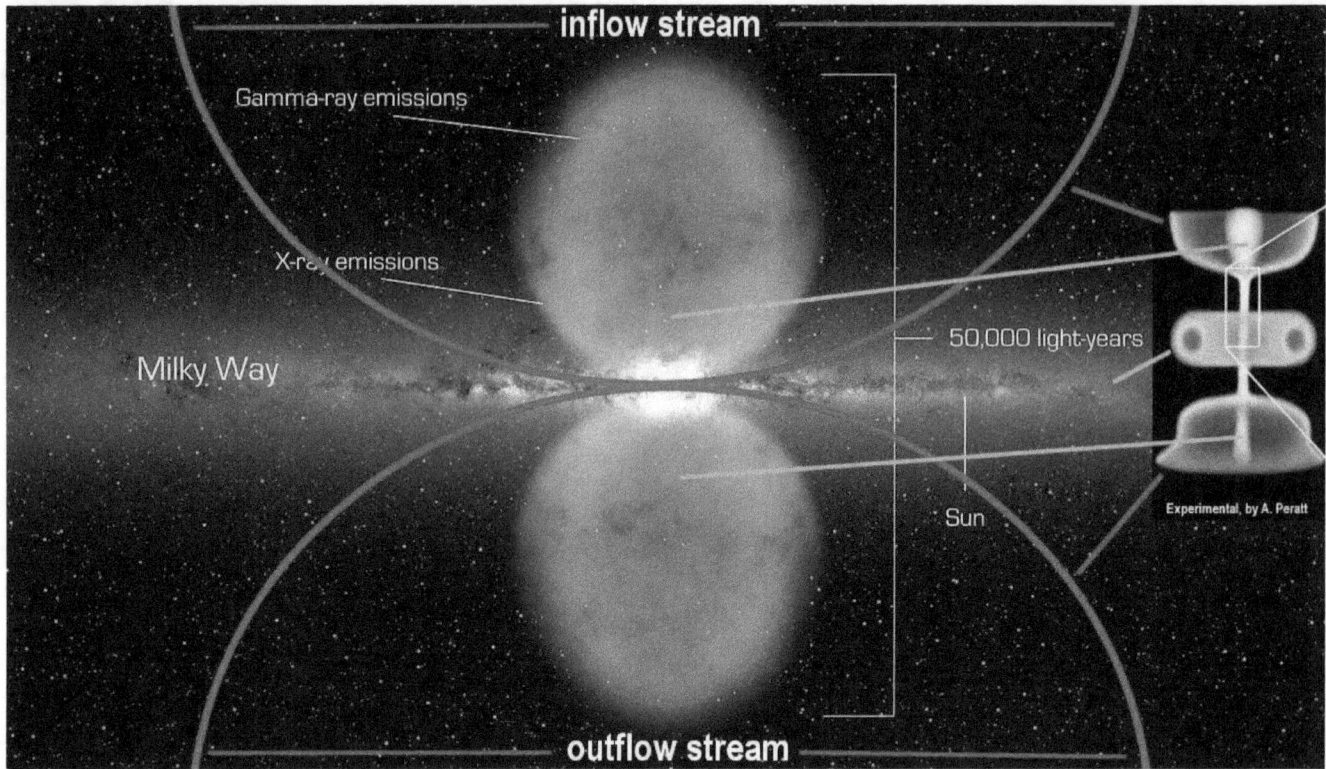

Image: Milky Way plasma domes

Part 1: The Sun is NOT a gas sphere. The Sun is plasma.

Image: The Milky Way in comparison

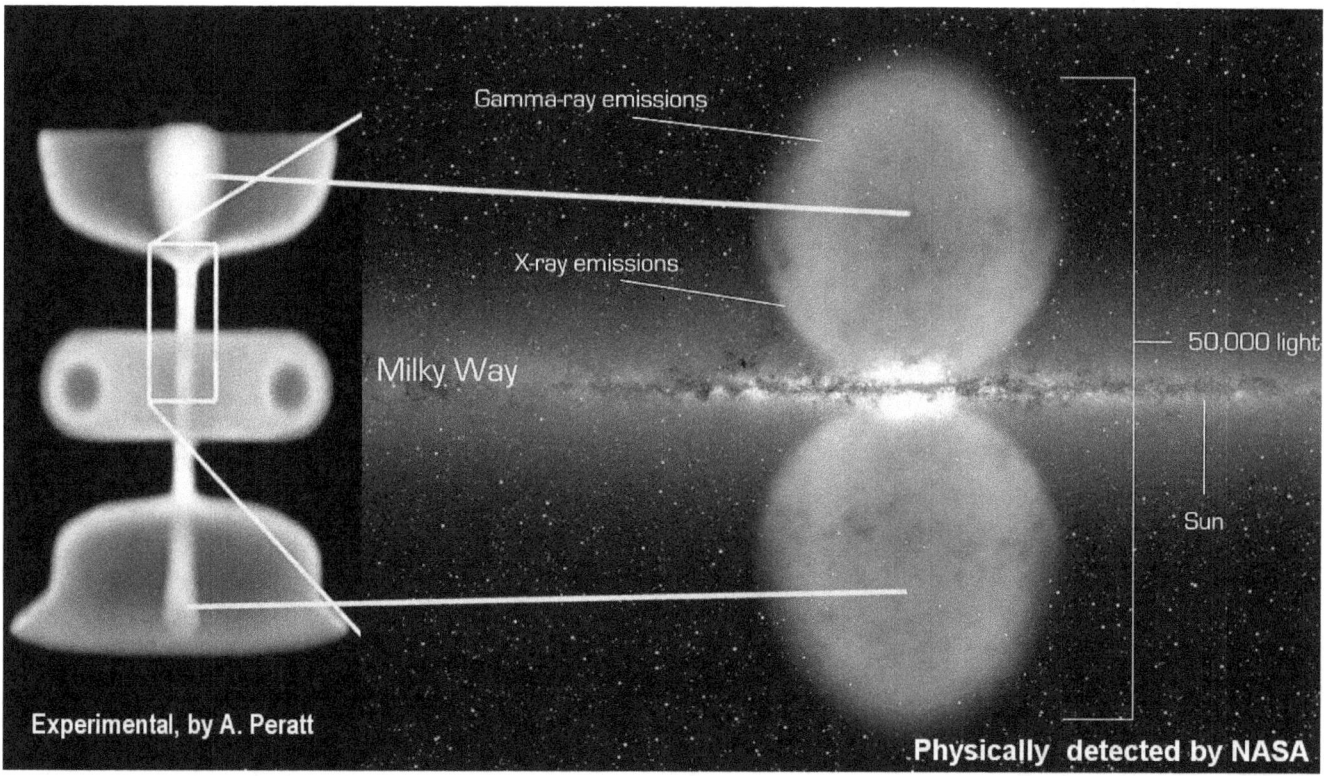

Image: The Milky Way in comparison

Image: Stars at node points

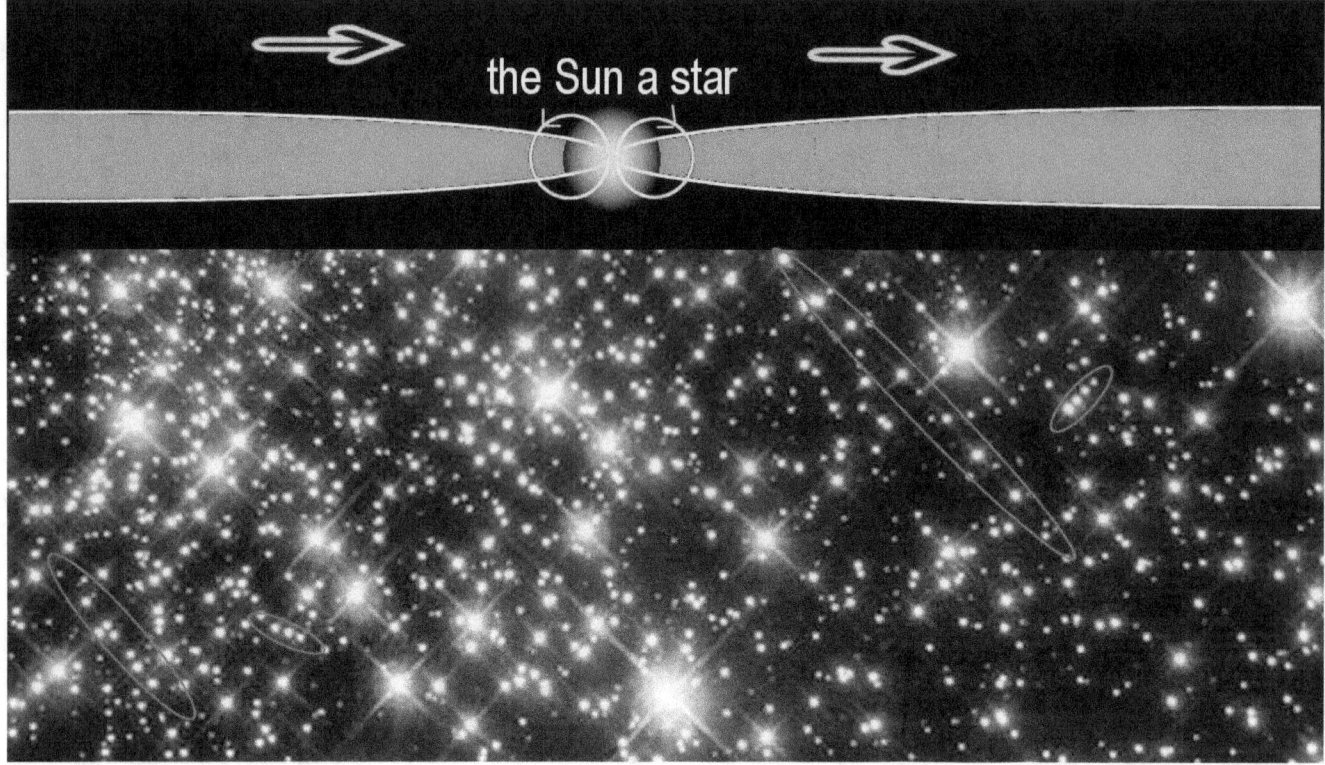

Image: Stars at node points

Image: Plasma focused onto the Sun

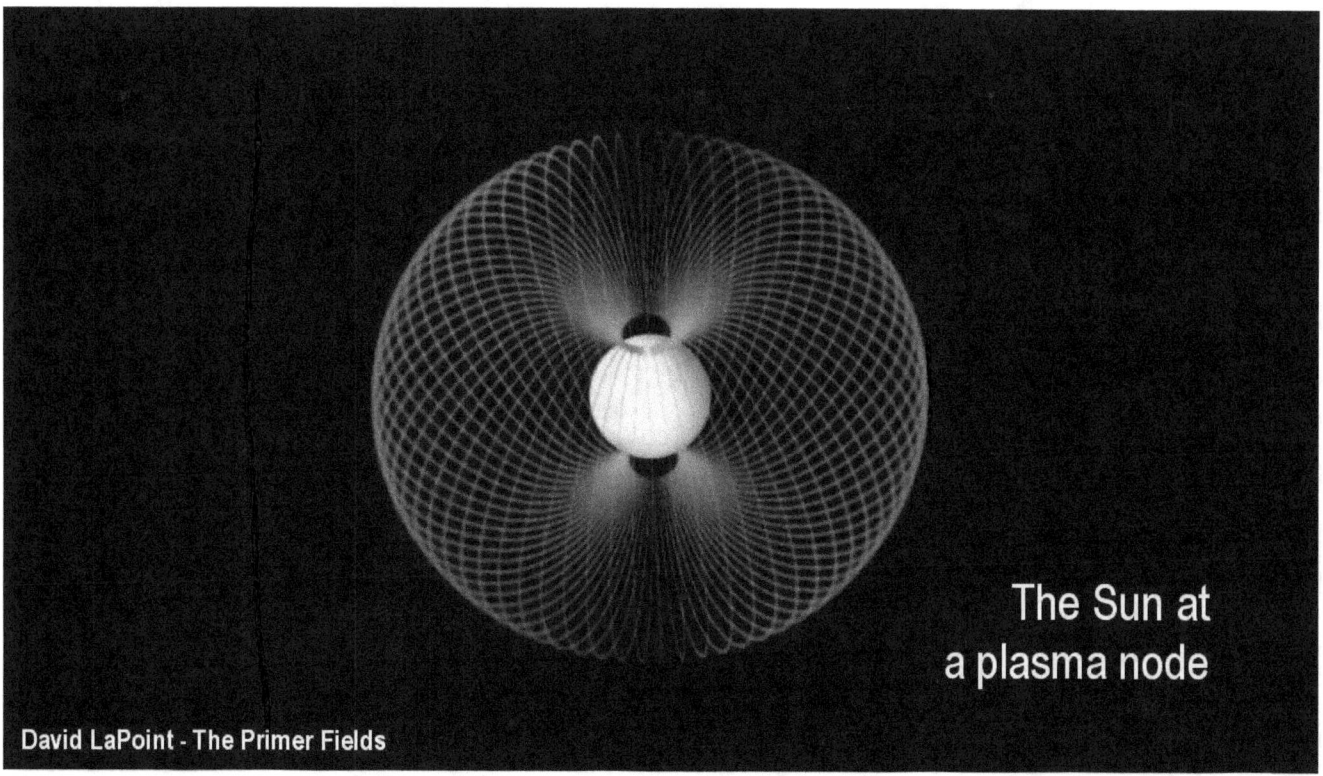

Image: Plasma focused onto the Sun

Image: Powered by interstellar plasma streams

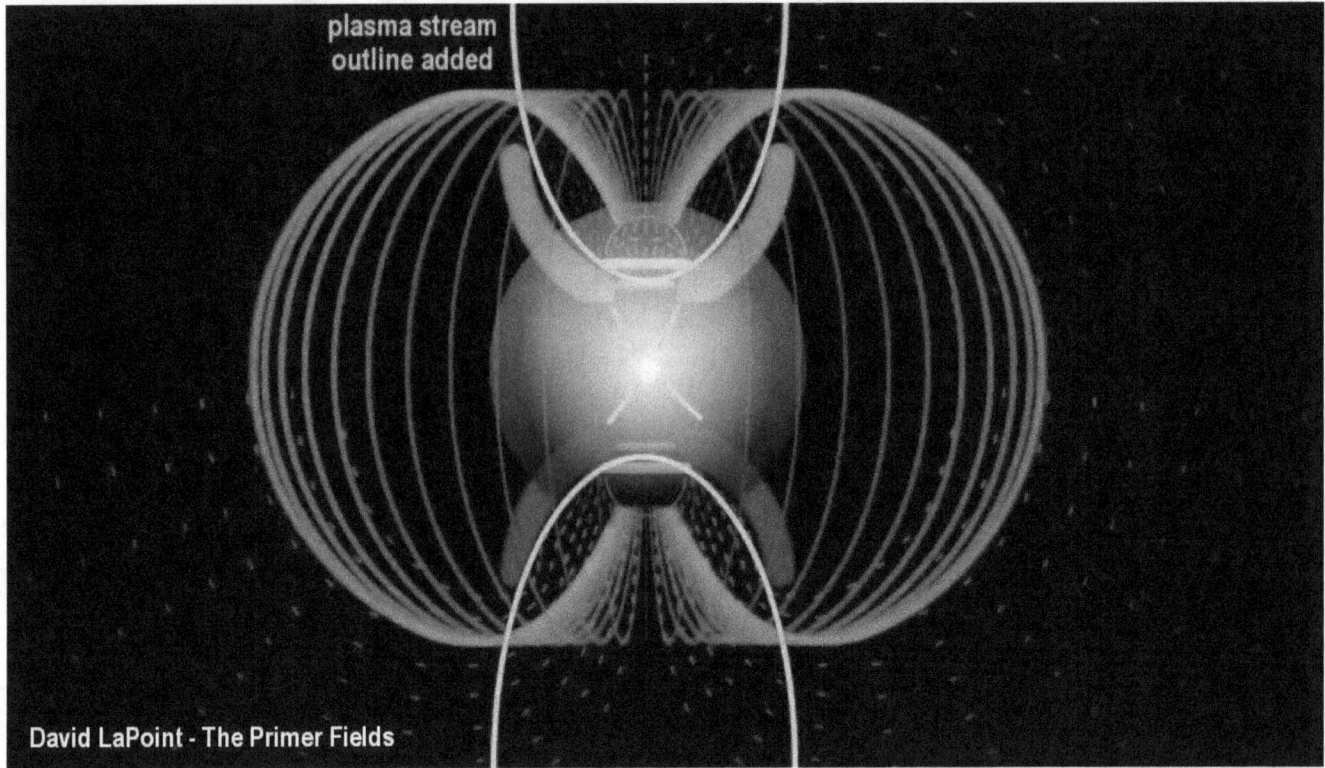

Image: Powered by interstellar plasma streams

Part 1: The Sun is NOT a gas sphere. The Sun is plasma.

Image: LaPoint and Peratt discoveries

Image: LaPoint and Peratt discoveries

Part 1: The Sun is NOT a gas sphere. The Sun is plasma.

Image: LaPoint, artificial Sun

Image: LaPoint, artificial Sun

Image: The Milky Way lookalike

Image: The Milky Way lookalike

Image: The Sun located in the Milky Way

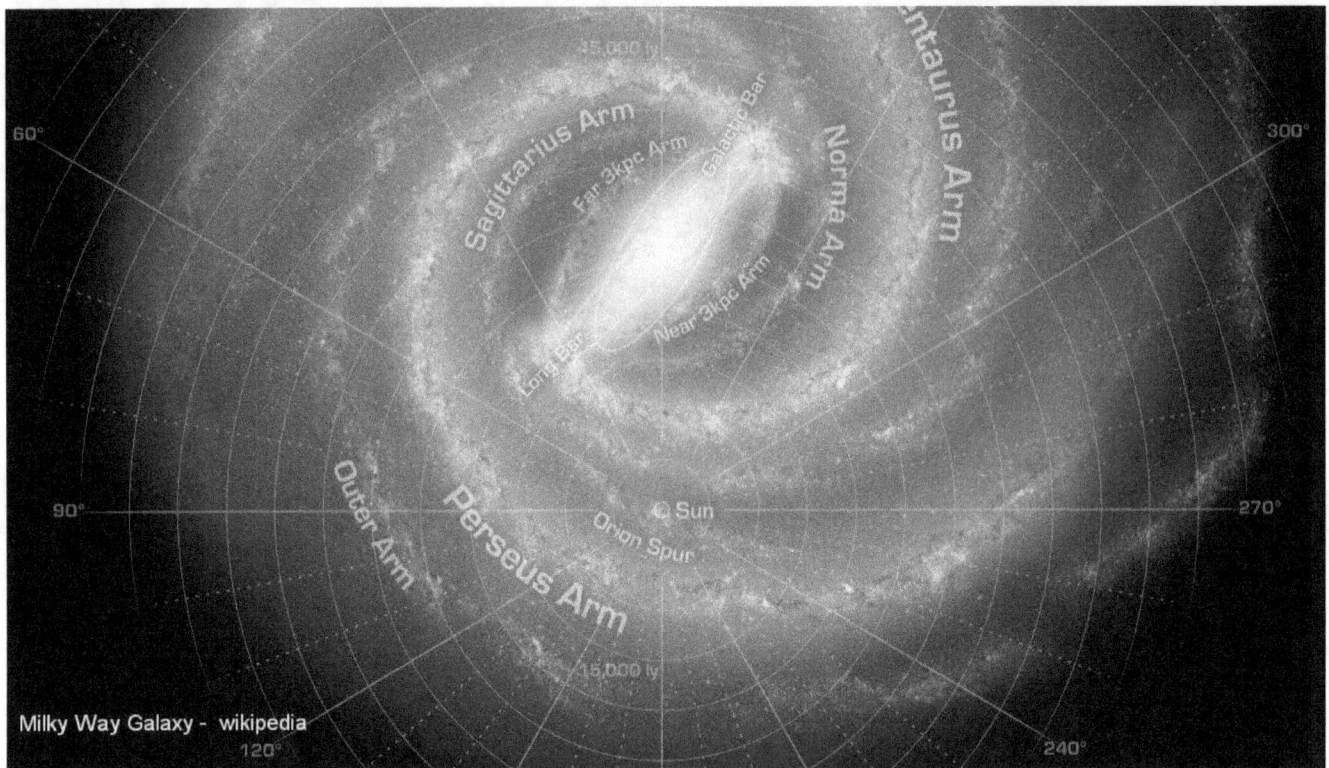

Image: The Sun located in the Milky Way

Part 1: The Sun is NOT a gas sphere. The Sun is plasma.

Image: The x-ray Sun

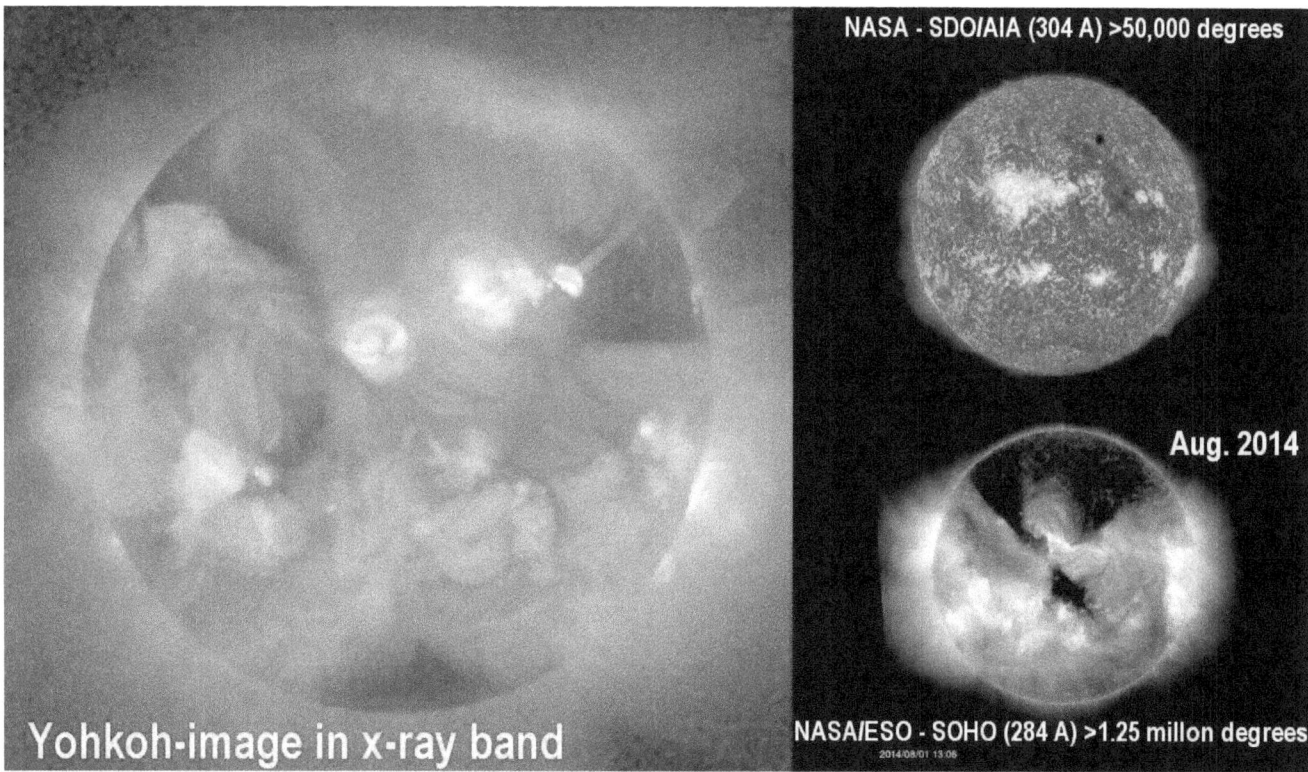

Image: The x-ray Sun

Part 1: The Sun is NOT a gas sphere. The Sun is plasma.

Image: Star-size comparison

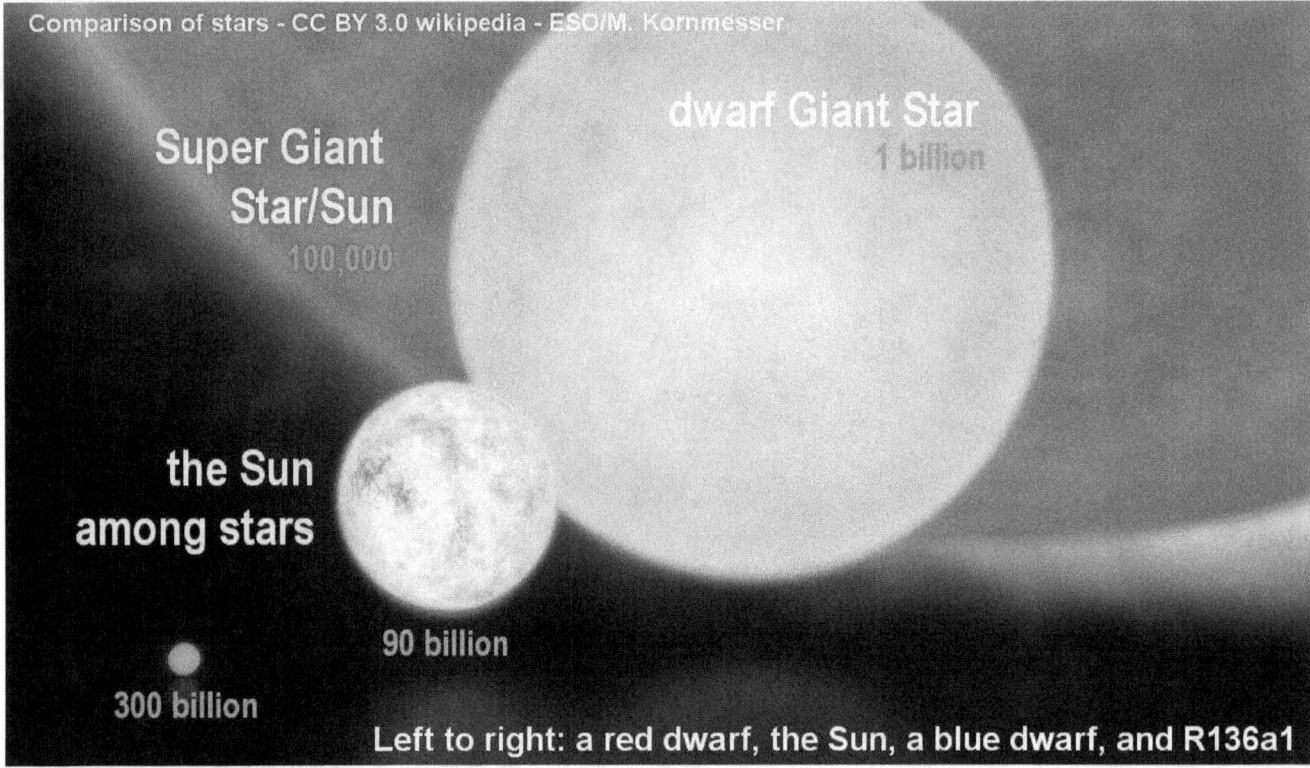

Image: Star-size comparison

The Sun cannot exist

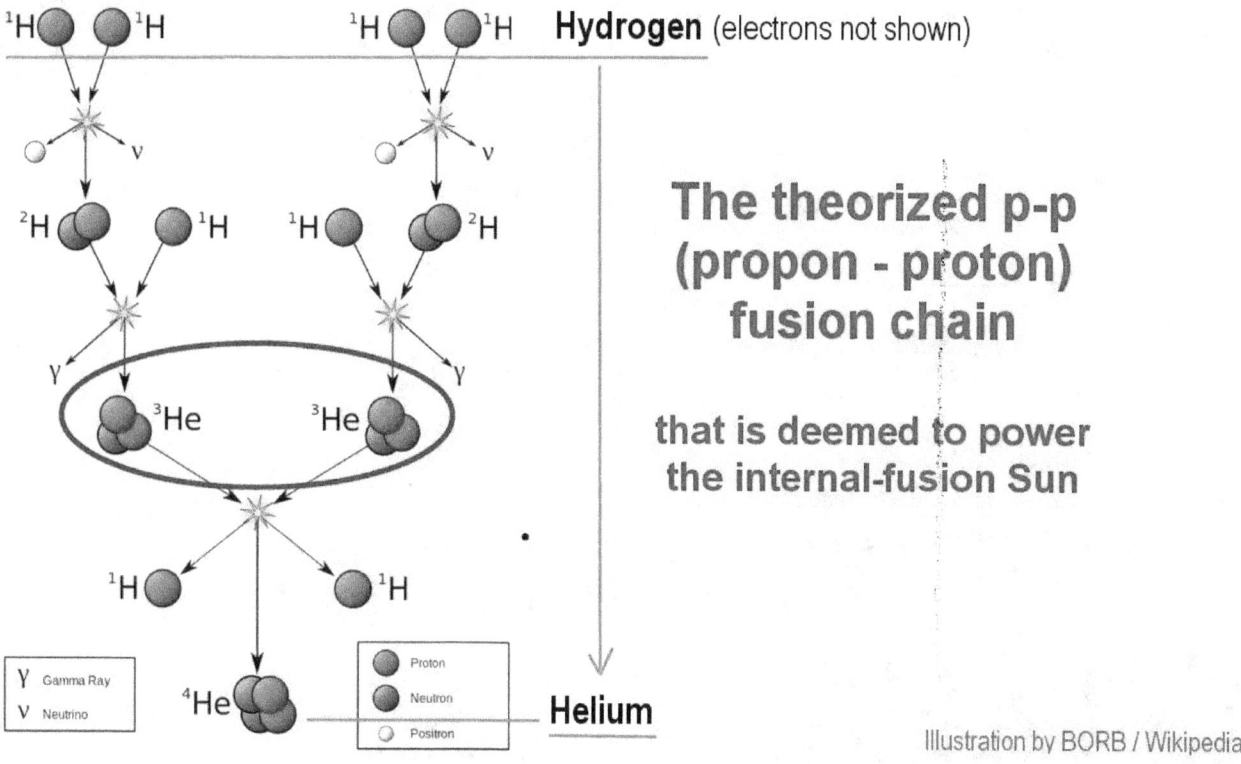

In real terms, the conditions for nuclear fusion to occur deep inside a sun, that force hydrogen atoms to fuse into helium atoms, which the theory is based on, do not exist at all. A sphere of hydrogen gas of the size of the Sun cannot exist. Its weight would crush all atoms in its core, much less enable the building of bigger ones. In addition, the theorized helium-3 fusion-stage is the hardest fusion to achieve, because of the much larger Coulomb Barrier of helium-3 that needs to be overcome for such proton-heavy atoms to fuse. In a NASA related helium-3 fusion experiment, it took a million times greater energy input to force the fusion than the energy that the fusion had generated. A sun cannot operate on this basis.

In real terms, there is no need for an internal nuclear-fusion power process to happen, for a sun to radiate light and heat. The real solar nuclear fusion process is much simpler; more certain; and more powerful; and in addition the fusion occurs right on the surface of the Sun, where it counts.

The solar-wind particles are plasma

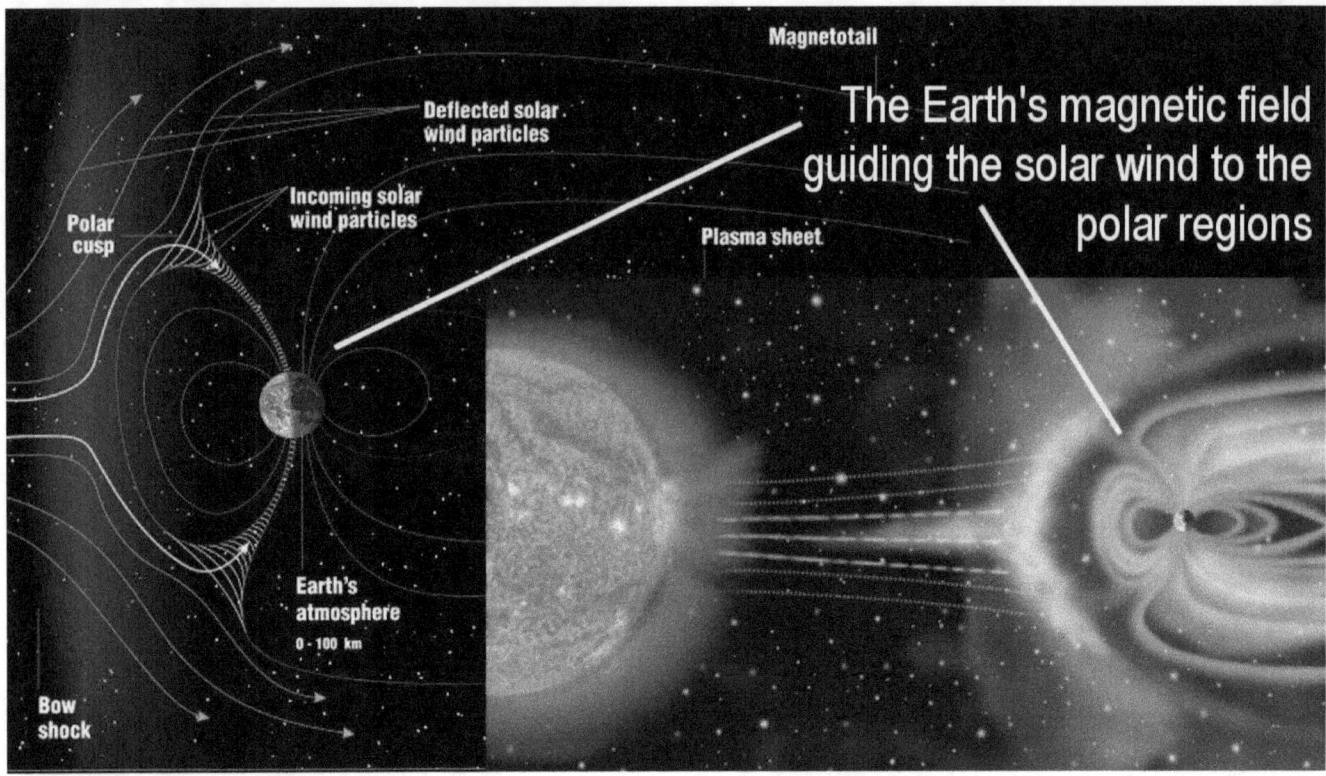

The solar-wind particles are plasma. They carry an electric charge by which they interact electrically with the Earth's magnetic fields. By this interaction, they are guided magnetically to the polar regions where they encounter the atmosphere and create their highly visible light show.

Guides interstellar plasma onto our Sun

The same type of process, only on a much larger scale, also guides interstellar plasma onto our Sun itself.

In the case of the Sun

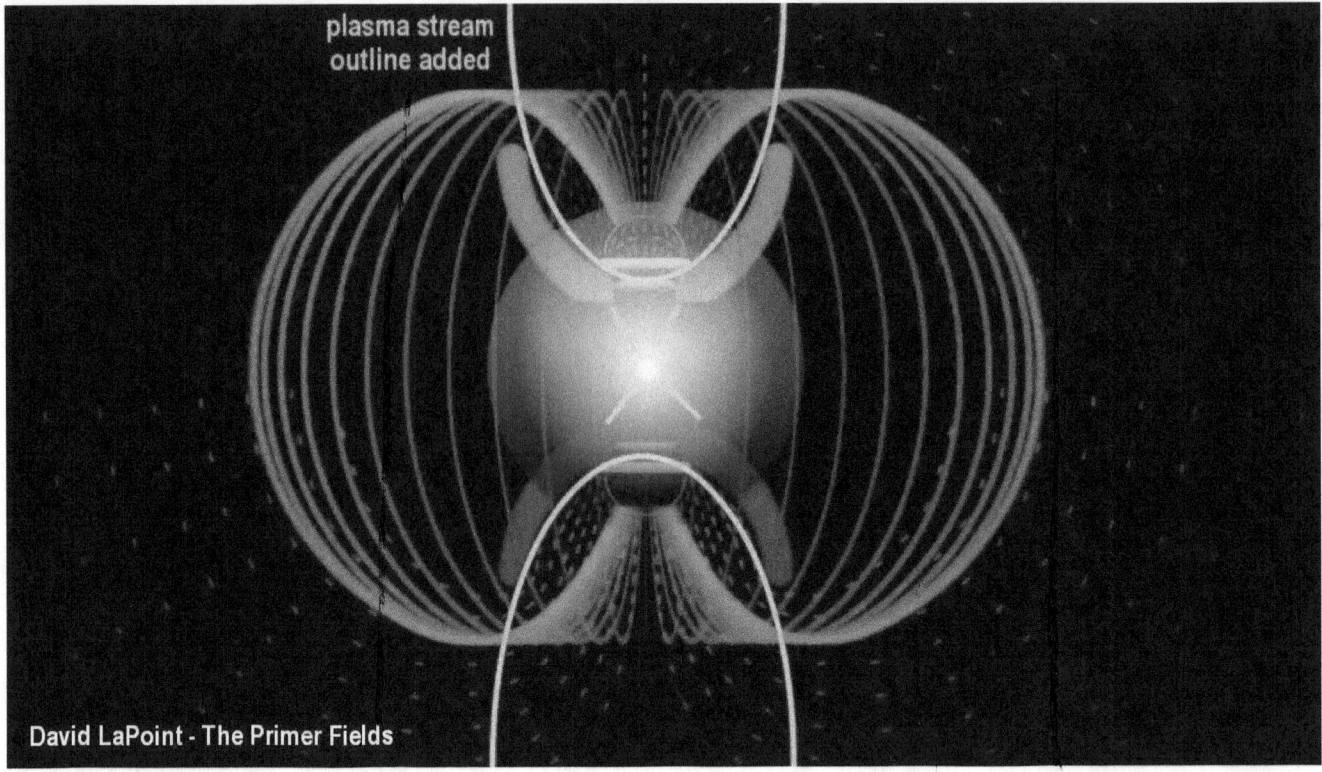

The difference is, that in the case of the Sun, the plasma comes from a far more-distant source. It is drawn to the Sun from interstellar space, in the form of long-distance plasma streams.

Plasma is electricity

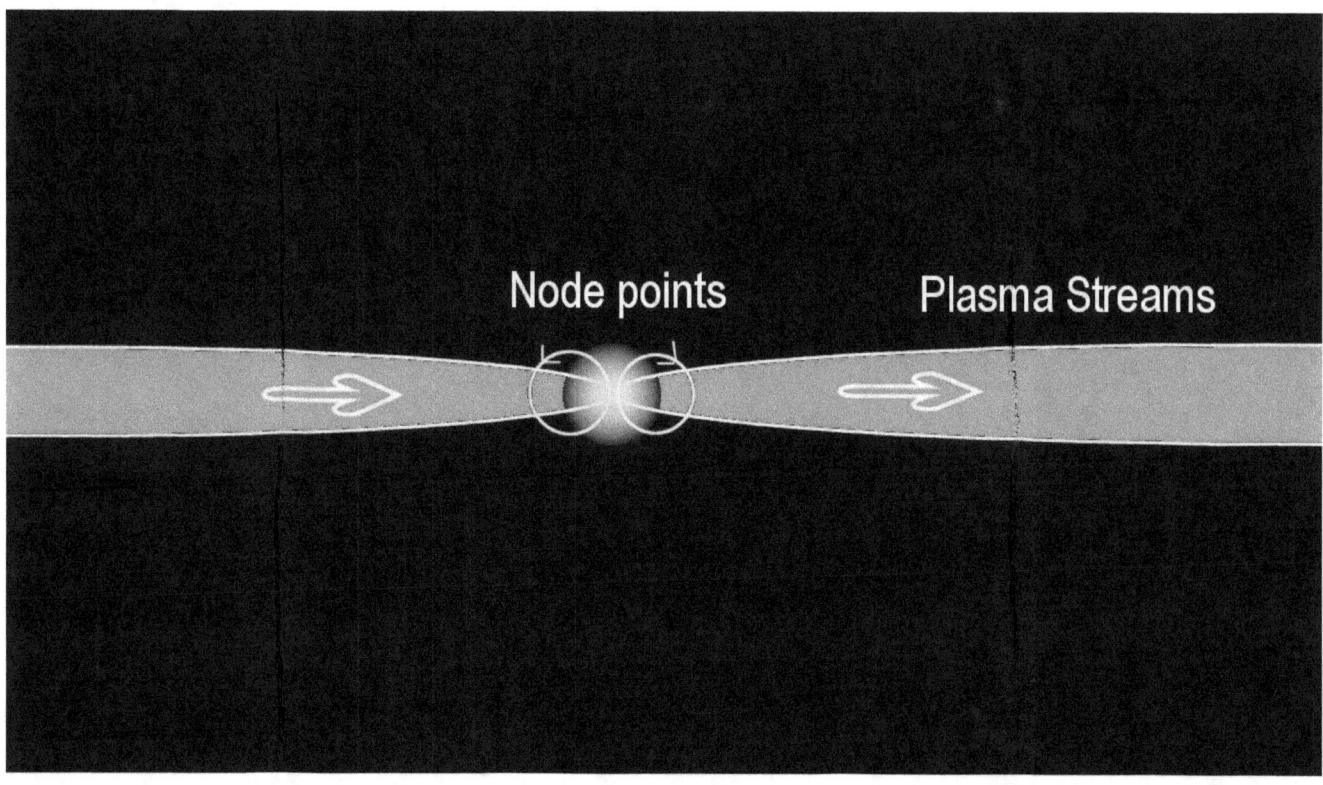

Since plasma is electricity, the flowing plasma currents create their own magnetic fields around them that pinch the flowing plasma into ever tighter confinement. The resulting concentrated magnetic fields, in turn, amplify the magnetic field of the Sun with their own magnetic structure.

The Sun becomes surrounded

The result is that the Sun becomes surrounded with a densely concentrated sphere of plasma. The plasma becomes so dense that the resulting 'plasma light-show' extends right around the Sun.

Part 1: The Sun is NOT a gas sphere. The Sun is plasma.

Fusion reactions on the surface of the Sun

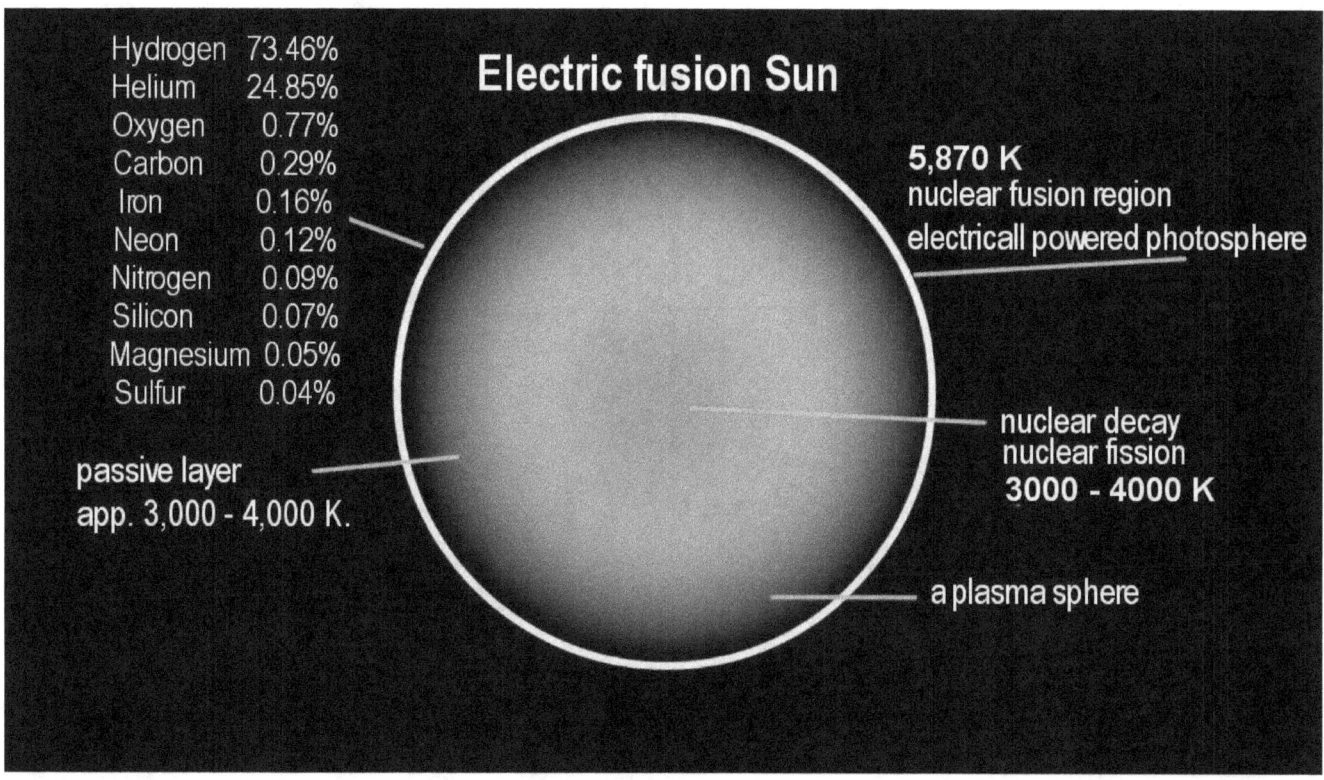

When plasma is motivated by the electric force, which is 39 orders of magnitude stronger than mass and gravity, chains of highly efficient nuclear fusion reactions occur. They occur right on the surface of the Sun, where the plasma meets the Sun's reaction cells. In the fusion reactions on the surface of the Sun, all the atomic elements are synthesized that the planets are made of. The synthesized atoms then flow away with the solar wind. During the early phase of the Sun, all the atoms for the planets were created and carried in the solar winds, till they fell out and condensed into planets.

This in short, is how the solar system was created near the center of the galaxy, and how it still functions fundamentally, though with lesser intensity.

Operational differences between the Sun and the Earth

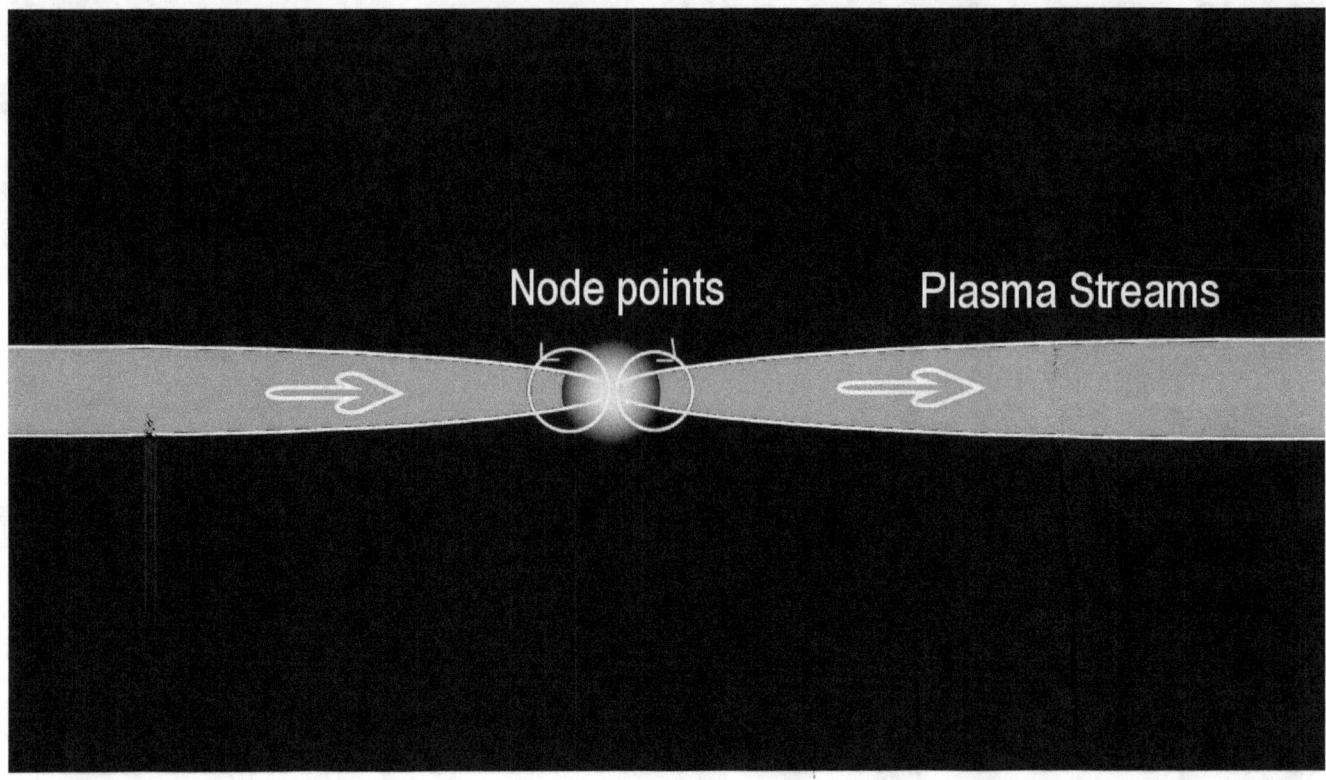

Of course, we also encounter some operational differences between the Sun and the Earth. For the intense plasma compression around the Sun to happen, which enables surface nuclear fusion to occur, large plasma streams are required to achieve the compression. They are needed to produce the necessary strong magnetic fields. Massive electric movement creates strong magnetic fields,

Plasma streams that connect stars

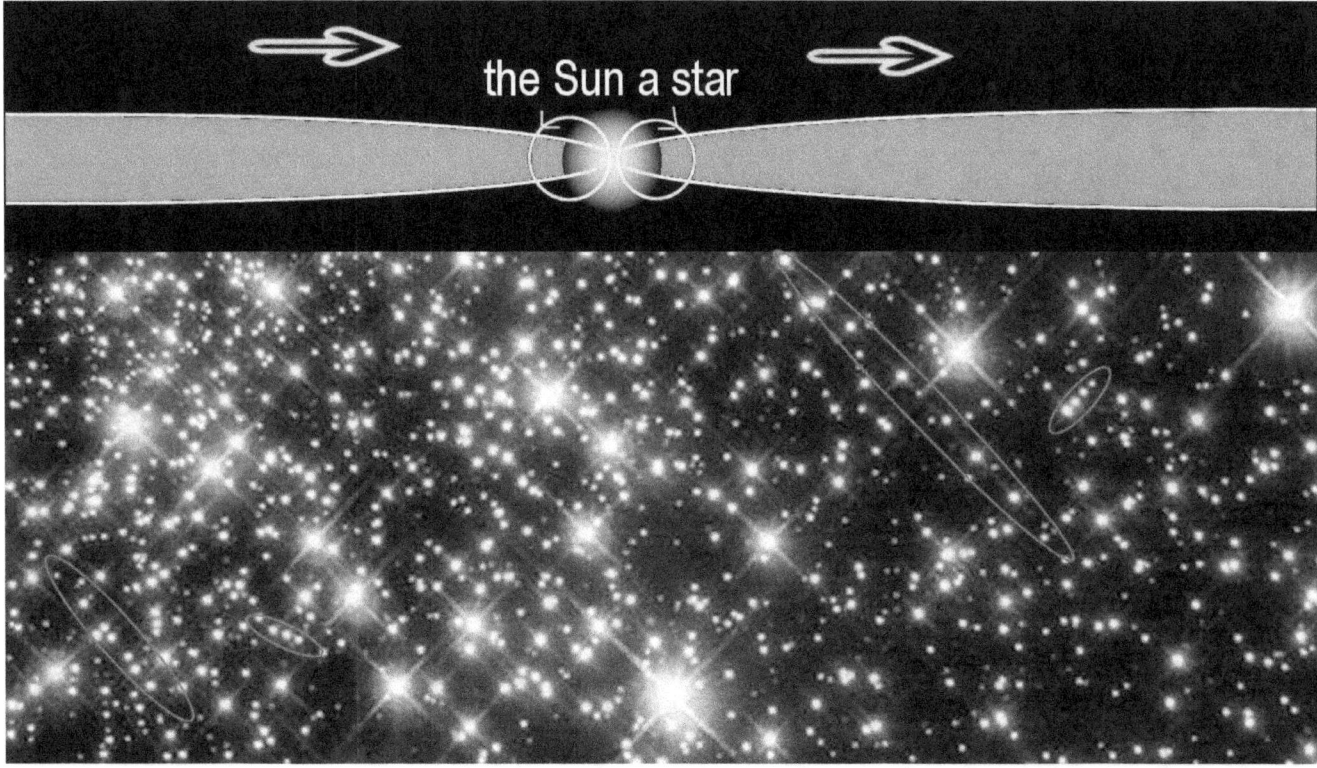

The interstellar plasma streams that connect stars in the galaxy, typically carry a greater volume of plasma than a Sun can consume with its nuclear-fusion synthesis. The excess plasma then simply flows on and away from the Sun in an outgoing plasma stream.

While the streams flowing towards the Sun become magnetically pinched into ever-tighter confinement, the weaker streams flowing away from the Sun, begin tightly, and then expand. They typically pick up plasma along the way. They continue to expand till the density becomes large enough again for them to contract by magnetic pinching as they flow towards the next star, at the next node point.

The evidence for this dynamic interconnection between the stars is found in the typical alignment of stars into short and long strings. This string-like alignment is also visible in the fields of galaxies.

An internally powered sun is not needed

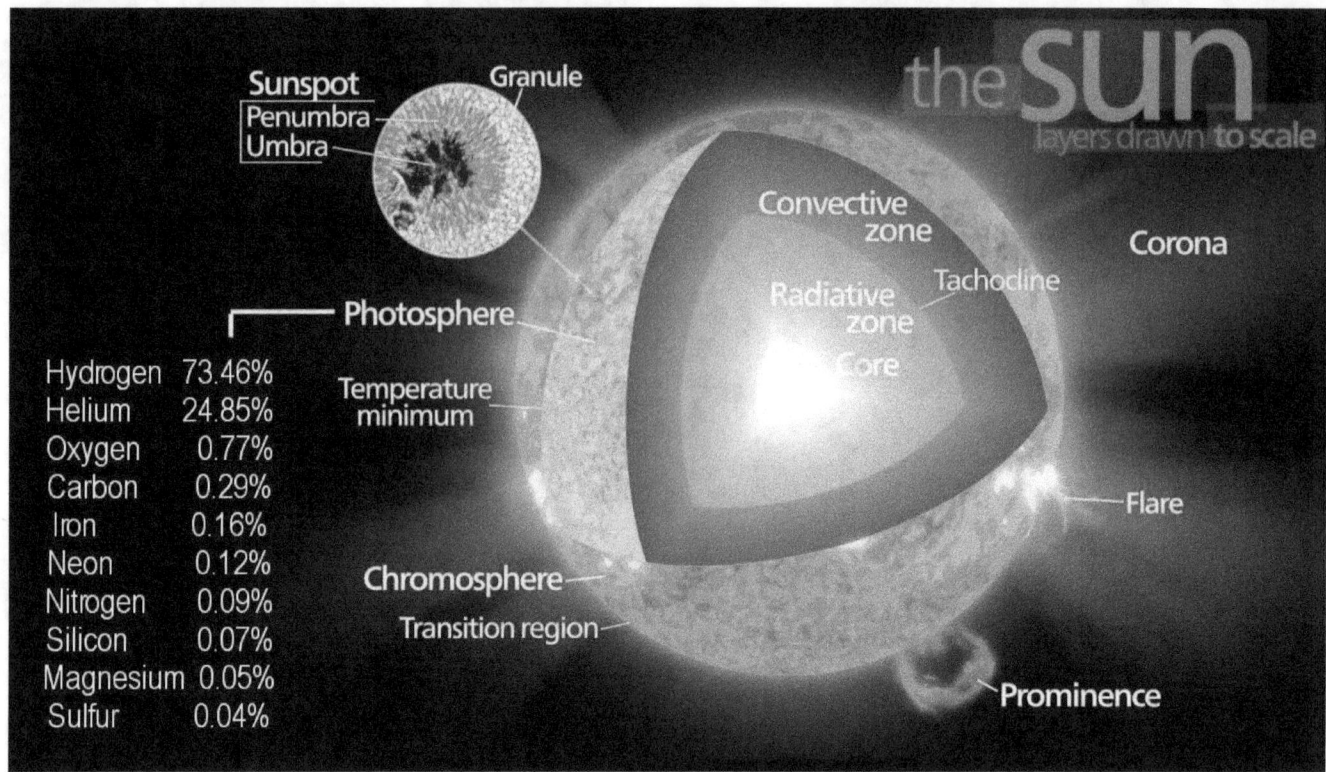

All this means that an internally powered sun is not needed. In fact, the internally powered Sun cannot perform the services for the universe that its dynamics require. The theory of the internally powered Sun simply doesn't work. It is so full of holes that it can't possibly work. Of course, it will likely take some time before the defective model of the Sun will be let go in the world of science and in public perception. Still, this needs to happen. The breakout to reality is urgently needed.

Part 1: The Sun is NOT a gas sphere. The Sun is plasma.

Why is the Sun not a sphere of hydrogen gas?

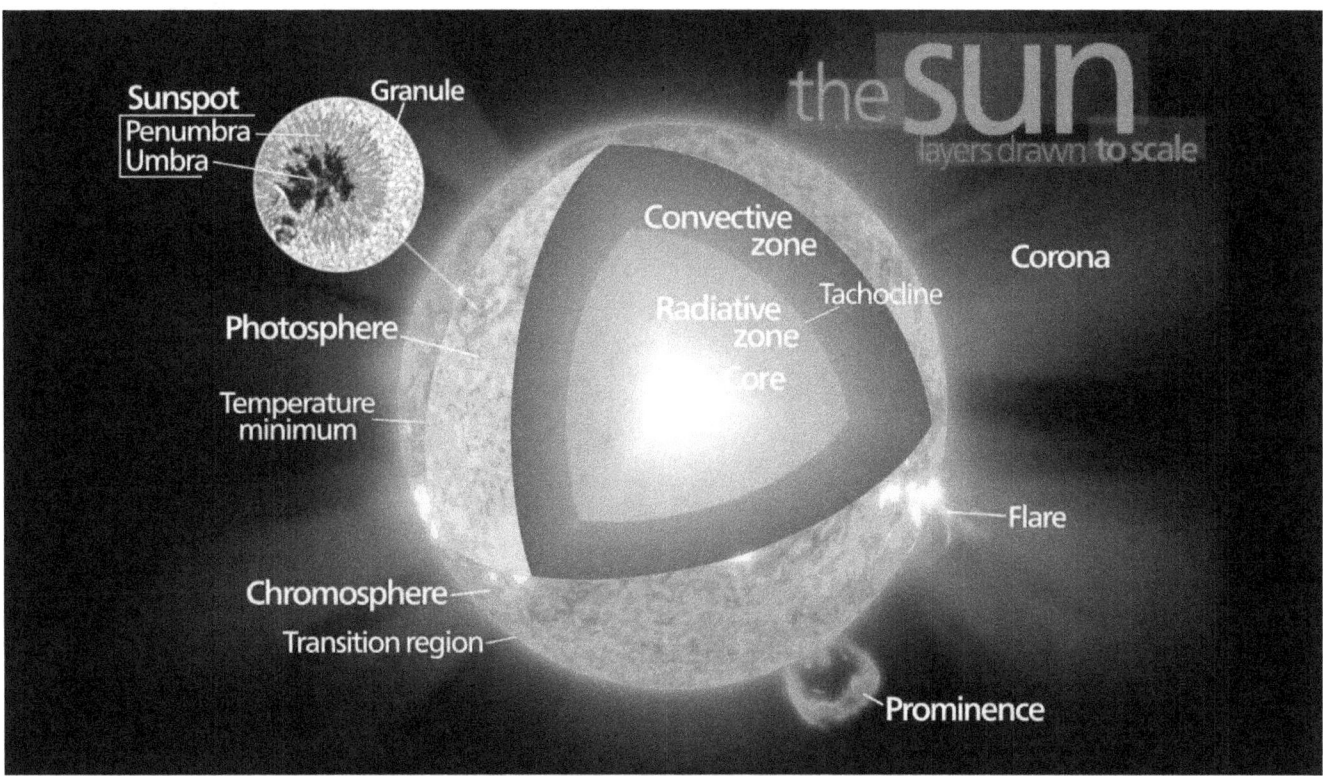

So what is wrong with the old theory about our Sun? Why is the Sun not a sphere of hydrogen gas? What is the generally accepted theory saying to us, that is wrong?

The general perception is that the Sun is a sphere of atomic gas, mostly hydrogen and some helium, and that within its core, at a calculated temperature of 15 million degrees Kelvin, at a gas density 150 times the density of water, nuclear fusion reactions do occur that generate vast amounts of energy by which the Sun becomes a brilliantly radiant sphere in the sky that has burnt for billions of years with unchanging intensity, and will keep on burning for a few more billion years to come.

The Sun is theorized to fuse hydrogen atoms together

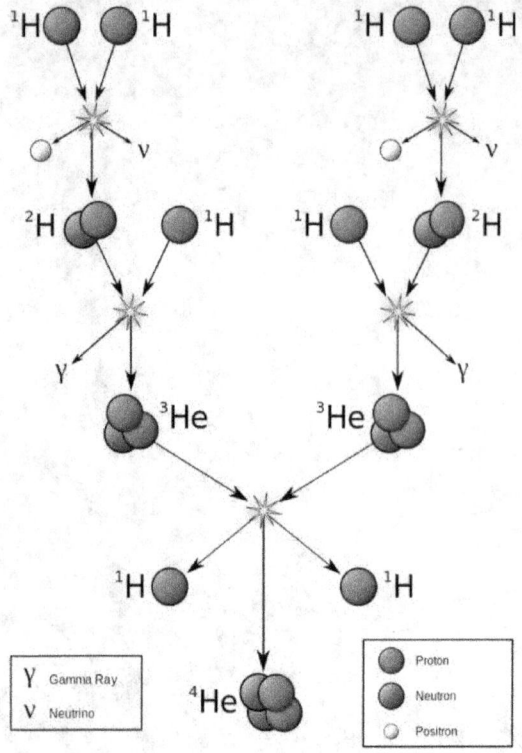

The theorized p-p (propon - proton) fusion chain

that is deemed to power the internal-fusion Sun

Illustration by BORB / Wikipedia

The Sun is theorized to fuse hydrogen atoms together in a chain of reaction by which helium atoms are produced and some energy. The produced energy-flux by mass is deemed to be equal to the energy generated by the metabolism of a human being. The extreme energy of the Sun is deemed to be the result of energy accumulation derived from its great volume.

That's the prevailing theory.

The Sun is anything but a constant star

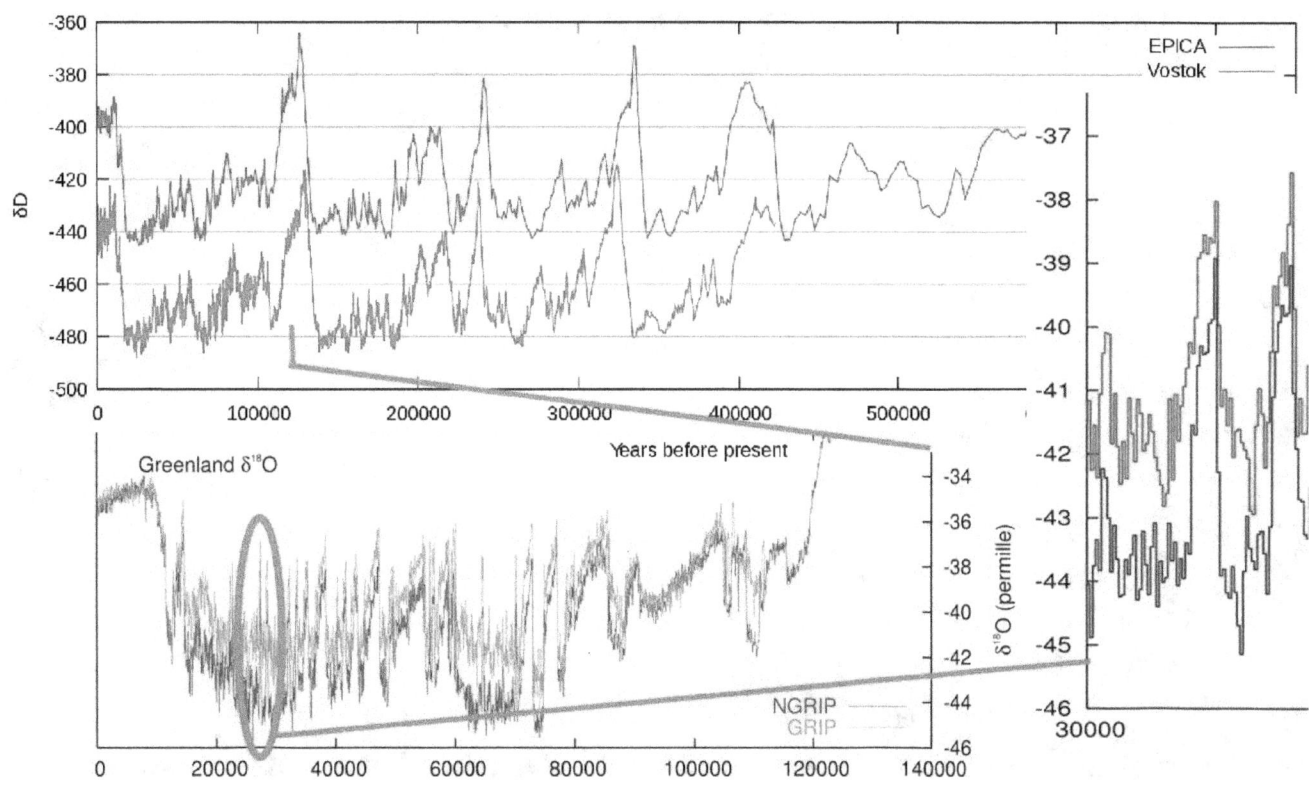

But the theory doesn't add up, does it? The long-term climate history that has been discovered, the big ice age cycles that we have evidence of, and even the ice core records of the last Ice Age with its gigantic short-term temperature oscillations that are evidenced in them, all tell us that the Sun is anything but a constant star. The the internal solar fusion theory does not allow for the enormous short and long-term fluctuations that we have evidence of.

The heat generated at the core of the Sun

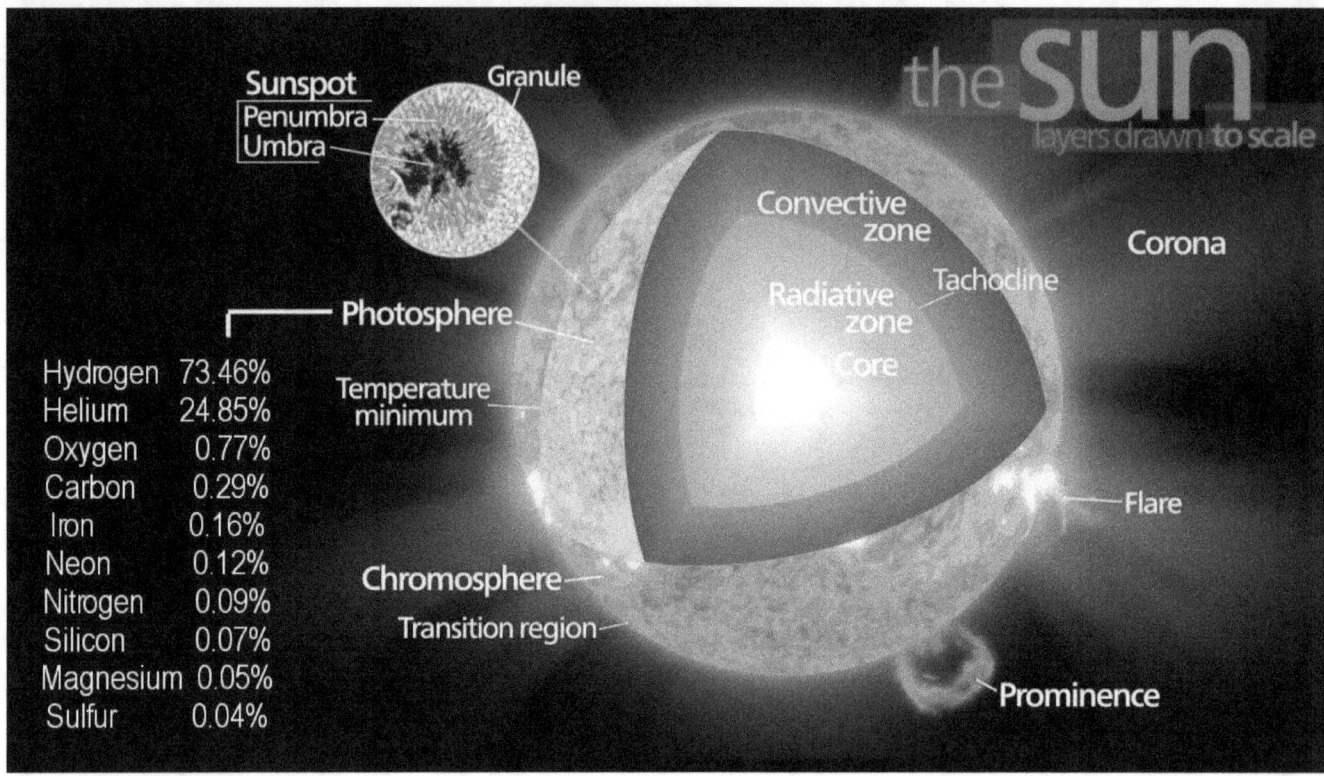

The theory tells us that the heat generated at the core of the Sun, by nuclear fusion, takes a path of 30 million years to slowly ooze to the surface by convection, and that even the heat transmission by photons takes 10,000 to 170,000 years to reach the solar surface. By this slow process, the Sun should be rock-solid, unvarying. But it isn't.

The Sun's energy cycle is oscillating

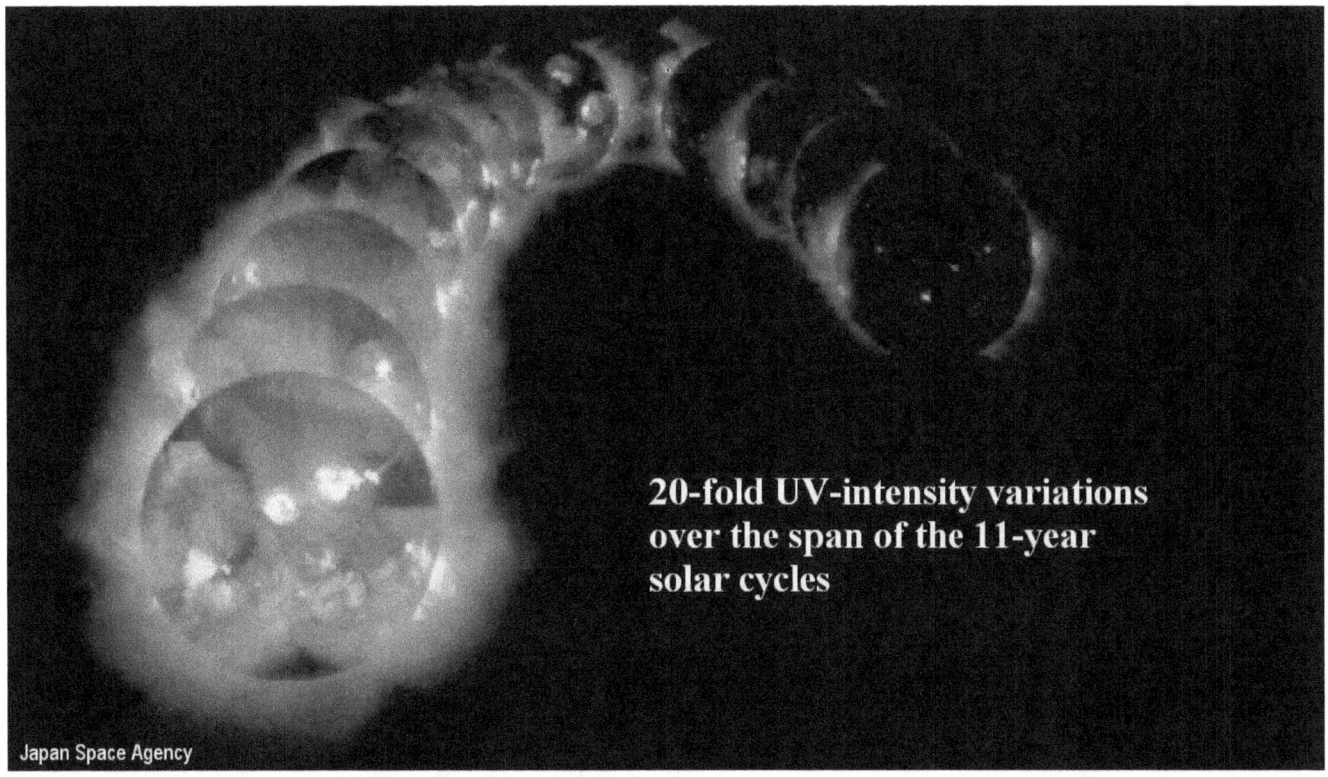

20-fold UV-intensity variations over the span of the 11-year solar cycles

Even now, the Sun's energy cycle is oscillating at an 11-year beat. Although its visible light output remains presently steady within seven-tenth of a percent, in other parts of the spectrum its energy radiation, varies by a factor of 20, that's a two thousand percent difference.

There are so many items of physical evidence coming to the surface that shouldn't happen under the internal-fusion theory, but which do happen, that one is forced to conclude that the widely accepted theory is evidently wrong.

It shouldn't be possible for the solar wind

- an example of the amazing solar eclipse photography of Milloslav Druckmueller
http://www.zam.fme.vutbr.cz/~druck/Eclipse/

For example, it shouldn't be possible for the solar wind to accelerate against the force of gravity, as the wind flows away from the Sun.

The corona around the Sun is hotter than the Sun

by Luc Viatour / www.Lucnix.be

Likewise, it shouldn't be possible that the corona around the Sun is hundreds of times hotter than the Sun itself. However, these impossibilities all happen. The theoretical impossibilities of self-evident facts, create paradoxes.

When we look at the sunspots on the Sun

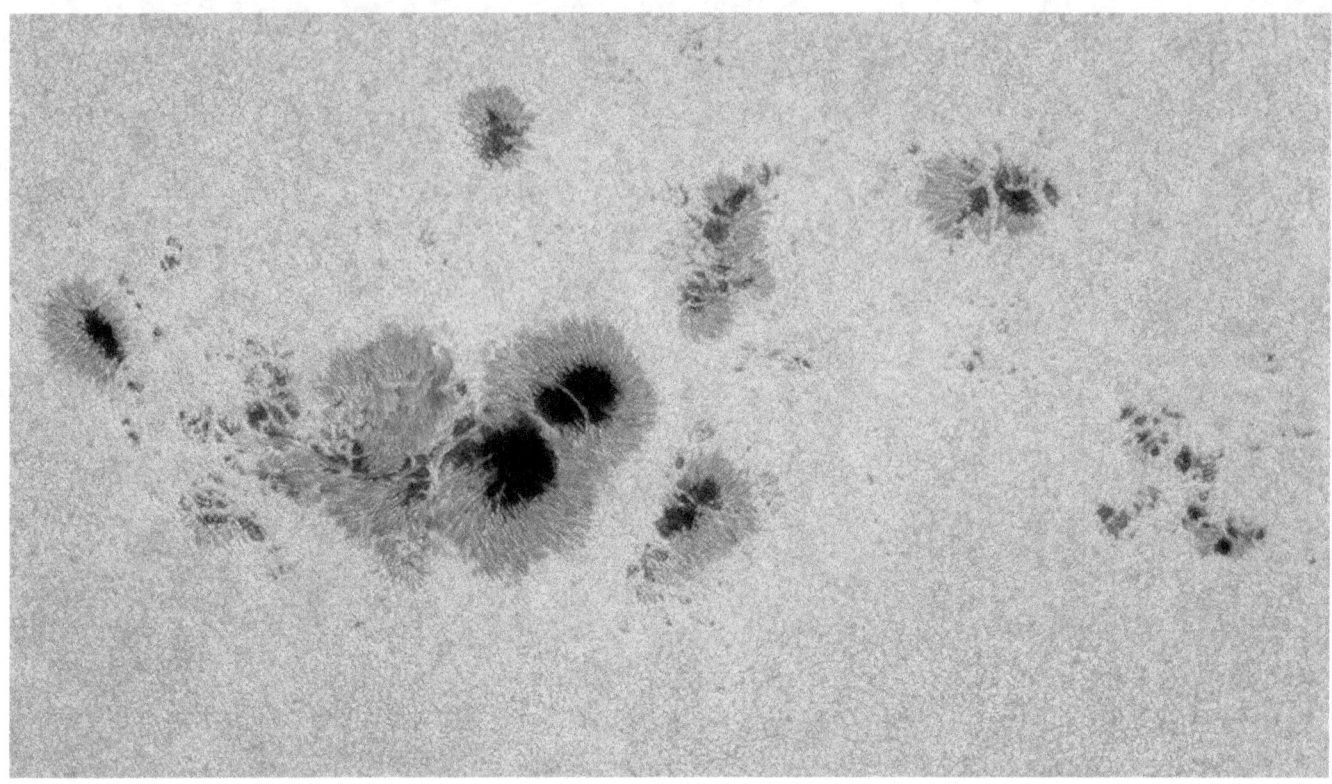

Neither should it be possible that when we look at the sunspots on the Sun, and look through the umbra below the surface, that the Sun is dark inside instead of being brilliantly radiant. However, as you can see for yourself, below its shiny skin, the Sun is dark.

Is this a paradox? No, it isn't. What we see is precisely what the Sun should be like, because nothing else is possible.

Impossible for the Sun to be a gas sphere

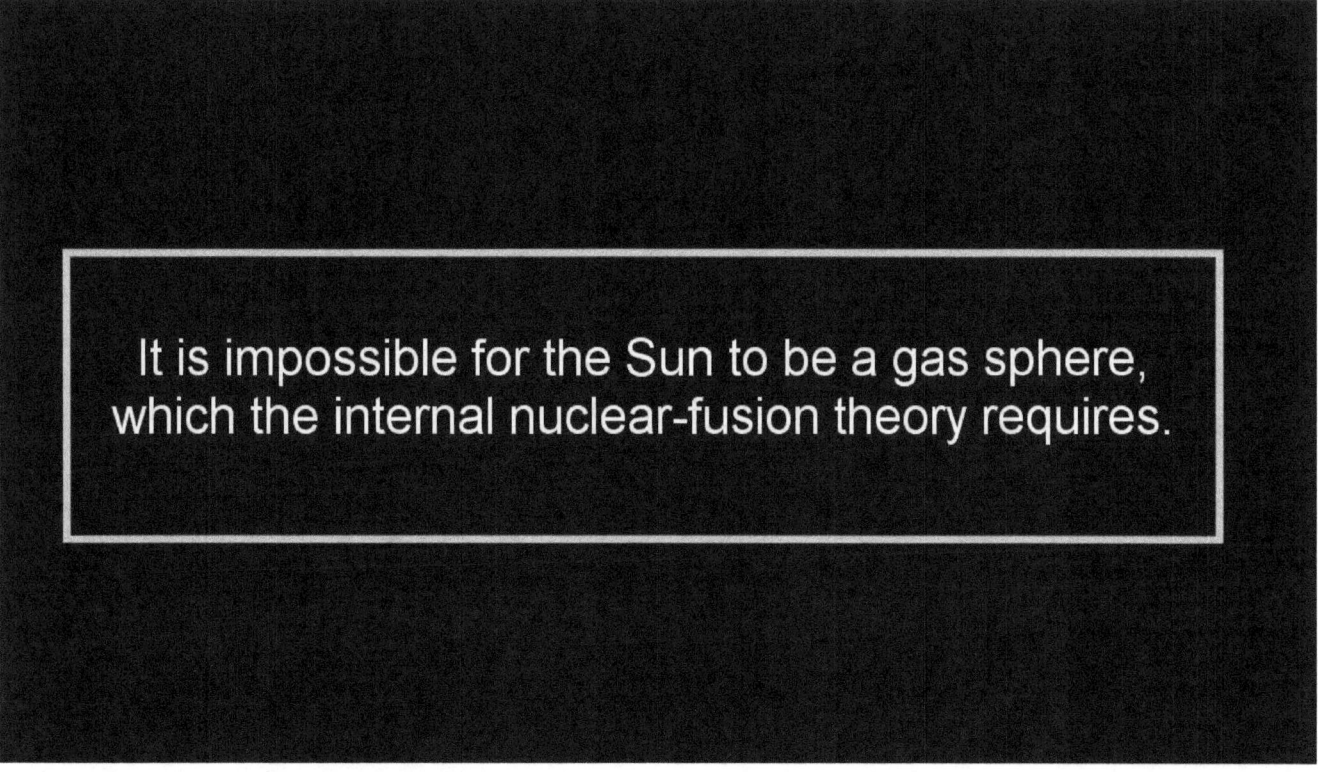

It is impossible for the Sun to be a gas sphere,

which the internal nuclear-fusion theory requires.

The mass-density of the Sun

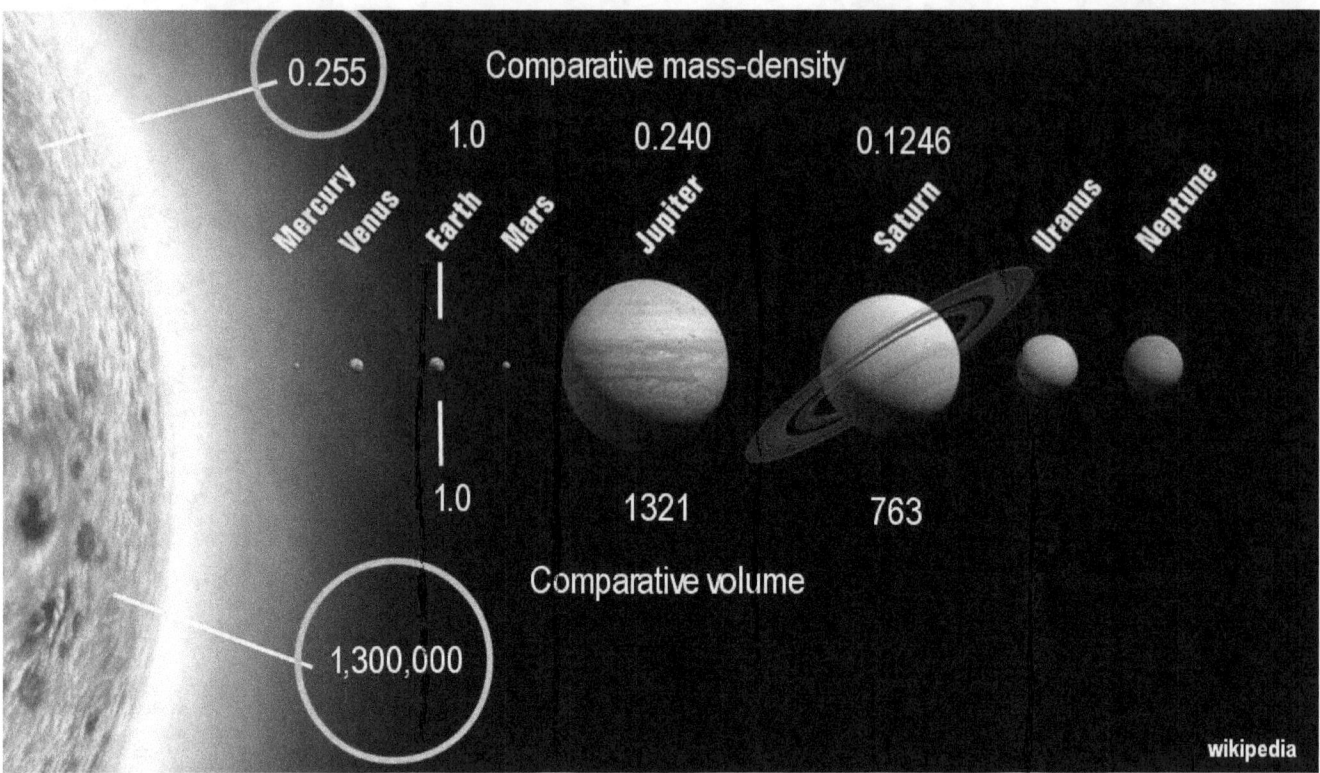

By what is required for the standard solar theory to function, the Sun should not exist. It is simply not possible for a gas-sphere of the size of the Sun, to exist. The gravity pressure in a gas sphere of such an immense size would be so great that all atoms in its core would be crushed, instead of atoms being fused into larger atoms. A gas sphere of the size of the Sun would also be a thousand times heavier than the Sun is known to be. No rational theory, no matter how exotic, can bridge this impossible paradox.

Just do some simple comparison. The gas planet Jupiter, at double the volume of Saturn, has double the mass density of Saturn. The doubling of the mass-density results from the greater compression of the gas by the greater gravity, which the larger gas-volume generates. This is what one would expect.

The Sun, in the same comparison, has a thousand times the volume of Jupiter, but it is known to have roughly the same mass-density as Jupiter. That's not possible.

The mass-density of the Sun would be more than a thousand times greater if it was a gas sphere. However, if the Sun was a plasma sphere, which is diffused by the repelling electric force that is inherent in plasma, the Sun's 'measured' mass-density is just about right.

The resulting gigantic paradox renders the internal fusion Sun theory, which depends on atomic hydrogen being fused into helium, to be totally wrong.

Great efforts have been made to explain the paradox away with exotic excuses, in order to rescue the false theory for which no evidence exists. Unfortunately, the process of shrouding paradoxes with exotic epicycles, for which no evidence exists either, is like saying to society, "we really don't know how the thing works. We are guessing. The paradoxical theory, impossible as it is, is the best we can come up with."

The premise that 99.999% of the universe does not exist

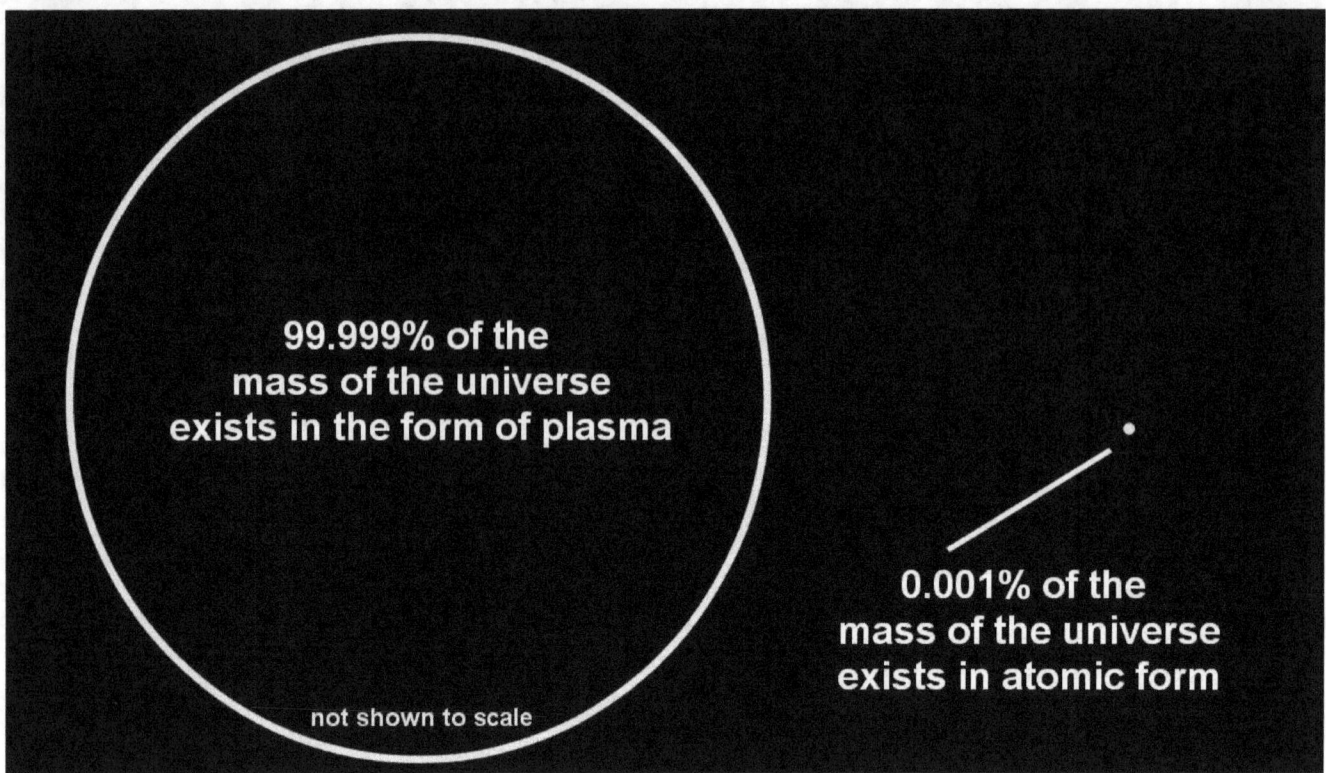

Of course, the paradoxical theory is the best theory possible when one is forced to proceed from the premise that 99.999% of the universe does not exist, so that the theory of the functioning of the Sun must be built on the one-thousandth of a percent of the universe that is allowed to be acknowledged in general perception. The result is necessarily, a hopelessly narrow perception, the kind that is characteristic of fairy tales that are told to children at bed time to put them asleep.

Plasma is the blood of the universe

Researchers at the Los Alamos National Laboratory have come to the conclusion that the Universe is not as empty as the fairy-tale scripts make it out to be.

The researchers have come to the recognition that 99.999% of the mass of the universe - which is the portion that the fairy tales do not include - does indeed exist, and exists in the form of plasma.

Plasma is the blood of the universe. It consist of the basic particles that all atoms in the universe are made of. The particles also exists in free-flowing form with a ratio way above 10,000 to 1.

While free-flowing plasma in space is invisible, as its particles are 100,000 times smaller than the smallest atom, the existence of giant plasma streams in cosmic space is discernable by their effects.

The effects of the plasma streams in space

For example, in cosmic space, we see the effects of the plasma streams in space in the alignment of stars and galaxies, often lined up in neat rows like beads on a string.

Secondary effects of flowing plasma

In the solar system we see the secondary effects of flowing plasma in the form of planets orbiting in an ecliptic around the Sun.

Plasma streams as solar wind

http://www.zam.fme.vutbr.cz/~druck/Eclipse/ - an example of the amazing solar eclipse photography of Milloslav Druckmueller

We also see the effect of plasma streams as solar wind, which is in part why plasma is called the lifeblood of the universe. Plasma is the lifeblood of the Sun. It enables electric nuclear fusion at its 'surface' layer. The solar wind is the result of it.

The Sun is a plasma star

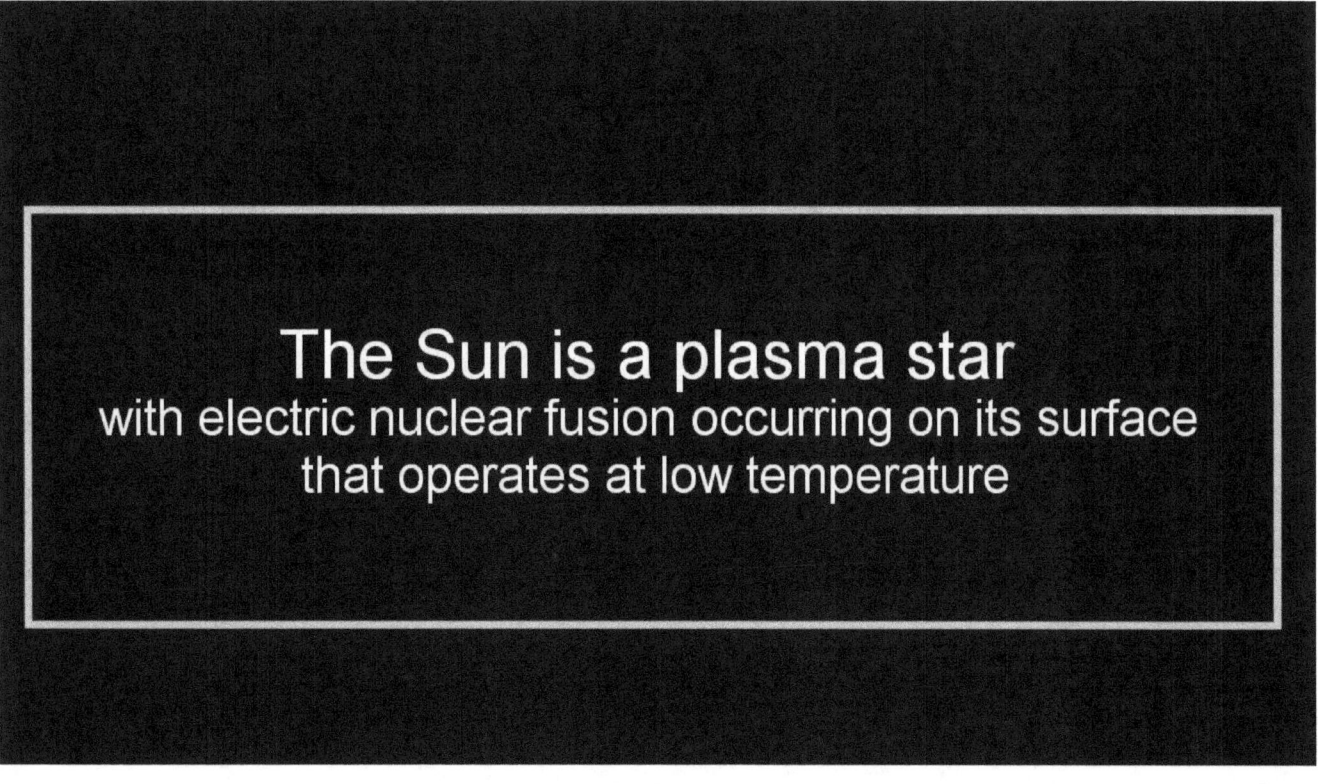

The Sun is a plasma star

with electric nuclear fusion occurring on its surface that operates at low temperature. No other types of solar energy are physically possible, nor would there be a need for other types, since electric plasma fusion on the surface of the Sun is the most efficient type of nuclear fusion possible. It is so efficient that it operates at such low temperature as the Sun's current 5,505 degrees Celsius.

The principle is efficient

The principle is efficient because its process is driven by electric interaction - by the interaction of one of the strongest forces in the universe.

Two main types of plasma particles

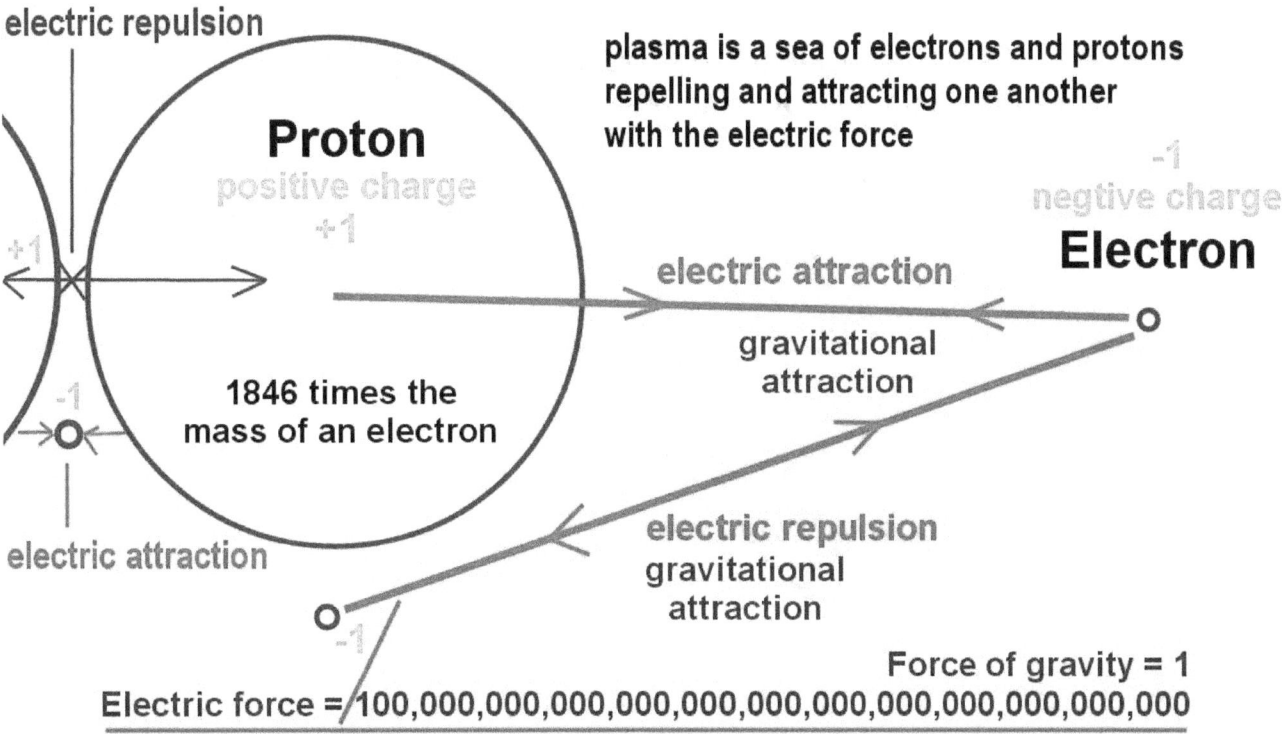

Plasma exists in the form of extremely small energized particles that are deemed to be themselves but constructs, of constructs of energy. They carry both a quantity of mass, and a specific electric charge.

There are two main types of plasma particles recognized, a large type and a small type. The small type is named an electron. It carries a negative electric charge. The large type is named a proton. It is a thousand times larger than the electron and has nearly 2,000 times more mass, and it carries a positive electric charge.

As particles of mass, plasma particles gravitate towards one another by the effect of their mass, called gravity. The particles also affect one-another by the electric force. Particles that are of complimentary polarity, meaning opposite polarity, such as protons and electrons, attract one another by the electric force that is 39 orders of magnitude stronger than the force of gravity. Unlike, by the effect of gravity, plasma particles do also repel one-another by the electric force. This happens between particles of like polarity.

Electrons rebound

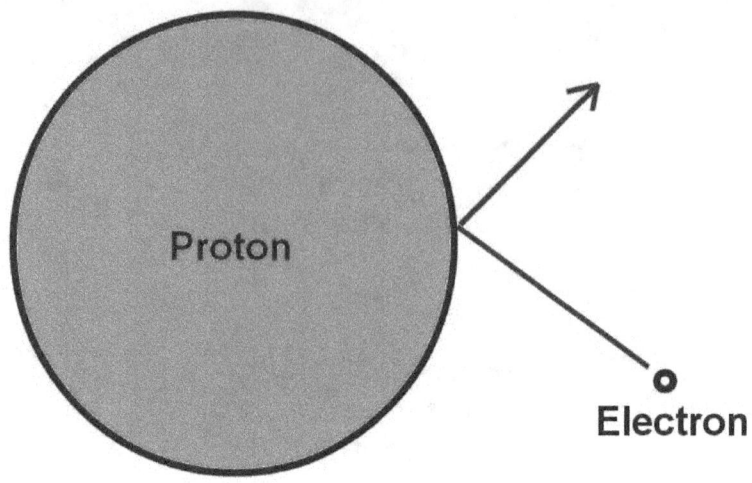

**electrons (-1) are attracted to protons (+1) by their unequal polarity
on contact, the electron is forced to rebound, only to be attracted anew**

However, the opposite happens when the attracting particles come extremely close to one another. The reason is, that the universe would not function without this additional principle.

For example, if electrons were merely attracted to protons, they would latch onto one-another. As a result, nothing would move. The electrons would simply remain stuck there. Nothing useful would result. But this doesn't happen. Nowhere in the universe can we find electrons latched onto protons. Before they would latch on, the electrons rebound.

Instead of getting stuck by being latched onto protons, at close distances, the electrons bounce away from the protons, like a rubber ball rebounds when it hits a wall. By this rebounding effect, electrons become drawn into an endless dance around the protons in plasma. They are attracted from afar, and repelled at close distances, only to become attracted anew.

A density determined zone

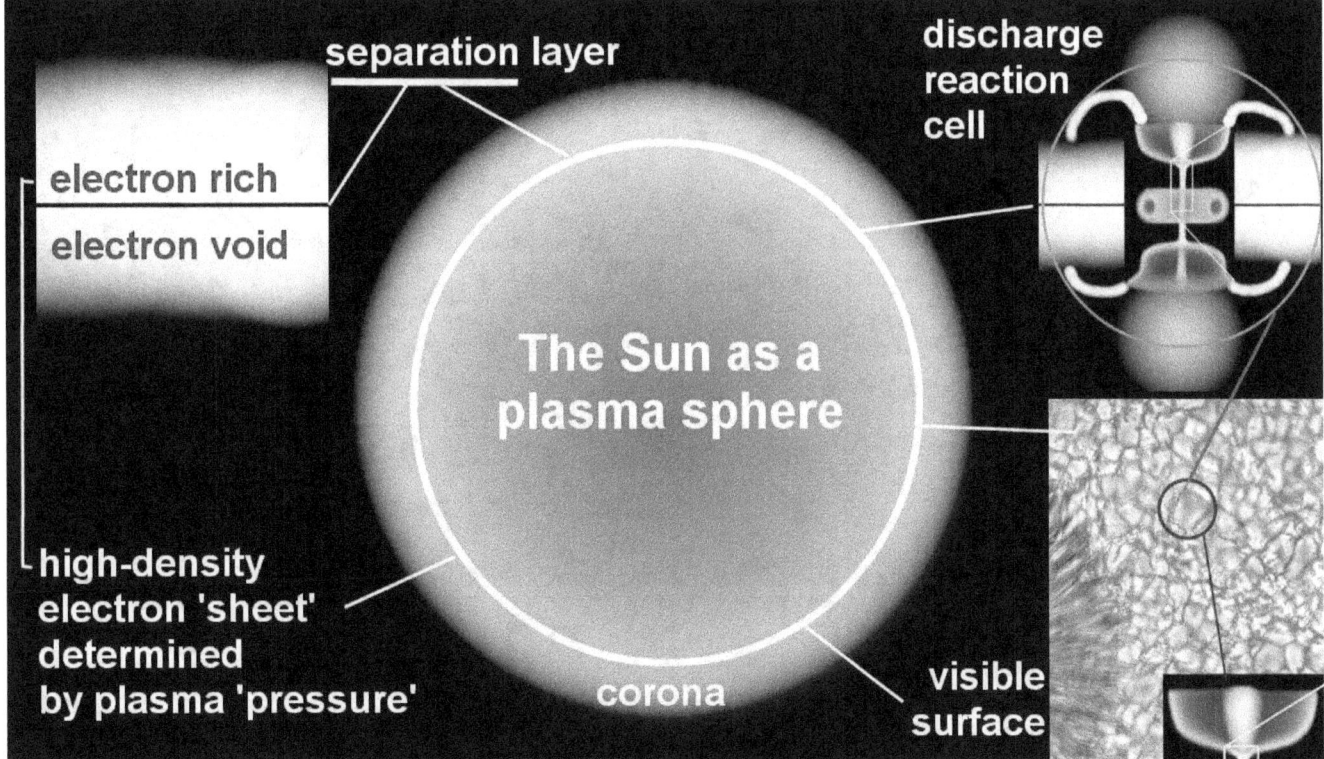

The same principle, may on the large scale, cause a repulsive barrier to form against the electron-dance, past a specific plasma density, which repels electrons out of the dense plasma region, somewhat like oil floating on water. This means that the visible surface of the Sun that we see, isn't actually a surface in the standard sense, but is merely a density determined zone where all the plasma interactions take place, including the fusion reactions that require a high electron-density.

A creative atom synthesizing 'engine'

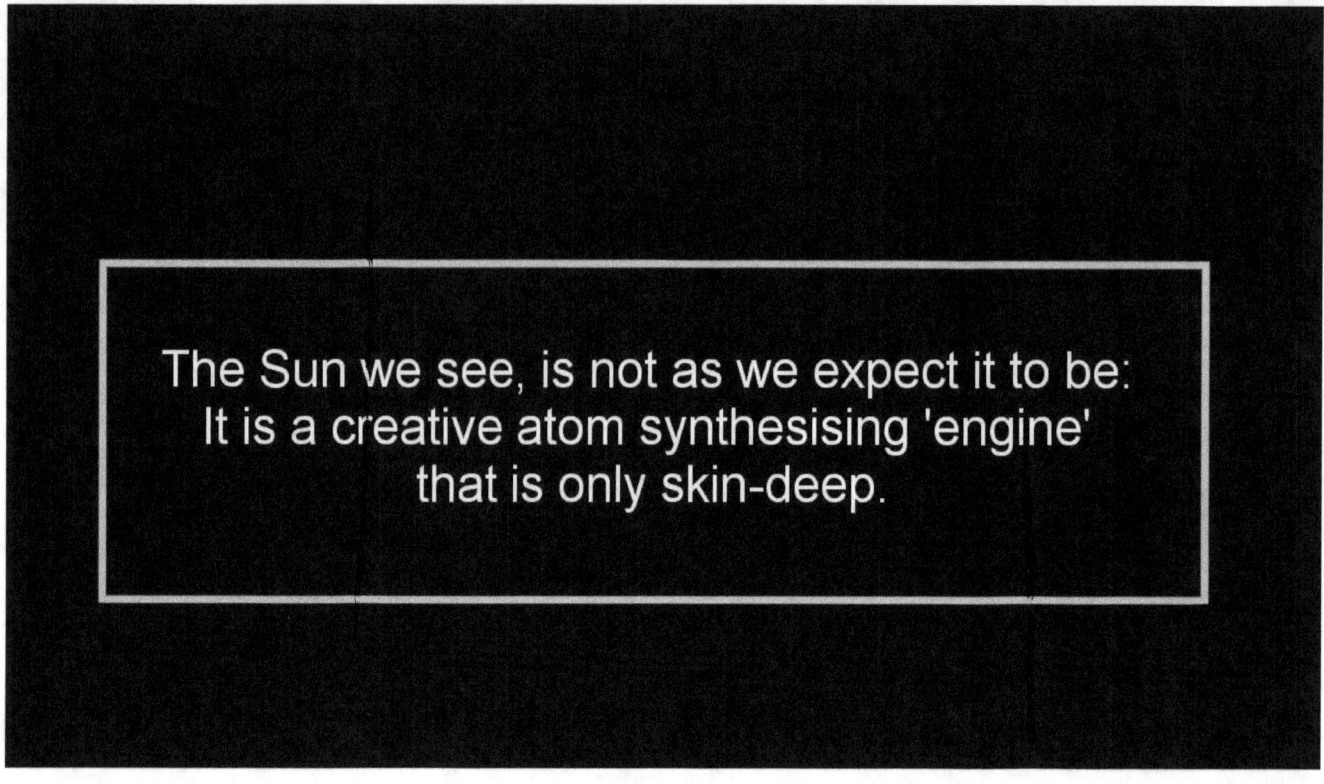

The Sun that we see, is not as we expect it to be: It is a creative atom synthesizing 'engine' that is only skin-deep.

When we look at the Sun

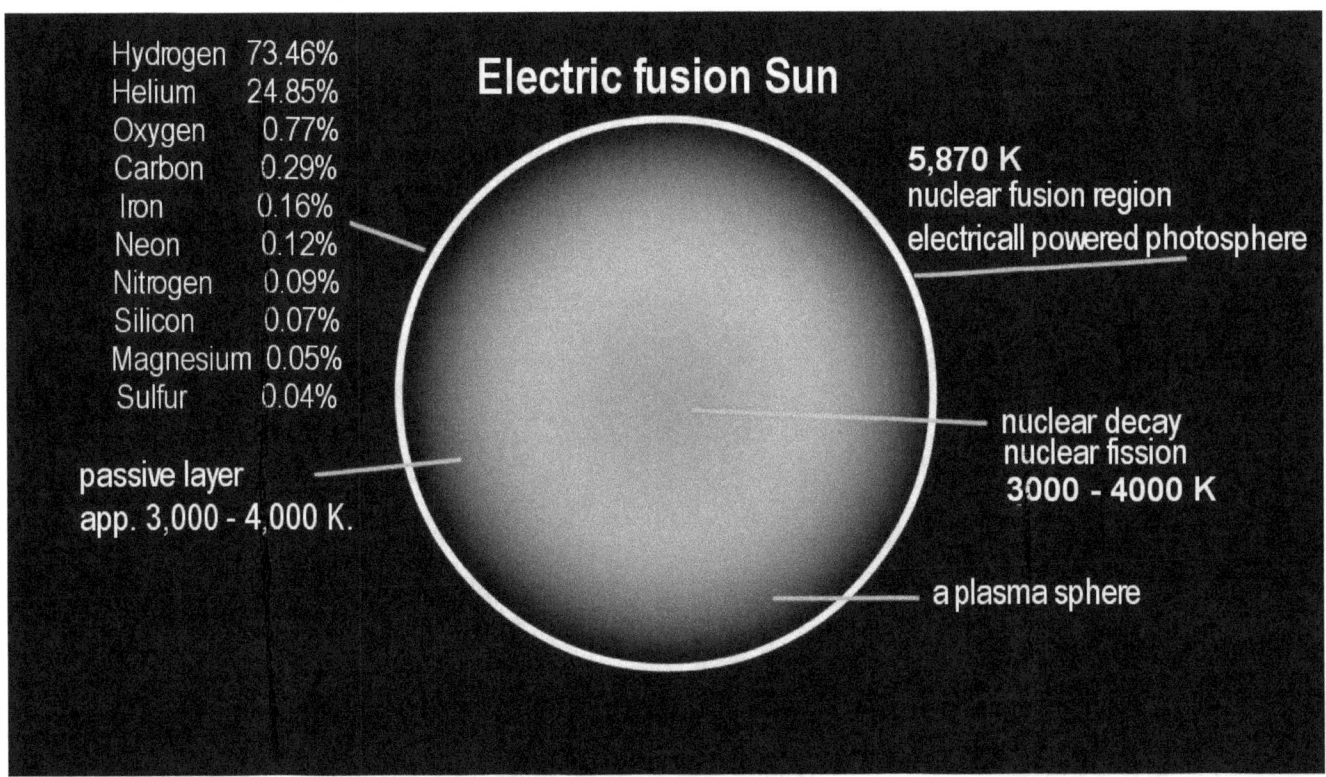

This means that when we look at the Sun, what we see, is in reality quite different than we imagine it to be. What we see is a thin active layer where electric interaction occurs, in which also nuclear fusion takes place, with an inactive ball of plasma below the active layer.

At the core of the plasma ball, some nuclear decay processes may occur that result when atomic structures drift down from the surface and become crushed by the plasma pressure, which then revert back to the plasma state from which they were created on the surface.

When the inflowing plasma stream becomes weak

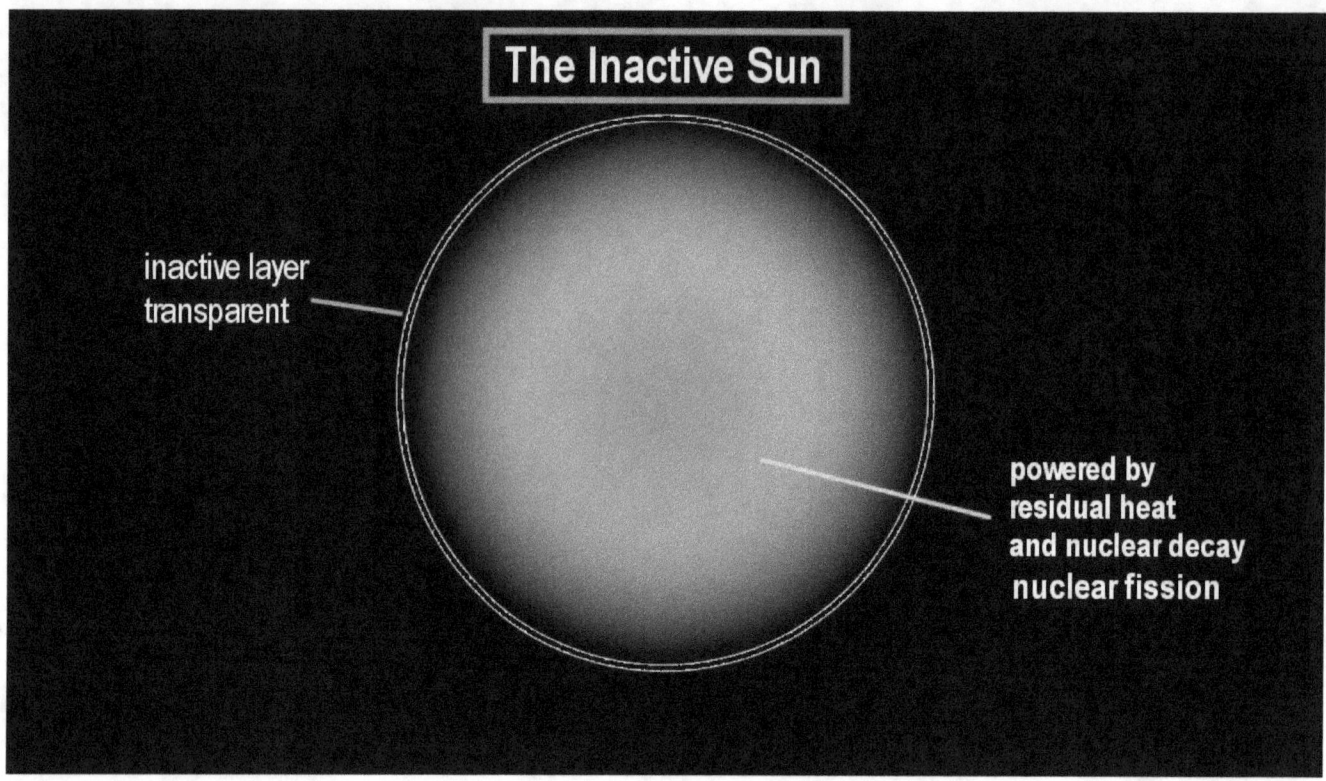

When the inflowing plasma stream becomes weak, and becomes insufficient to maintain the required electron density for the electric nuclear fusion to occur, the reactive layer will cease to function. It will simply cease to exist. A faint glow may remain as some weak reactions may continue to occur. And at the deep center a bright zone may be found, caused by the fission reactions of nuclear decay. In its inactive state, our Sun will likely become a 'white dwarf.'

Part 1: The Sun is NOT a gas sphere. The Sun is plasma.

Nuclear fusion happens on the surface of the Sun

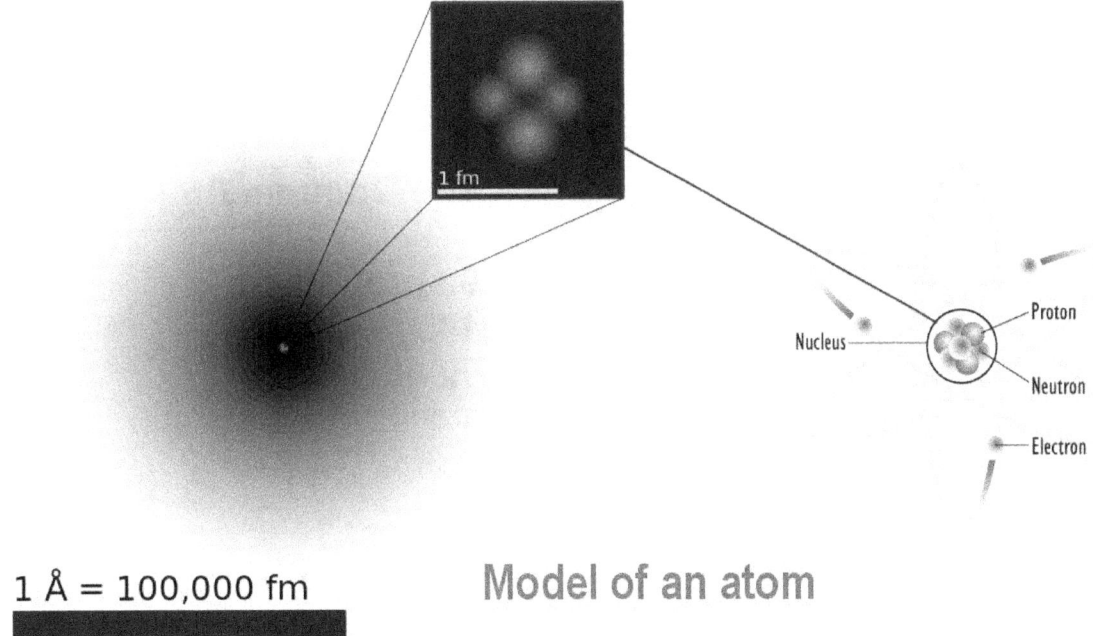

Model of an atom

wikipedia

Nuclear fusion happens on the surface of the Sun, for as long as the Sun operates in a healthy, electron rich, environment. When electrons become highly energized, the rapid movements of their dance around the proton has the effect that the electrons seem to be everywhere. When this dance happens with a great energy, an atom is being born. A hydrogen atom results from a single electron 'swarming' a single proton. The atom that is formed by this highly energetic dance, is typically 100,000 times larger than the proton at the center of the dance, and is a million times larger than the swarming electron itself that gives the atom its effective form.

The dynamic dance of electrons

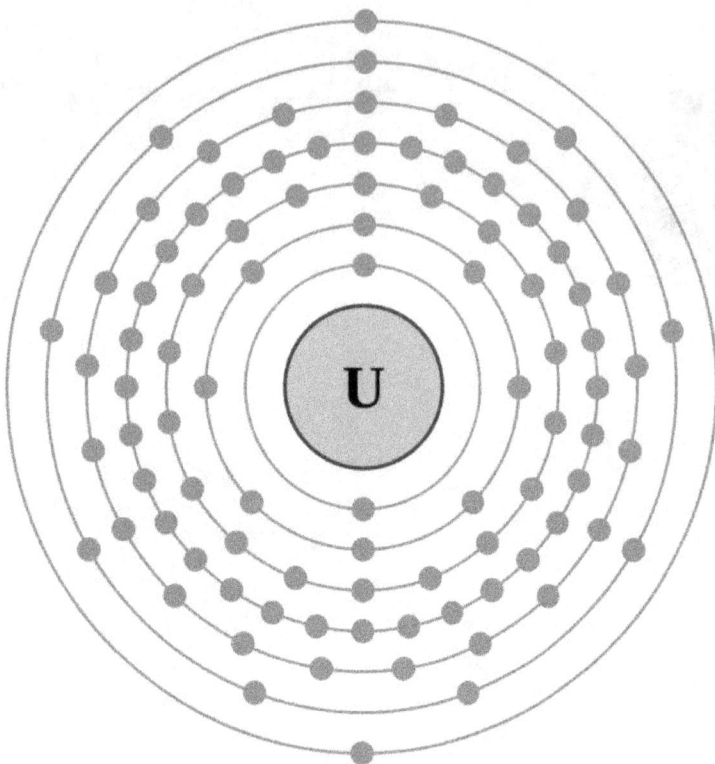

The Uranuim Atom

92 protons
146 neutrons
92 electrons
aranged in layers of
2, 8, 18, 32, 21, 9, 2

The dynamic dance of electrons, can be a swarm of one, or of more that a hundred swarms of electrons arranged into 'rooms' and layers of rooms, up to 7 levels deep, all contained within a single atom, which, altogether give the atom its form.

It may be the powerful electromagnetic effect in flowing plasma, accelerated by magnetic fields, which accelerates the protons to the velocity needed for them to fuse. The fusion cells on the surface of the Sun are typically up to 1000 kilometers wide. They operate as very large particle accelerators that fuse plasma into such large structures as the uranium nucleus that required the fusion of 238 protons, some of which become converted into neutrons in the process of fusing.

By the intense electric interactions of plasma in the fusion process and with its synthesized atoms, electromagnetic energy, both in the form of light and thermal energy, is being emitted.

When two protons are forced closely to each other

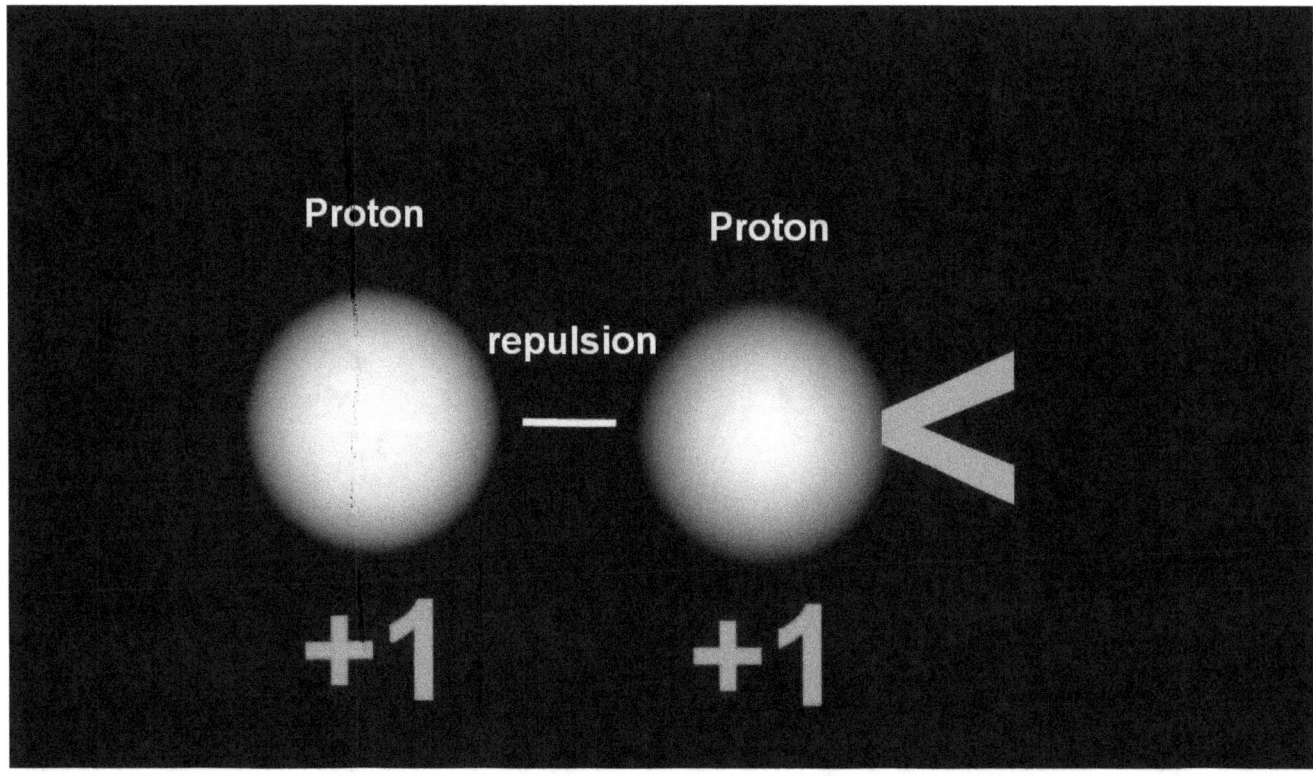

A different type of close-encounter effect occurs when two protons are forced closely to each other, with a greater force than their repelling electric force. In this case, at an extremely close distance, the protons' electric repelling force is reversed, and becomes an attractive force instead, which snaps the two protons together.

The neutron, essentially acts like a glue

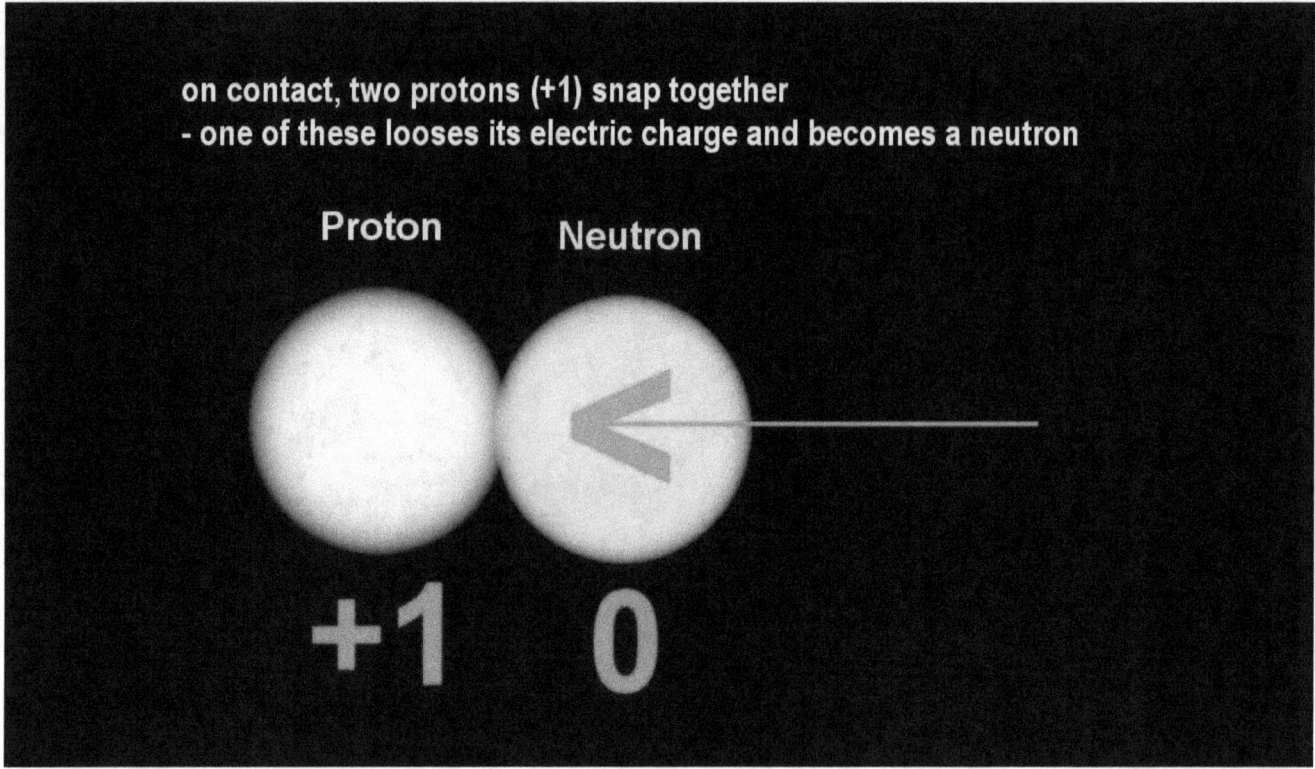

The energy that overcomes the repulsion for this to happen, of course, has to be absorbed. It becomes absorbed by one of the protons, that thereby looses its electric polarity. The joined proton becomes electrically neutral in the process. It becomes a neutron. By this radical transformation of the previous proton into a neutron, the fused particles form a larger unit in which the neutron, essentially acts like a glue between protons in the resulting nucleus that becomes the center of an atom.

By the process of protons becoming bundled

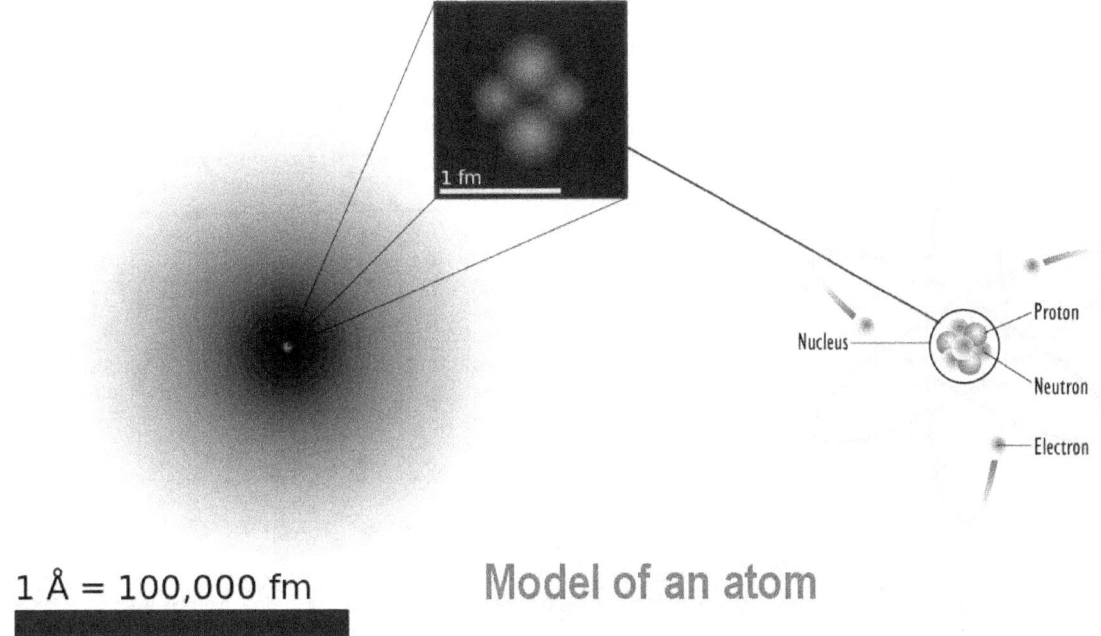

Model of an atom

wikipedia

By the process of protons becoming bundled together, it becomes possible for large atoms to be formed that have numerous protons at their center, forming large clusters for large atomic structures, as in the case of uranium, where the entire large nucleus is being latched together with more neutrons that serve as glue for the giant package, than protons in the package.

Part 1: The Sun is NOT a gas sphere. The Sun is plasma.

A hundred different elementary atoms

Group → Period ↓	1	2	3	4	5	6	7	8	9	10	11	12	13	14	15	16	17	18
1	1 H																	2 He
2	3 Li	4 Be											5 B	6 C	7 N	8 O	9 F	10 Ne
3	11 Na	12 Mg											13 Al	14 Si	15 P	16 S	17 Cl	18 Ar
4	19 K	20 Ca	21 Sc	22 Ti	23 V	24 Cr	25 Mn	26 Fe	27 Co	28 Ni	29 Cu	30 Zn	31 Ga	32 Ge	33 As	34 Se	35 Br	36 Kr
5	37 Rb	38 Sr	39 Y	40 Zr	41 Nb	42 Mo	43 Tc	44 Ru	45 Rh	46 Pd	47 Ag	48 Cd	49 In	50 Sn	51 Sb	52 Te	53 I	54 Xe
6	55 Cs	56 Ba		72 Hf	73 Ta	74 W	75 Re	76 Os	77 Ir	78 Pt	79 Au	80 Hg	81 Tl	82 Pb	83 Bi	84 Po	85 At	86 Rn
7	87 Fr	88 Ra		104 Rf	105 Db	106 Sg	107 Bh	108 Hs	109 Mt	110 Ds	111 Rg	112 Cn	113 Uut	114 Fl	115 Uup	116 Lv	117 Uus	118 Uuo

Lanthanides	57 La	58 Ce	59 Pr	60 Nd	61 Pm	62 Sm	63 Eu	64 Gd	65 Tb	66 Dy	67 Ho	68 Er	69 Tm	70 Yb	71 Lu
Actinides	89 Ac	90 Th	91 Pa	92 U	93 Np	94 Pu	95 Am	96 Cm	97 Bk	98 Cf	99 Es	100 Fm	101 Md	102 No	103 Lr

In this manner, more than a hundred different elementary atoms are being created on the solar surface. We find them arranged in the periodic table of elements, according to the number of protons contained in each package.

Typically the ratio of protons to neutrons is 1 to 1. In larger atoms, a greater ratio of neutrons is required to hold the big nuclei together. As I said, more glue is needed there. However, in every case, the number of electrons in an atom, matches the number of protons, whereby the electric fields inside an atom balance each other out. By this perfect equality, every resulting atom becomes electrically neutral. While it is possible for the swarming electrons of closely spaced atoms to share each other's space, which latch atoms together into molecules and so on, the electric neutrality of the atomic structures remains always intact.

The balancing act that creates electrically neutral packages, which all atoms and molecules are, is of critical importance for the universe to exist, and for us to exist. This electric balancing act is the 'heart' that motivates everything. The electric balancing that creates electrically neutral atoms is one of the main factors why the cold-fusion Sun theory is possible, or for that matter, any atomic synthesizing fusion is possible.

Even the hydrogen atom is, heavy

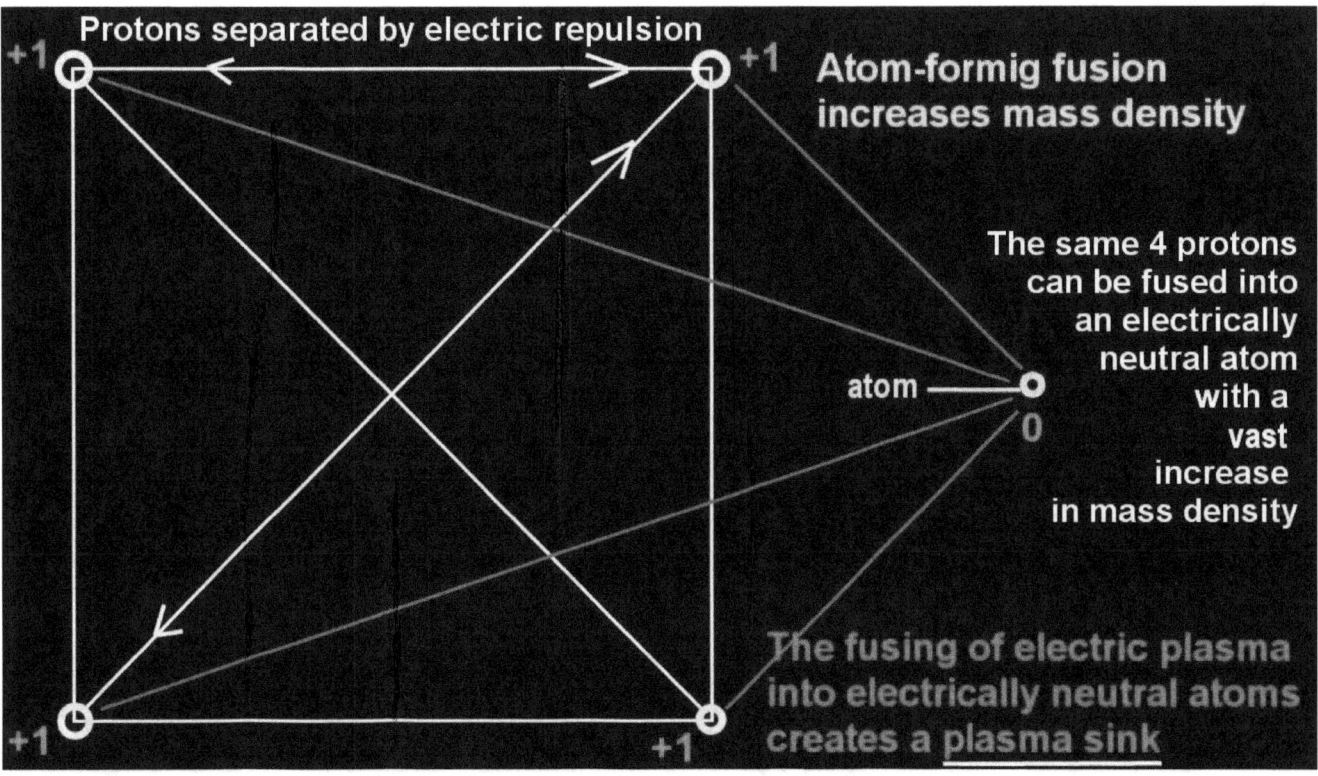

When an atom is formed that is electrically neutral, the repelling force that would keep the protons repelled from each other in plasma, is suddenly neutralized. This means that far-more protons, when joined to form atoms, can be packed into a given space than would otherwise be able to exist in that space.

In the unbound state, electric repulsion makes plasma extremely light and thinly diffused. In comparison, plasma bound into atoms is heavy, because the resulting package is dramatically smaller. Even the hydrogen atom is, heavy, in comparison with plasma, and of course, the helium atom is four-times heavier still.

The dense packaging of protons into atoms renders the internal solar nuclear-fusion theory, a fundamental impossibility.

Because of the tight packing of atoms, a gas sphere the size of the Sun, filled with atoms of hydrogen and helium, would likely be a thousand times heavier, if not more so, than the Sun actually is.

Hydrogen origin paradox

The Sun synthesises all the atoms in the solar system. What created the hydrogen atoms it is deemed to be made of?

As if the mass-density paradox was not enough to disprove the theory of the Sun being a hydrogen star, one may further consider that an even greater paradox exists. Where would the hydrogen have been produced that under the internal fusion theory, makes up a Sun? That's a paradox.

It is a paradox, because a sun is the only operating platform in the universe that causes the synthesis of atoms from plasma, including the hydrogen atom. The paradox is that the hydrogen could not have come from the Sun. A sun cannot produce itself. It cannot operate before it exists. Nor would the supposed accretion of dust separate out the light hydrogen for the Sun, and the heavy atoms for the planets. Accretion doesn't work that way.

The resulting built-in hydrogen origin paradox is so great that it closes the door on any possibility for the Sun to be a gas star with internal, atomic fusion happening in its core. This renders the theory as but a dream - a dream enabled by another dream, the Big Bang creation dream.

Part 2: The universe of plasma refuting the Big Bang

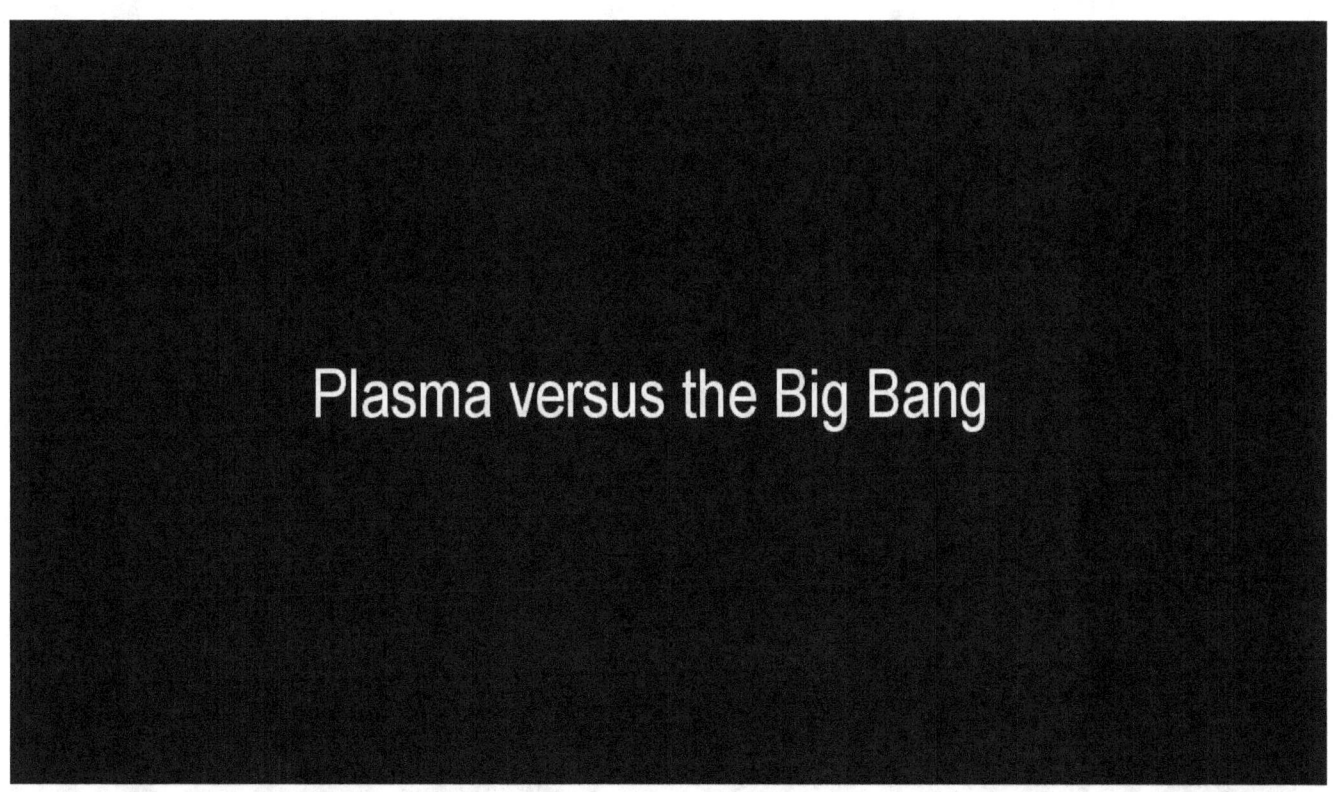

Our Electric Cold Fusion Sun (Part 2) Plasma vs. the Big Bang

The Big Bang creation is more a myth than a theory

The Big Bang creation is more a myth than a theory. It may be termed a science-dream on the order of fairy tales.

The dream begins with nothing. Before the Big Bang happened, there was nothing. The universe did not exist. Reality was an inconceivable dark void.

The entire universe exploded into being

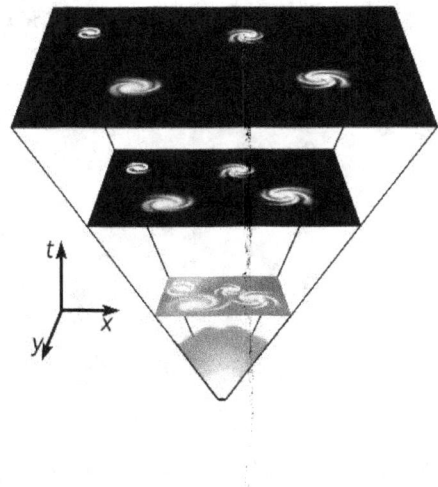

The Big Bang creation myth refuted by the electric solar fusion model

Suddenly in this infinite nothing, 13.8 billion years ago, the entire universe exploded into being.

All mater, all energy

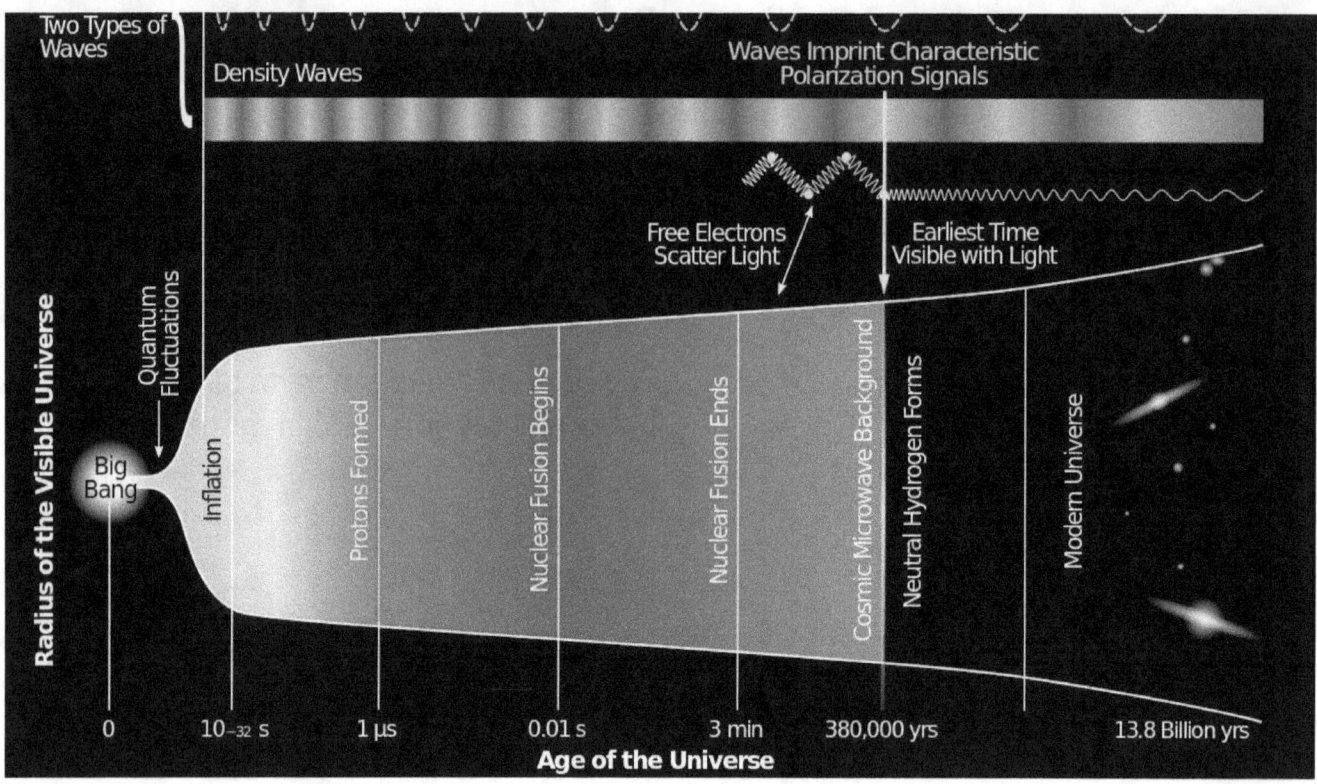

All mater, all energy, all that there is and ever will be, is deemed to have emerged in the space of three minutes out of the 'whirlpool' of the giant explosion that expanded into all directions from its central point as a primordial cloud of dust that is still expanding outward with ever-increasing speed.

The dust is deemed to have condensed

The dust is deemed to have condensed into galaxies and stars and planets. From within the dust, the hydrogen atoms are said to have grouped together into giant spheres of gas that became stars.

The tale is full of holes

A deemed star-forming region in the Large Magellanic Cloud. NASA/ESA image

It is deemed that each star, by its accretion of weight, became so hot under the resulting intense gravitational compression that the hydrogen atoms in its core began to fuse into larger helium atoms. The building of larger atoms is deemed to create energy, from which the stars derive their light. Thus each gas star became a Sun, powered from within by its own substance.

All this comes together as a nice tale that almost makes sense. But the tale is full of holes, holes created by paradoxes. Numerous impossible paradoxes, supported by physical evidence, solidly refute the impossible theory.

The Earth itself, is a paradox under the Big Bang theory

The existence of the Earth itself, is a paradox under the Big Bang theory. It shouldn't be what it is. Its composition disproves the theory that the universe was created in a primordial explosion 13.8 billion years ago.

Light disproves the explosion theory of the Big Bang

The center of the Milky Way, at the center of the Big Bang explosion of the universe

Even the nature of light disproves the explosion theory of the Big Bang, in which the formed dust condensed it into galaxies that are deemed to have expanded across the cosmos, and are still expanding, and racing away from us into all directions with accelerating motion.

The red-shift of light from distant galaxies

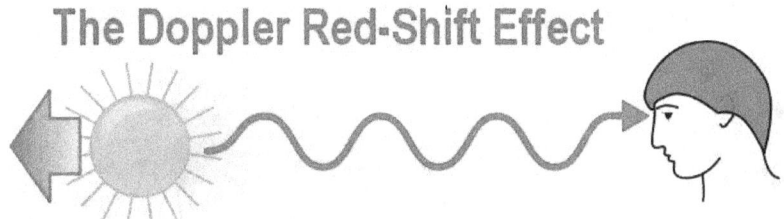

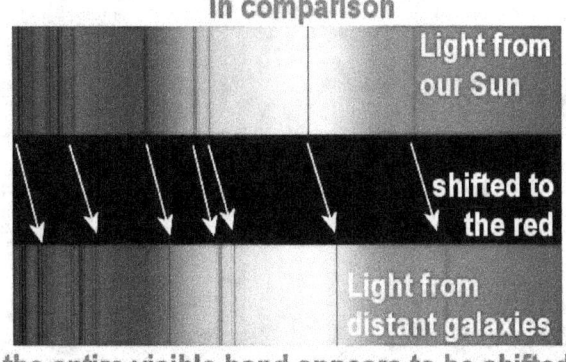

As proof for the accelerating motion, a theory has been invented that interprets the spectral red-shift in light from distant galaxies as being caused by the source of the light racing away from us, the observer.

The red-shift of light from distant galaxies is said to prove the Big Bang theory of the accelerating expansion of the universe. Light is said to be stretched out as the most distant galaxies are deemed to be racing away faster and faster, the further they are located from us. That's the tale that is being told. It is almost believable.

Red shift is the result of energy depletion

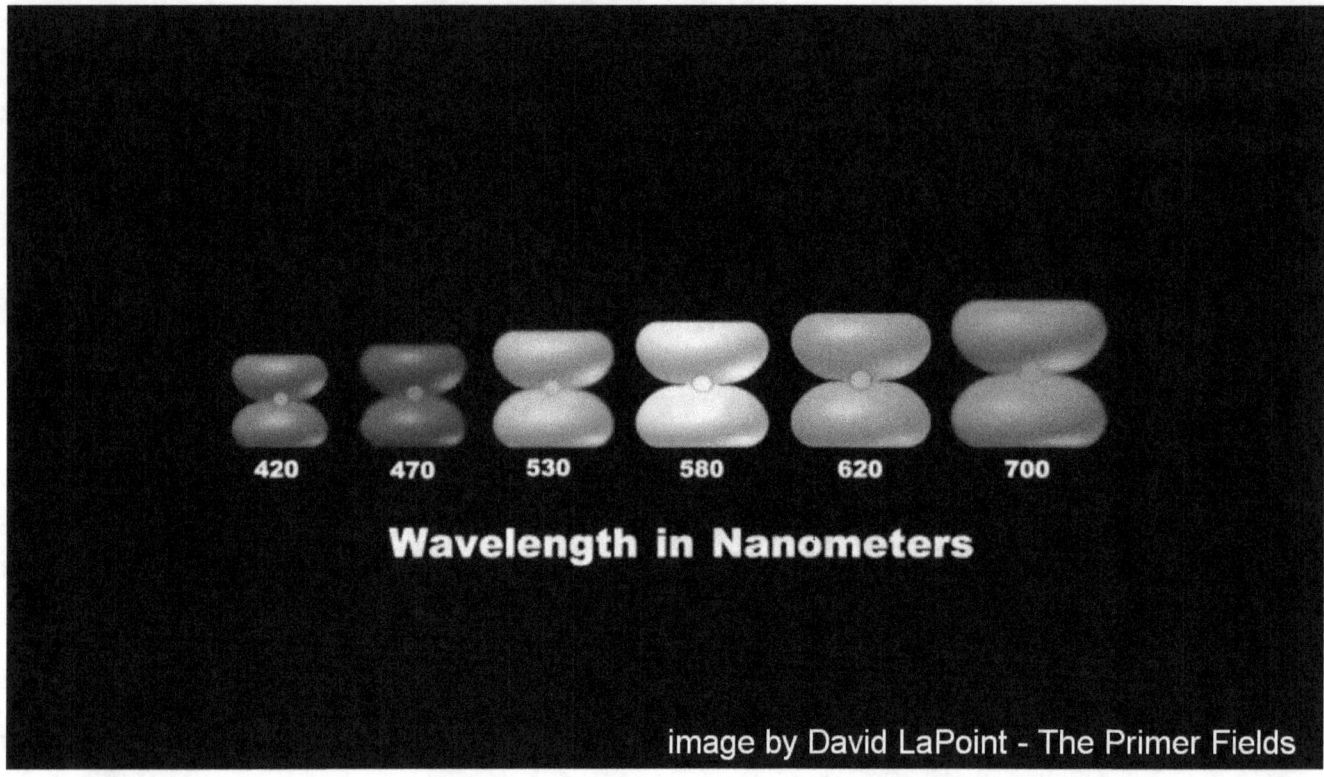

image by David LaPoint - The Primer Fields

In the real world the theory breaks down when the measured red shift is more rationally perceived as the result of energy depletion of the photons over long distances in their encountering plasma and atomic gases in intergalactic space. The energy depletion causes the entire electromagnetic spectrum to shift towards a lower energy state, which means a shift towards the red. The photons of light are said to be packages of energy. The more energy they contain, the tighter the packages are packed. When energy is dissipated, The packages become larger. The violet expand into blue, the blue into green, the green into yellow. The entire spectrum shifts towards the red. This simple reality, disproves the Big Bang creation theory.

A gigantic piece of evidence that disproves the Big Bang theory

The Earth from 98,000 miles during the Apollo 11 translunar injection on July 16, 1969

We also have a gigantic piece of evidence that disproves the Big Bang theory even more, and proves at the same time that every sun is a plasma-powered electric nuclear-fusion engine that synthesizes all atomic elements that are known to exist in the universe. This item of evidence is the Earth itself. The proof for this statement was delivered only recently in a massive scientific effort to determine the age of the Earth.

The Earth is proof that synthesizing atomic fusion

> The proof that lies in the Earth is proof
> that synthesizing atomic fusion of plasma
> is occurring on the Sun and always has been
> as the ONLY source for atomic elements
> in the universe
>
> **But what is plasma fusion?**

The proof that lies in the Earth is proof that synthesizing atomic fusion of plasma is occurring on the Sun and always has been as the ONLY source for atomic elements. But what is plasma fusion?

Plasma is a sea of electrons and protons

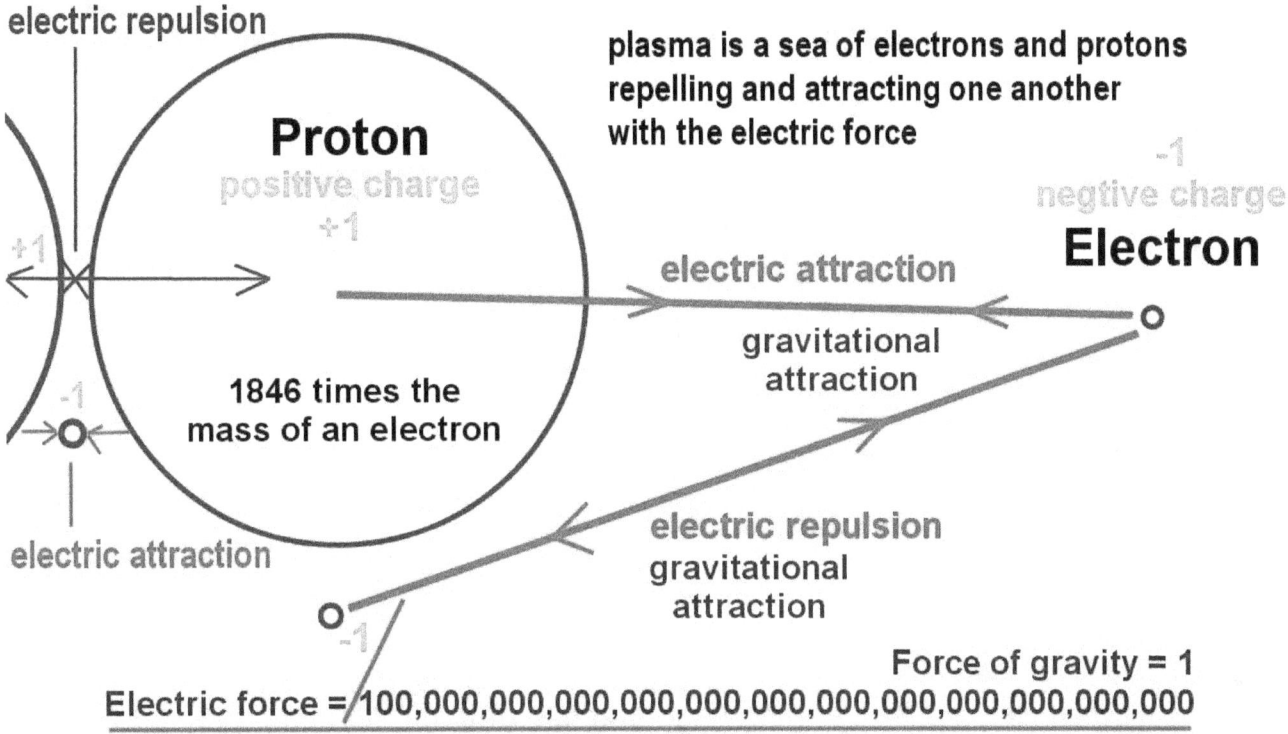

Plasma is a sea of electrons and protons repelling and attracting one another with the electric force.

Plasma in space

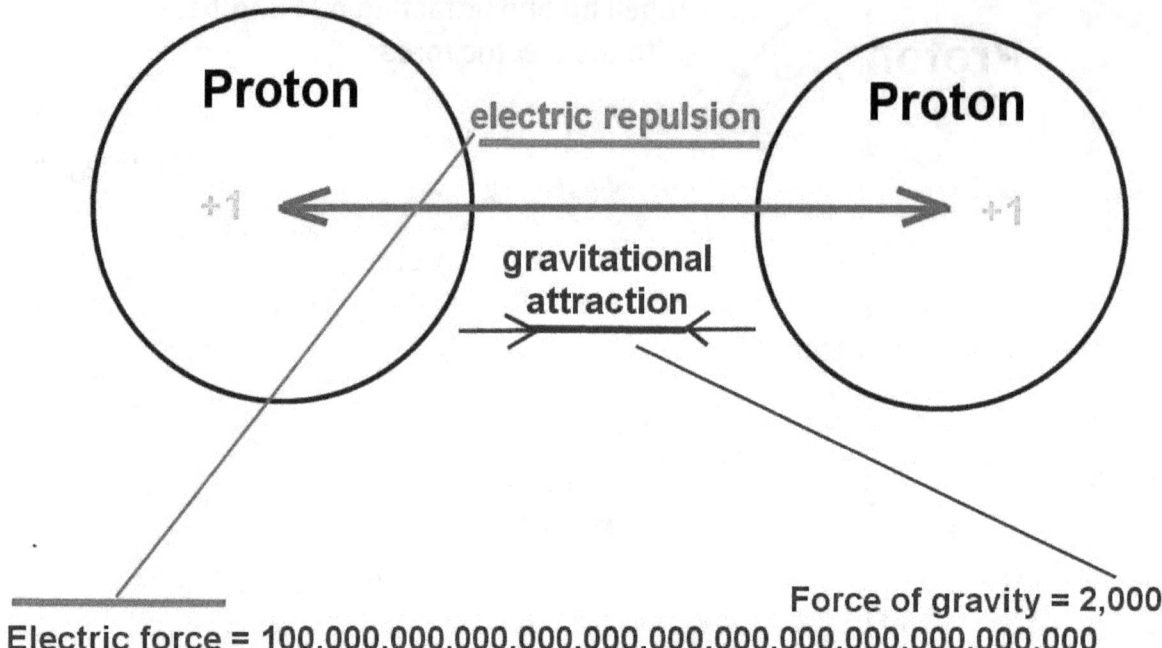

Plasma in space is thinly defused by the force of electric repulsion.

On contact, two protons snap together

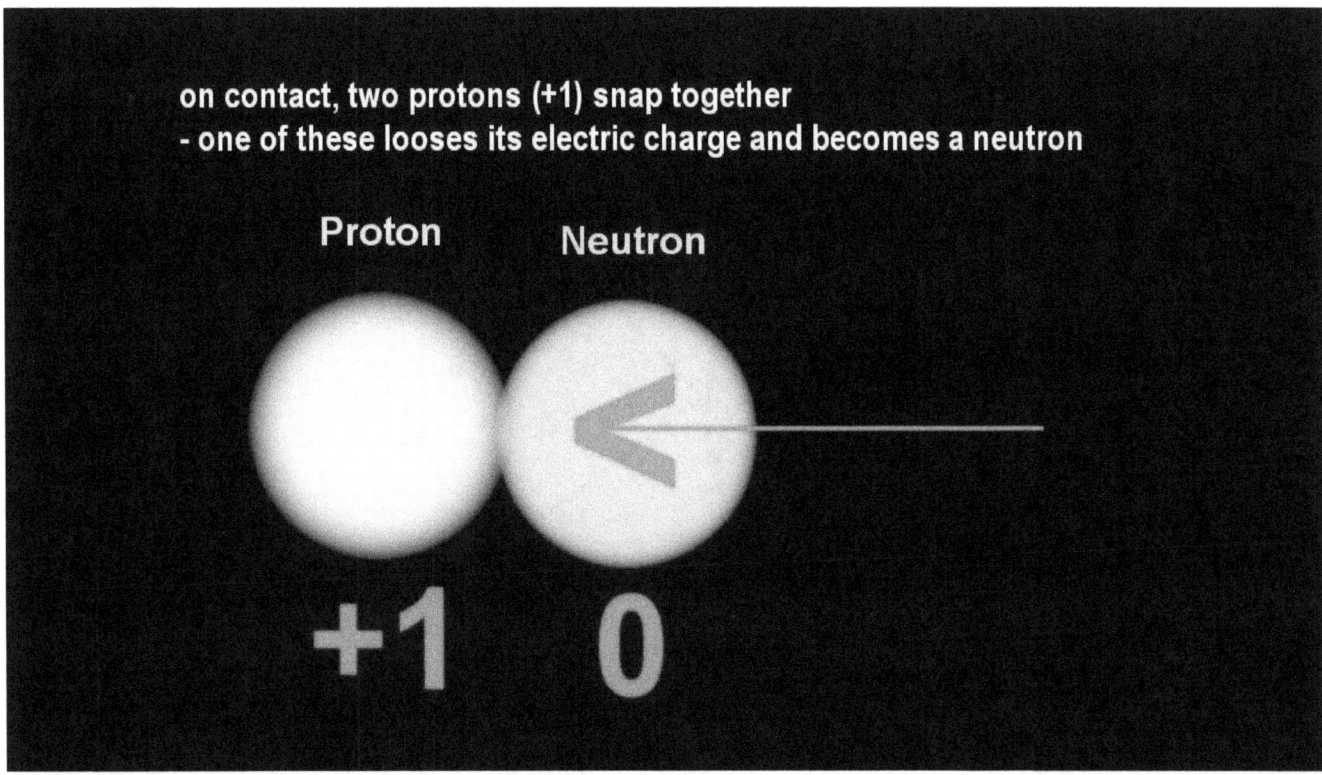

On contact, two protons snap together. 'Thereby, one of the protons looses its electric charge and becomes a neutron.,

On contact, the electron is forced to rebound

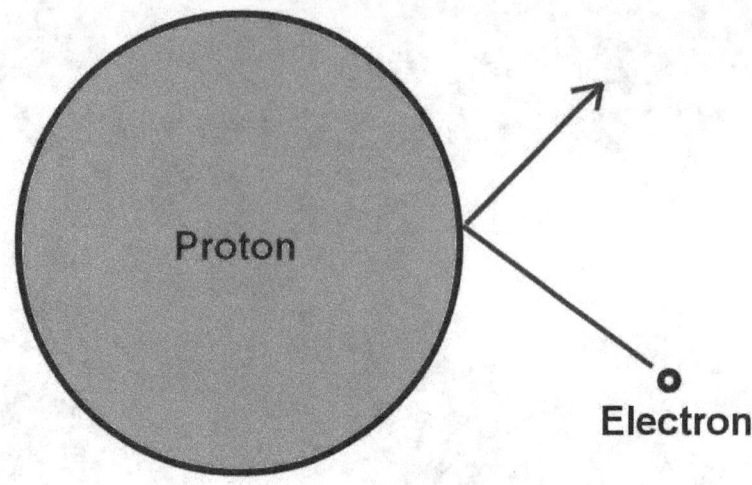

Electrons (-1) are attracted to protons (+1) by their unequal polarity. On contact, the electron is forced to rebound, only to be attracted anew.

Atoms are formed by the dynamic 'dance'

**atoms are formed by the dynamic 'dance'
of electrons being attracted and forced to rebound**

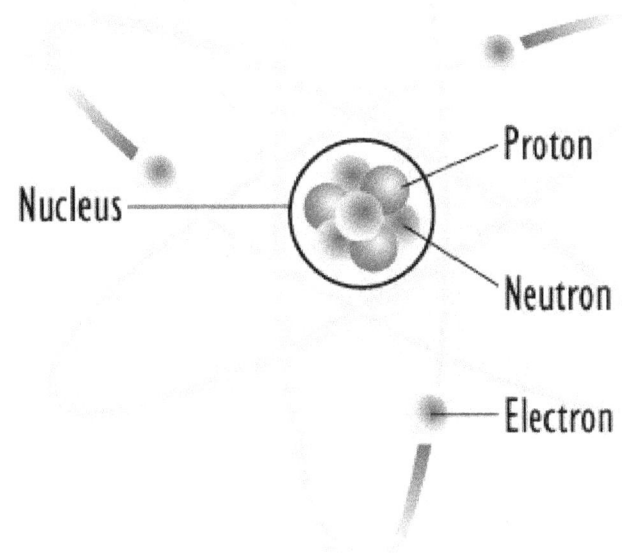

wikipedia (image)

Atoms are formed by the dynamic 'dance' of electrons being attracted and forced to rebound.

Atoms are electrically neutral plasma structures

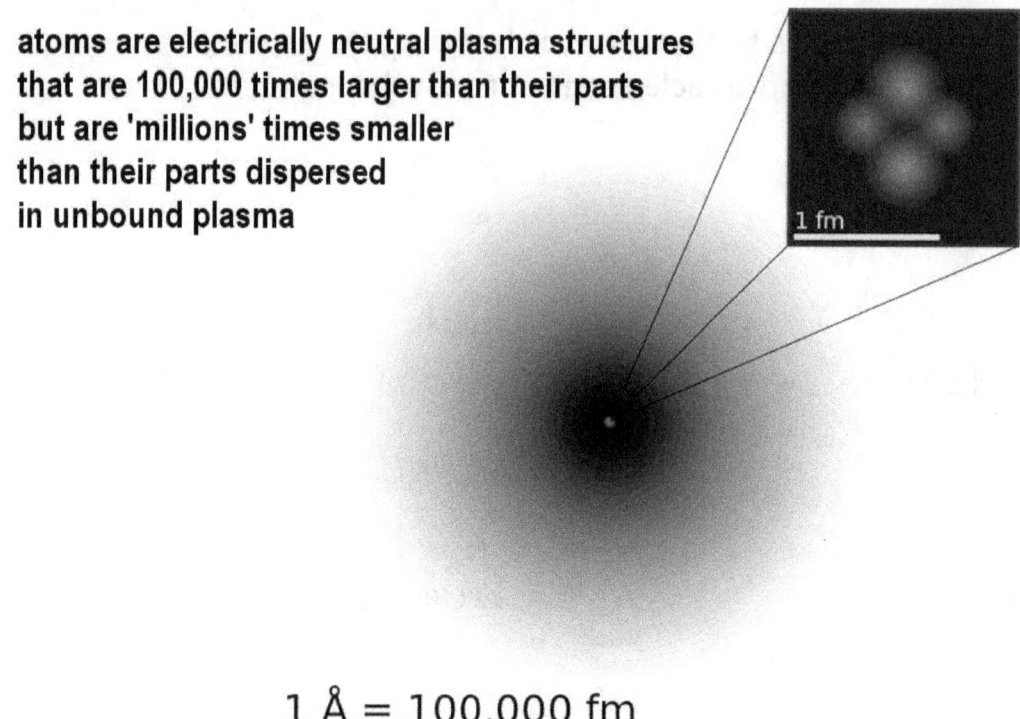

atoms are electrically neutral plasma structures that are 100,000 times larger than their parts but are 'millions' times smaller than their parts dispersed in unbound plasma

1 Å = 100,000 fm

wikipedia (image)

Atoms are electrically neutral plasma structures that are 100,000 times larger than their parts, but are 'millions' times smaller than their parts dispersed in unbound plasma.

Atom-forming fusion increases mass density

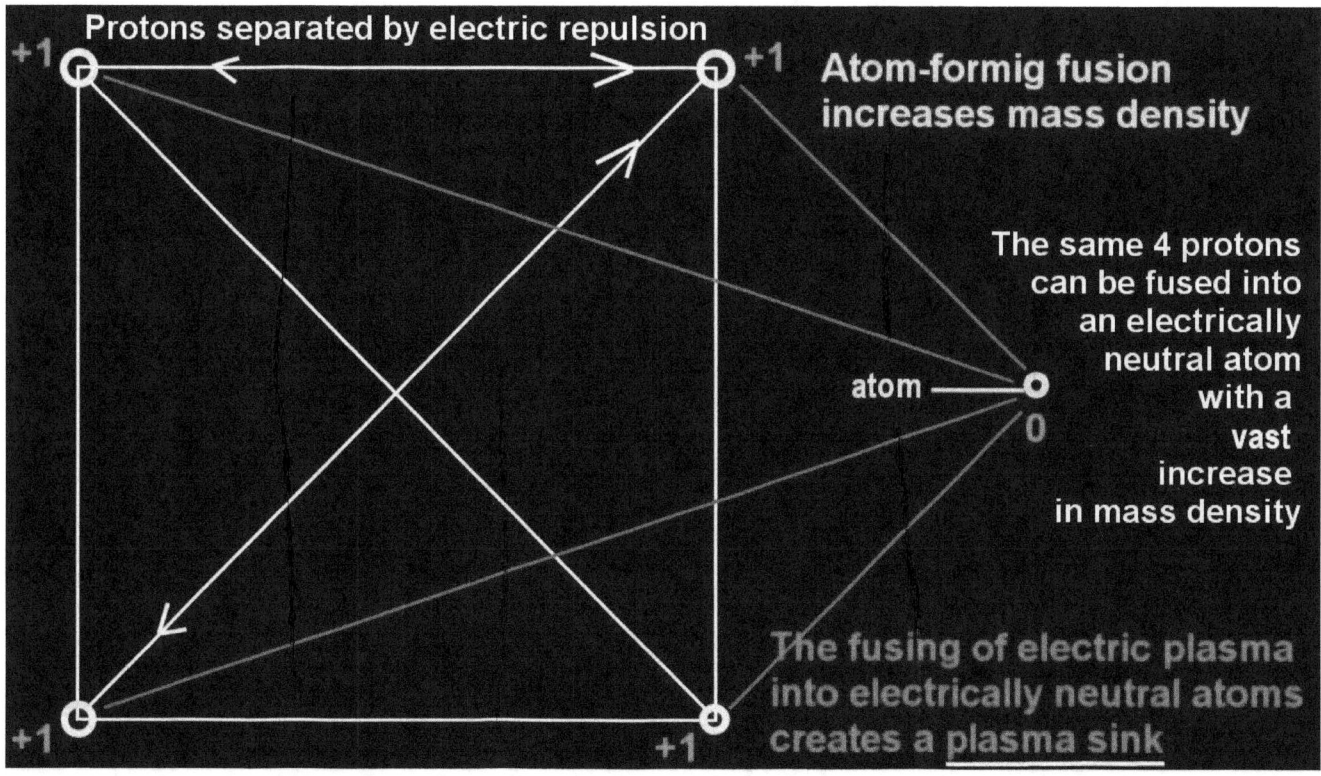

Atom-forming fusion increases mass density, while the forming of electrically neutral atoms creates a sink in the electrically charged landscape.

Electric nuclear fusion happens naturally

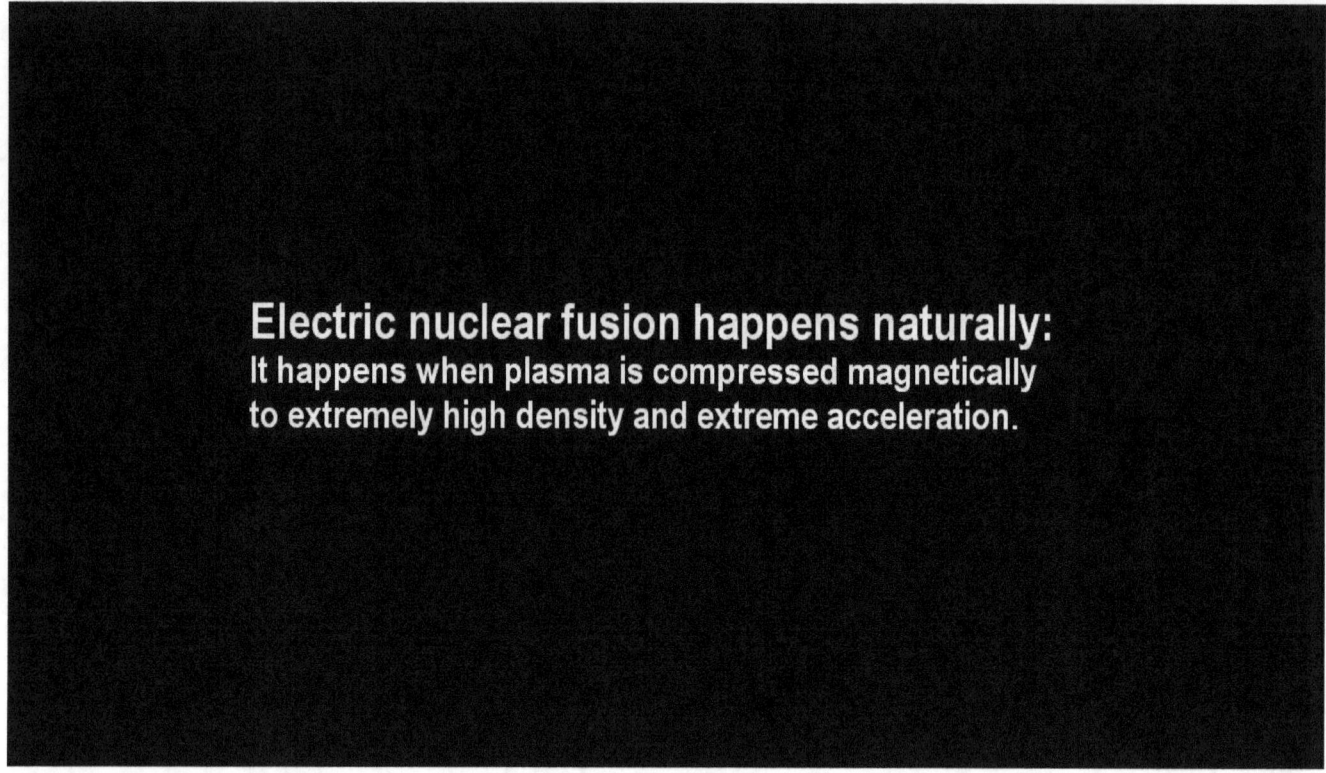

Electric nuclear fusion happens naturally: It happens when plasma is compressed magnetically to extremely high density and extreme acceleration.

Experiments at the Los Alamos National Laboratory

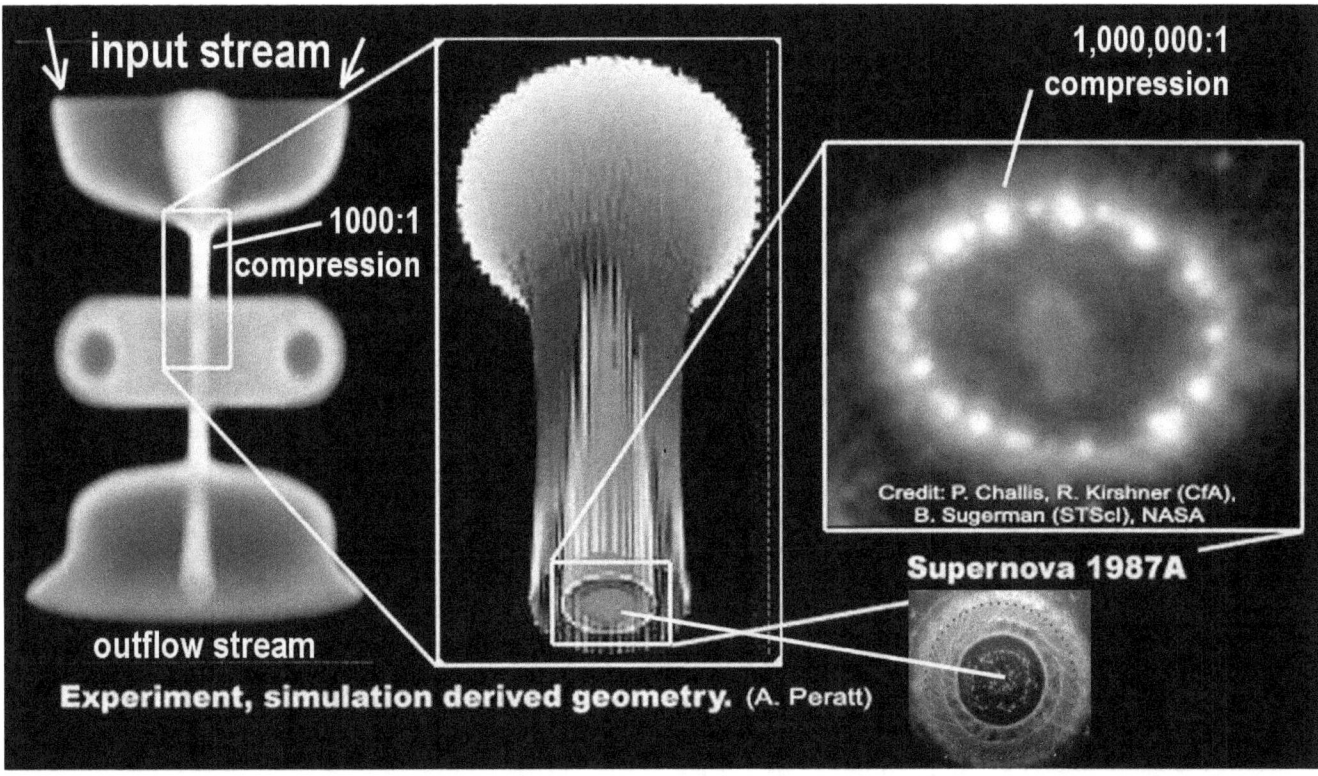

Experiments at the Los Alamos National Laboratory illustrate that a million to one plasma compression is natural in high-energy, magnetically focused plasma flow dynamics.

Plasma compression may be a billion-fold

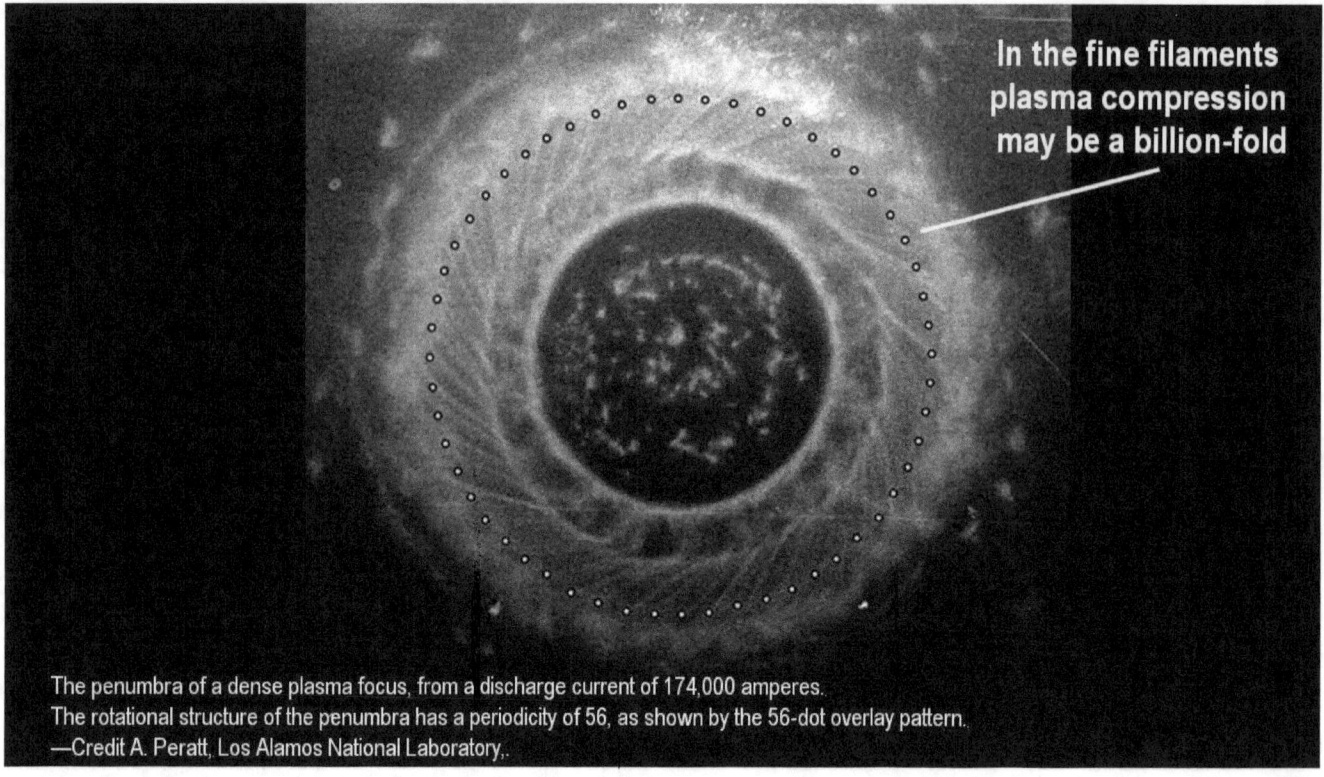

The penumbra of a dense plasma focus, from a discharge current of 174,000 amperes.
The rotational structure of the penumbra has a periodicity of 56, as shown by the 56-dot overlay pattern.
—Credit A. Peratt, Los Alamos National Laboratory,.

In the fine filaments, plasma compression may be a billion-fold.

Very large cosmic 'primer fields'

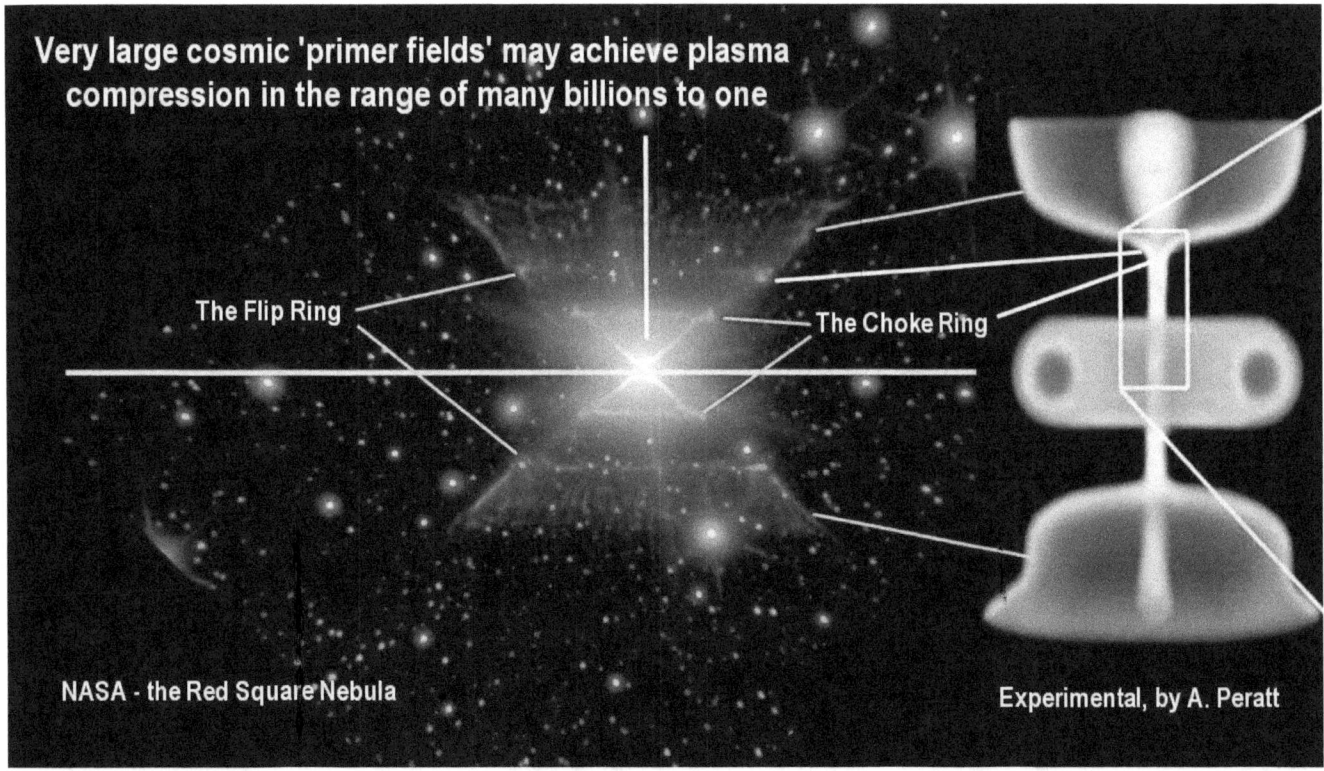

Very large cosmic 'primer fields' may achieve plasma compression in the range of many billions to one.

Plasma compression may exceed trillions to 1

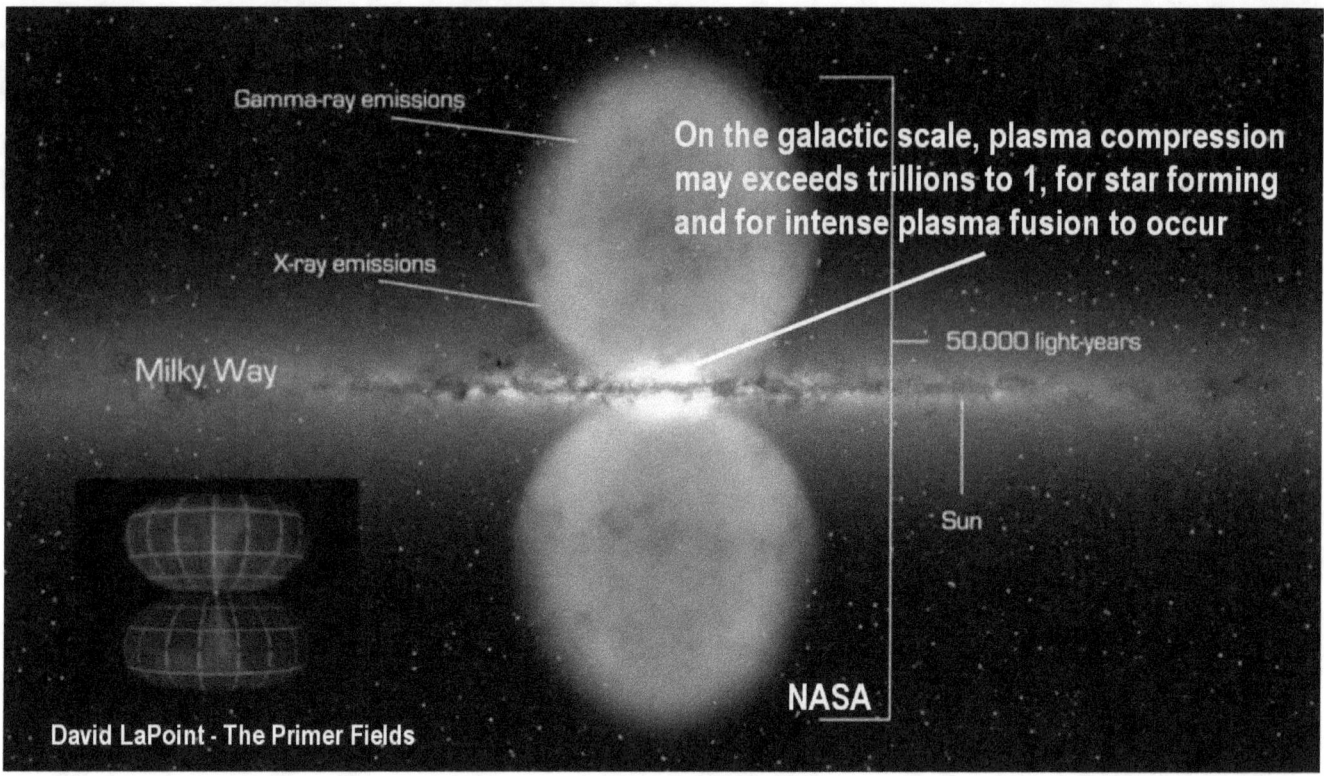

On the galactic scale, plasma compression may exceed trillions to 1, for star forming and for intense plasma fusion to occur.

The Earth is our witness

The Earth is our witness that atom-synthesizing electric nuclear fusion is continuously occurring.

Large atomic elements decay over time

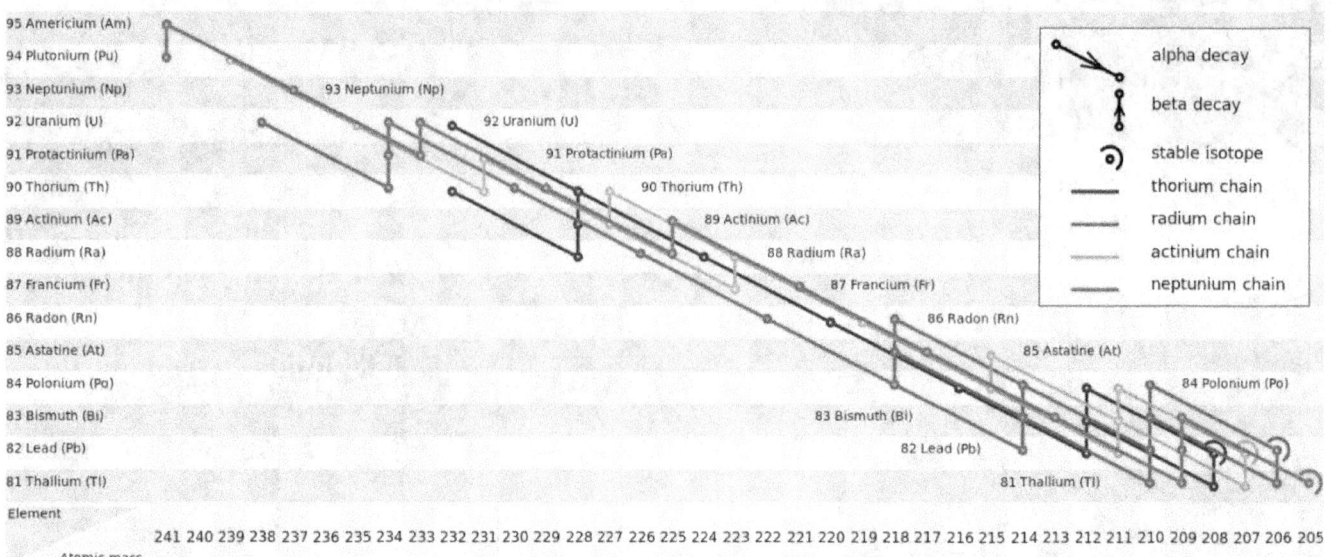

Artwork by JohanTheGhost, CC-BY-SA-2.5,2.0,1.0, - wikipedia

It had been discovered previously that large atomic elements, such as uranium, decay over time in a chain of mutation that ends in lead. Lead is the heaviest stable element that exists. It has been further discovered that the rate of decay is knowable for the different decaying elements.

The ratio of lead in uranium-containing rocks

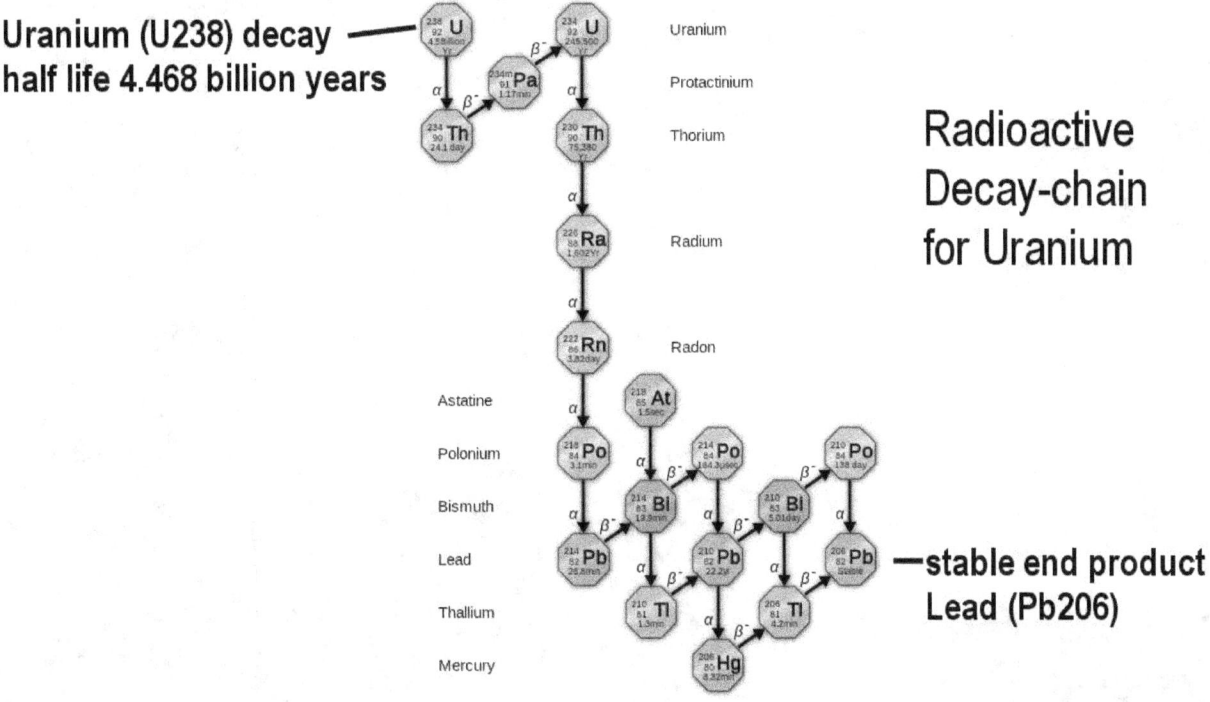

Uranium, for example is known to have a half-life of 4.5 billion years. This means that over the span of 4.5 billion years, half of the volume of uranium that existed in the beginning, has become transmuted into lead.

It has been reasoned that the measurable ratio of lead in uranium-containing rocks, can be used to accurately determine the age of the Earth. On this basis the age of the Earth, and of the solar system as a whole, has become known to be in the range of 4.54 billion years.

It has been discovered that the lead-to-uranium ratio is always the same, wherever uranium is found, including in cases were uranium is found in the remains of asteroids.

The measured ratio disproves the Big Bang theory

Milky Way look-alike NGC 6744
ESA - Wide Field Imager view - CC BY 3.0

The measured ratio disproves the Big Bang theory. The ratio proves that all the atomic elements on Earth and in the solar system nearby, were NEWLY created, at the time the Earth and the solar system was formed, which occurred likely near the center of our galaxy where the plasma pressure is strong and the fusion process on the surface of the Sun is immensely productive, so much so, that even the largest atoms were synthesized in substantial quantities, such as uranium.

The dating of the Earth, with the atomic clock

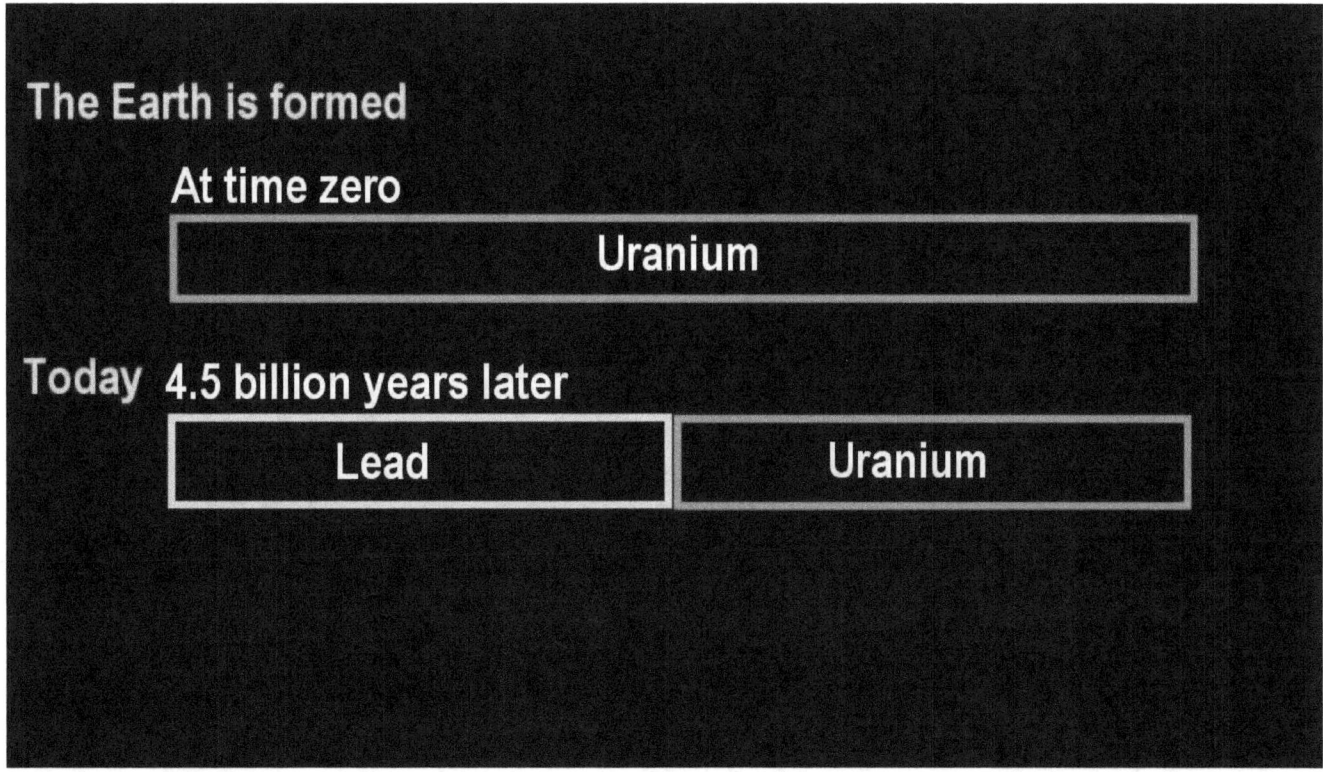

The scientific dating of the Earth, that has been timed with the atomic clock of the universe, proves that all the atomic elements that the Earth is made of, did not originate as materials that were formed in the Big Bang, more than 13 billion years ago. The radioactive decay would have produced a much greater ratio of led in uranium-containing rocks. The measured ratio proves that all the atoms for the planets in the solar system were synthesized NEW at the time of the forming of the solar system.

This means that the synthesis could only have occurred on the surface of the Sun, in the time of its initial intense state. While atomic-fusion synthesis still occurs on the surface of the Sun today, it does so with lesser intensity.

The dating of the Earth, with the atomic clock of the universe, doesn't match the Big Bang creation theory by a long way. It thereby disproves it. It leaves the external-fusion Sun standing as the only possible contender for the synthesis of the atomic elements in the solar system.

So it is that the scenario that solves the paradox, is a case of actual historic evidence.

Gravitational accretion of cosmic dust from the Big Bang

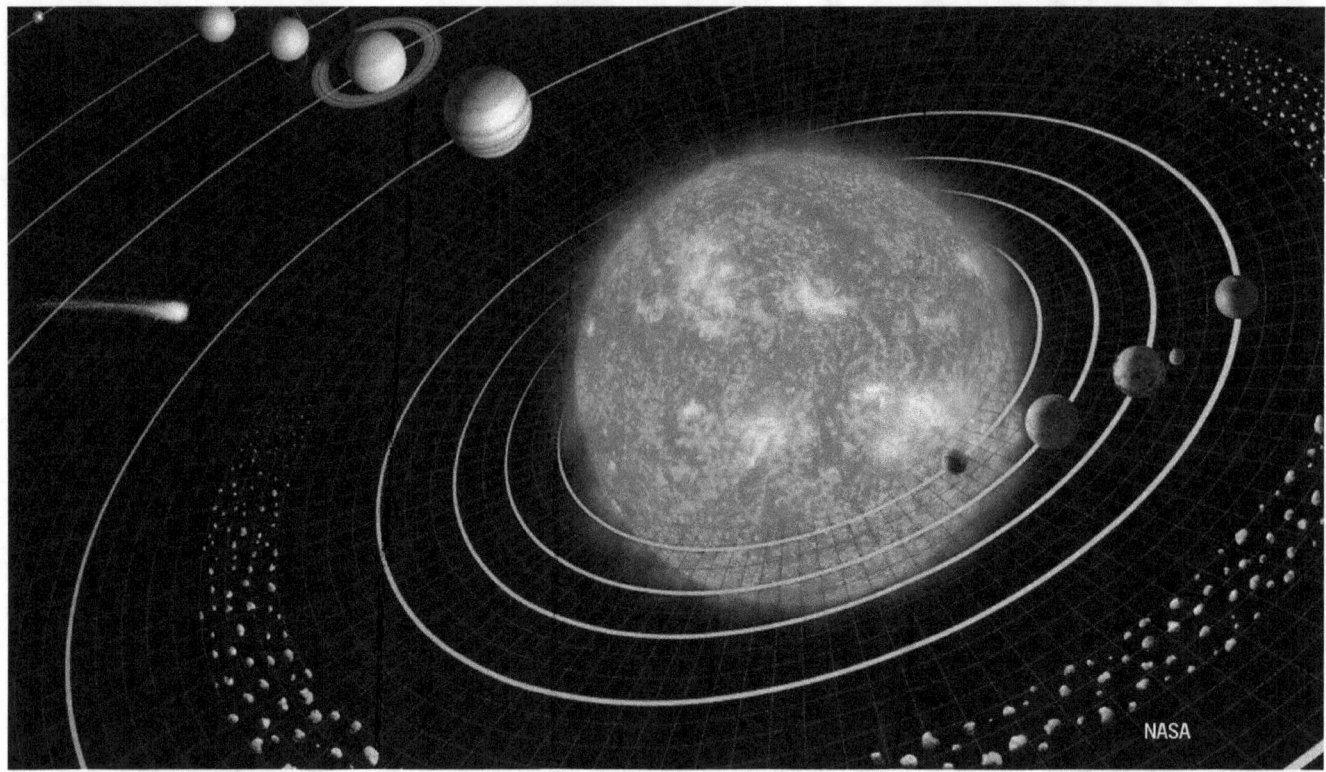

The fabled theory that all the stars and planets were drawn together by the gravitational accretion of cosmic dust from the Big Bang explosion, doesn't make any sense anyway. The theory is a paradox in itself. No principle exists that would single out the hydrogen atoms from the cosmic dust to gravitationally form a star. Hydrogen atoms have the least gravitational attraction of all the atoms in the universe. Even helium asserts a 4 times greater gravitational attraction than hydrogen. Shouldn't all the stars be made up of helium then? And what about the really heavy elements with huge gravitational attraction, such as iron, lead, or even uranium? Shouldn't they form the core of the stars? The entire hydrogen-sun theory is a paradox. It doesn't have a basis to stand on that makes any sense.

The hydrogen-sun theory

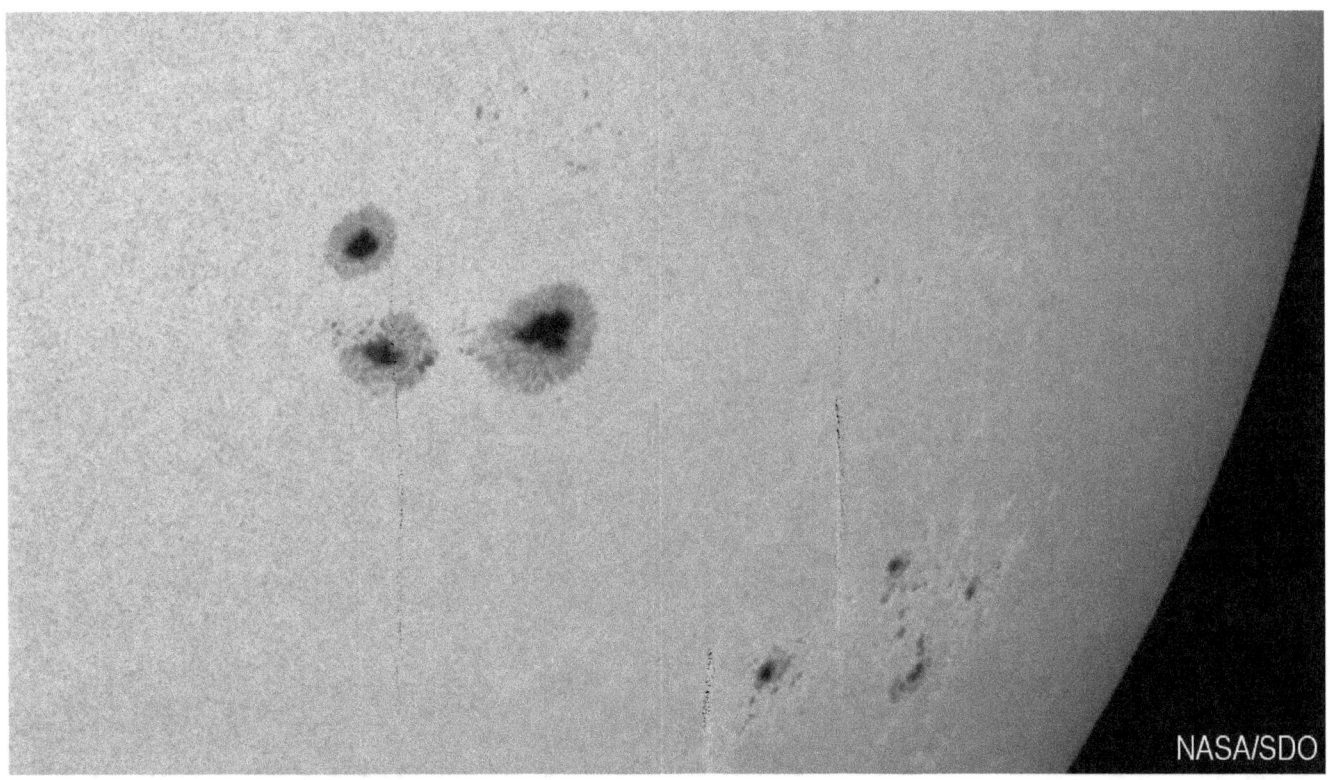

The hydrogen-sun theory is so evidently false that it is surprising that it is still maintained in the empty box of the gravity-only universe represented by the Big Bang dream.

Look at the volume of hydrogen that is needed

Just look at the volume of hydrogen that is needed to fill up the Sun, and also the four gas planets that are almost entirely made up of hydrogen. This enormously huge volume of hydrogen does not exist, according to the cosmic abundance ratio that has been measured on the Sun and throughout the universe.

According to the false theory

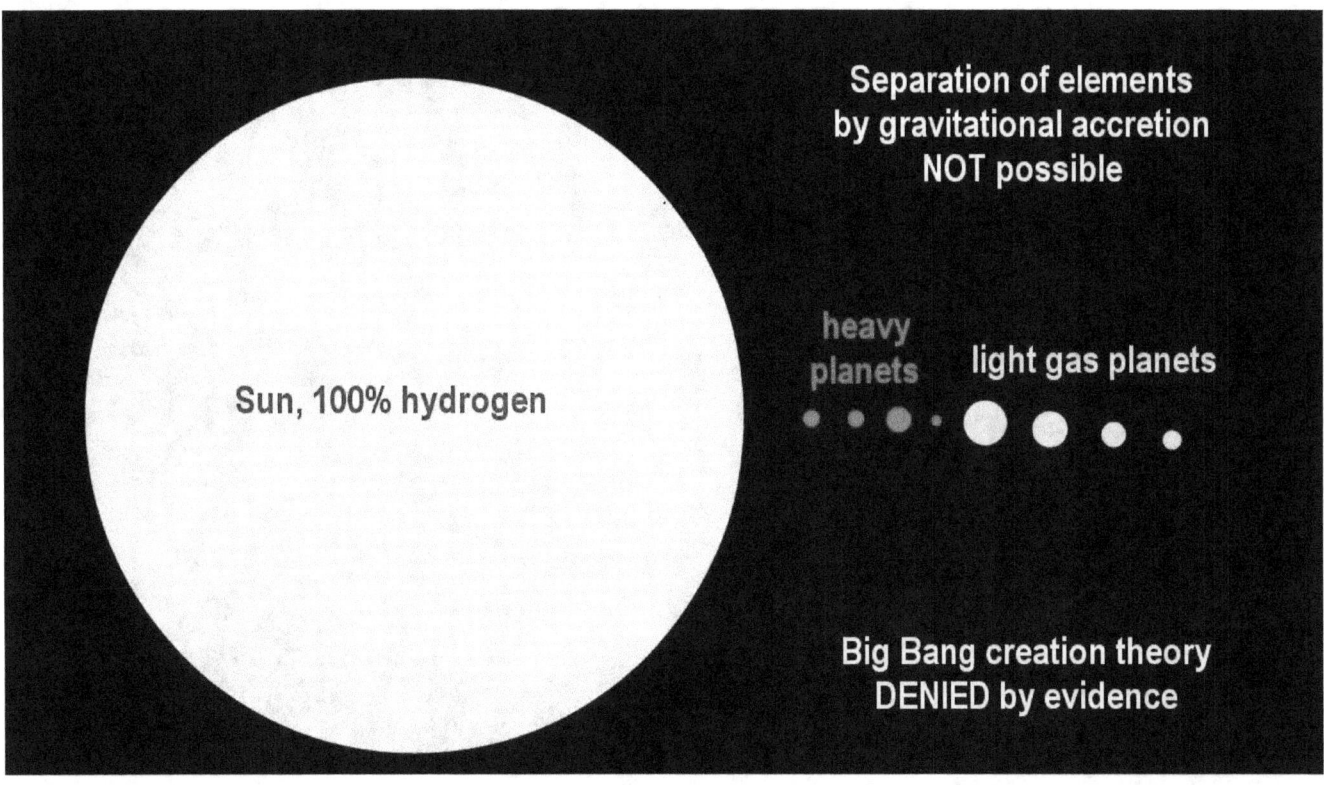

According to the false theory the Sun started as a hydrogen star, which has converted hydrogen into helium for more than 4 and a half billion years, towards its present ratio.

Where did the huge volume of hydrogen come from

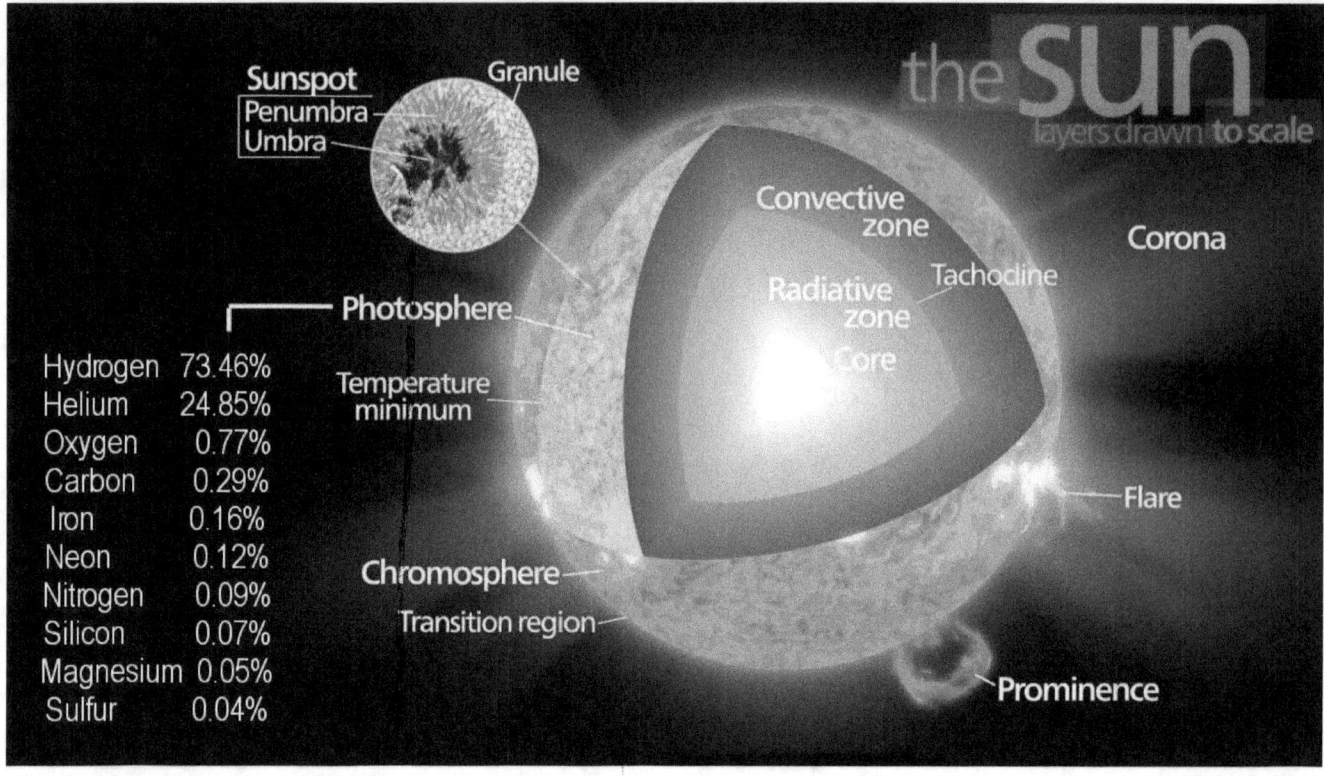

Where did the huge volume of hydrogen come from that the Sun, supposedly, has converted into helium over those billions of years of its operation at an estimated hydrogen conversion rate of 620 million metric tonnes per second?

Out of the range of the cosmic abundance ratio

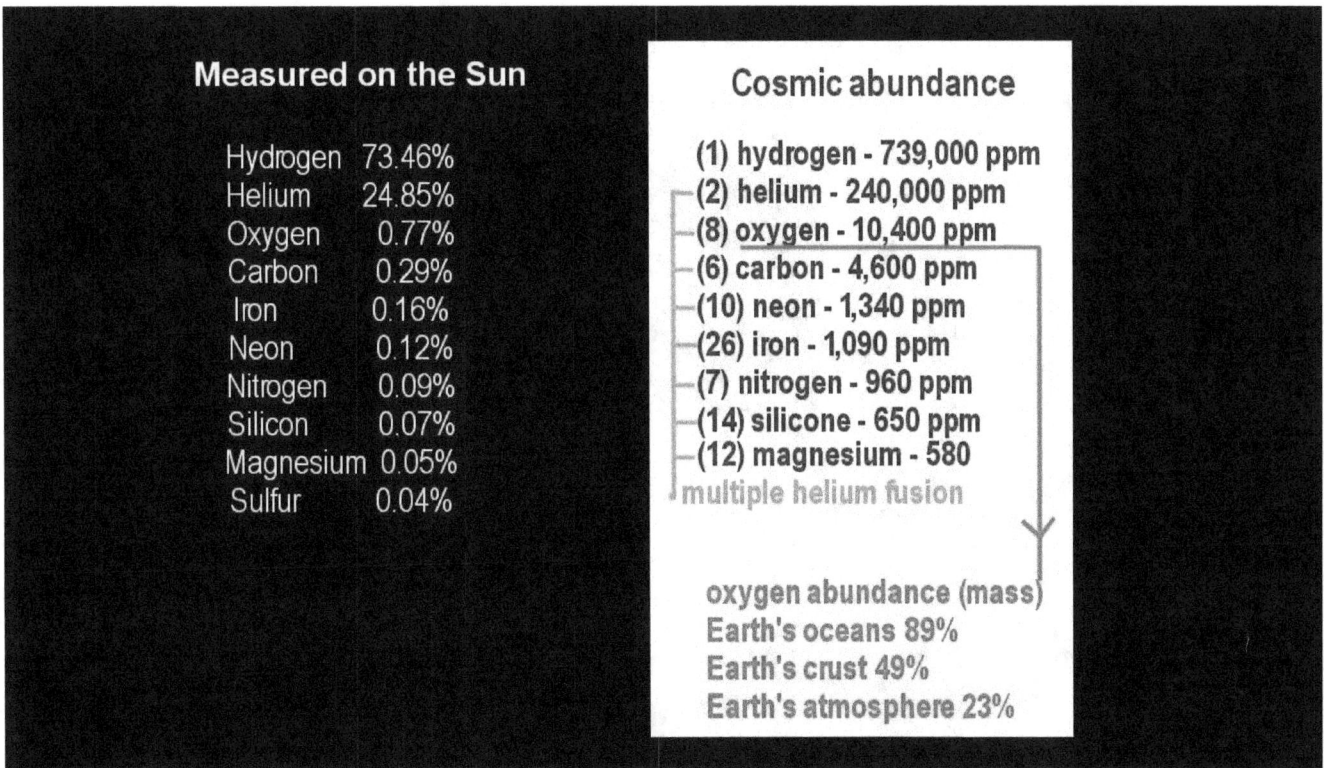

The resulting dynamic consideration places the hydrogen ratio that would have had to exist at the birth of the Sun, way out of the range of the measured cosmic abundance ratio. This impossible paradox can only be resolved on the basis of recognizing that the Sun is not, and never has been, a sphere of hydrogen and helium gas, but exists as a sphere of plasma. With this fundamental correction made in the theory, the gas ratio in the solar system closely reflects the measured cosmic abundance ratio.

The case of comparison also illustrates what it is that we actually see, when we look at the Sun through the umbra of the sunspots.

We see a Sun that is dark inside

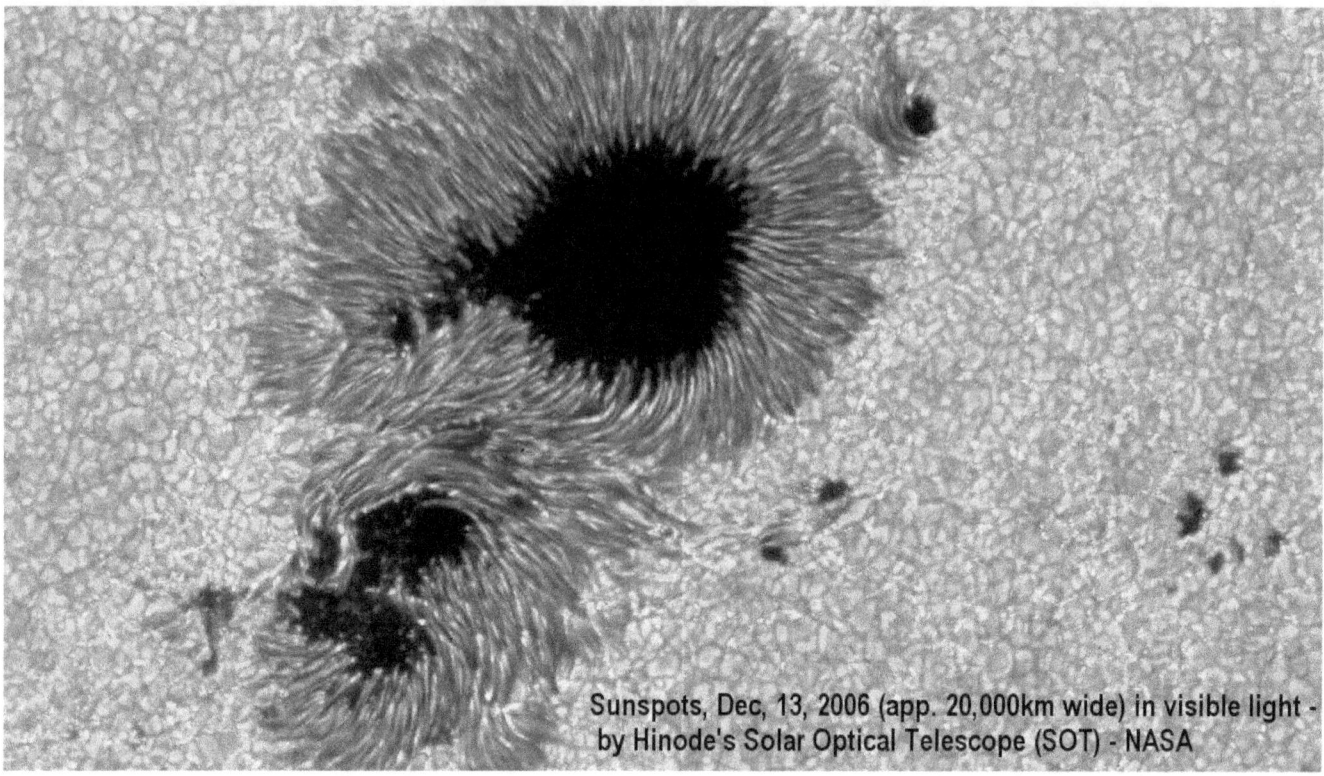

Sunspots, Dec, 13, 2006 (app. 20,000km wide) in visible light - by Hinode's Solar Optical Telescope (SOT) - NASA

When we look at the Sun, and look through the umbra of the sunspots, we see a Sun that is dark inside. We can see plainly that internal nuclear fusion isn't happening in the Sun. The Sun would be brighter inside if nuclear fusion would be happening internally, as the solar power source.

A Sun that is a sphere of thinly dispersed plasma

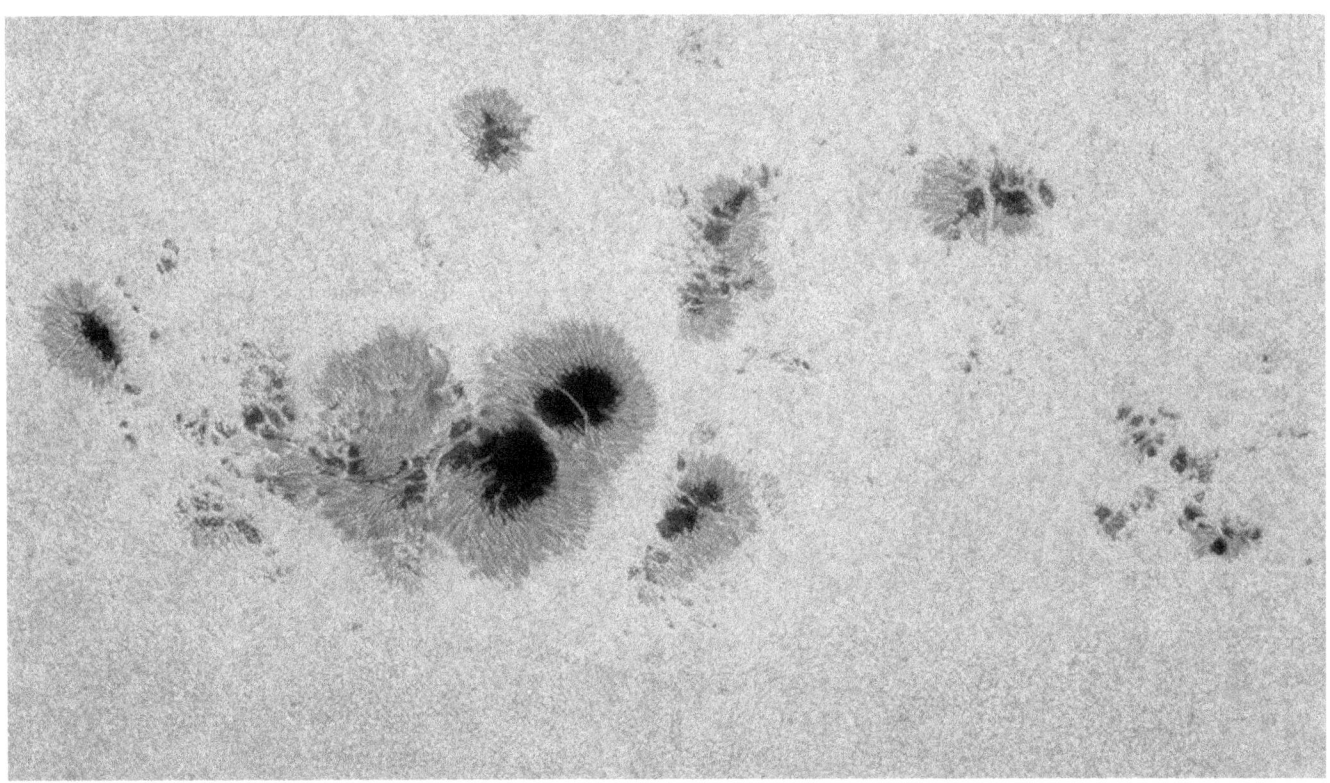

We see instead that everything happens on the surface of the Sun. This means that what we see in the Sun, is exactly what we should see for a Sun that is a sphere of thinly dispersed plasma with electric fusion occurring on its surface. We see that nothing exists there to be seen, below the surface. Plasma particles, which are 100,000 times smaller than an atom is, are invisible. They emit no light. Plasma is dark. It is a type of black hole. That's what we see.

Of course, by looking at the Sun, we can also see that the Sun is intensely energetic on its surface.

From where does the Sun derive its energy

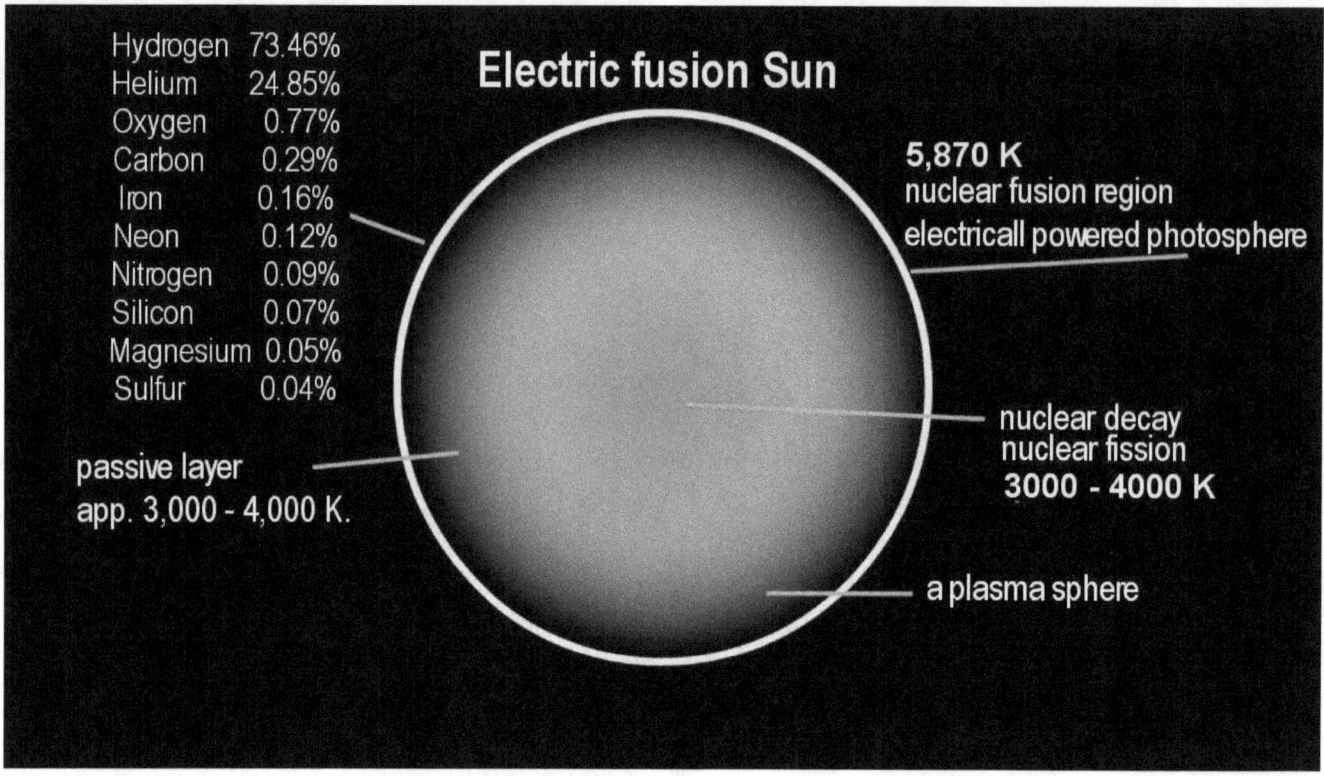

From where then, does the Sun derive its energy and its substance for synthesizing fusion? Where would the energy come from, if it didn't come from external plasma flowing into the Sun?

From plasma our world was formed!

The answer is simple. The Sun's energy and its fusion-input are both exclusively derived from plasma.

From plasma our world was formed!

But where does the plasma come from? Is it the product of the Big Bang? Evidently it isn't since it is well demonstrated that the Big Bang theory is a science fairytale. So, where did the plasma come from, and still does come from?

The radiometric dating of the Earth

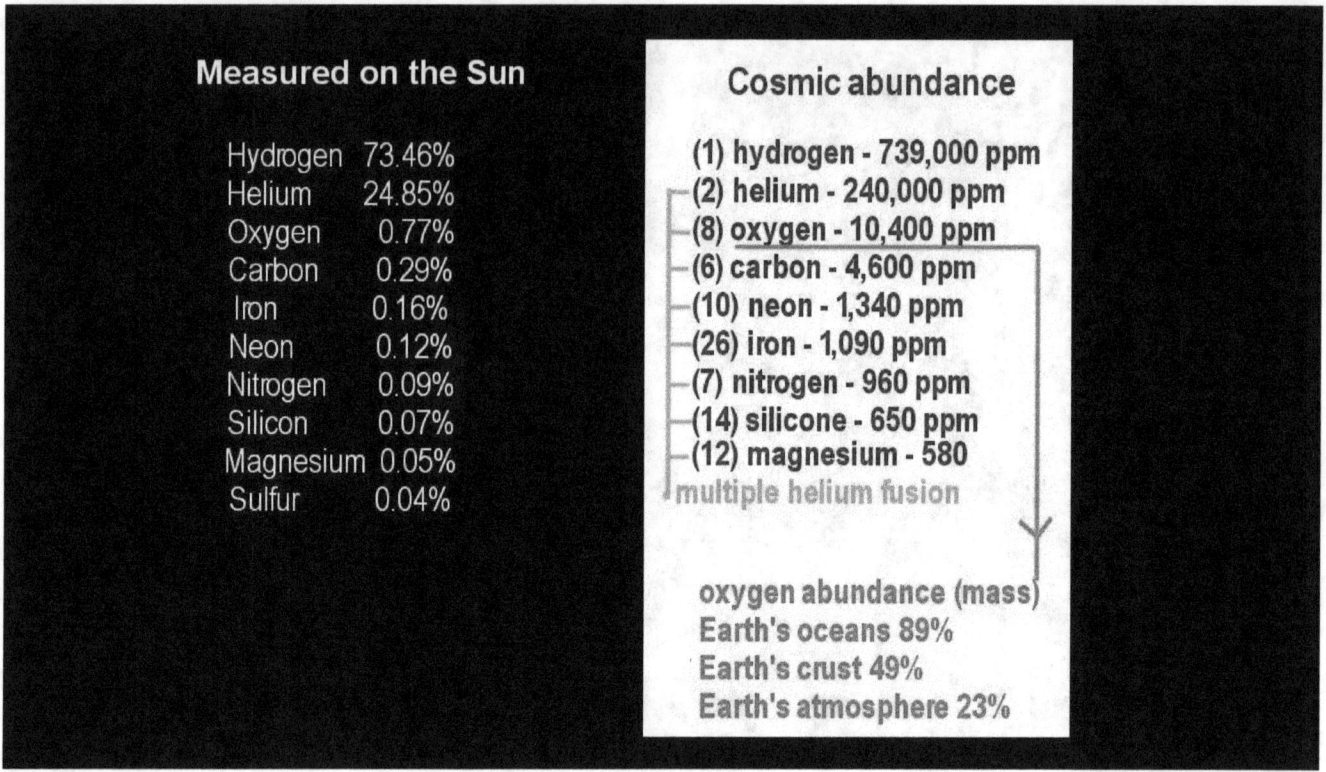

The radiometric dating of the Earth with the atomic clock of the universe, proves the solar synthesized creation of the Earth to be real, as a construct of fused plasma. We are looking at enormous volumes of plasma flowing into the Sun for this gigantic creative process to be possible.

Plasma gets 'consumed' by the solar electric-fusion process

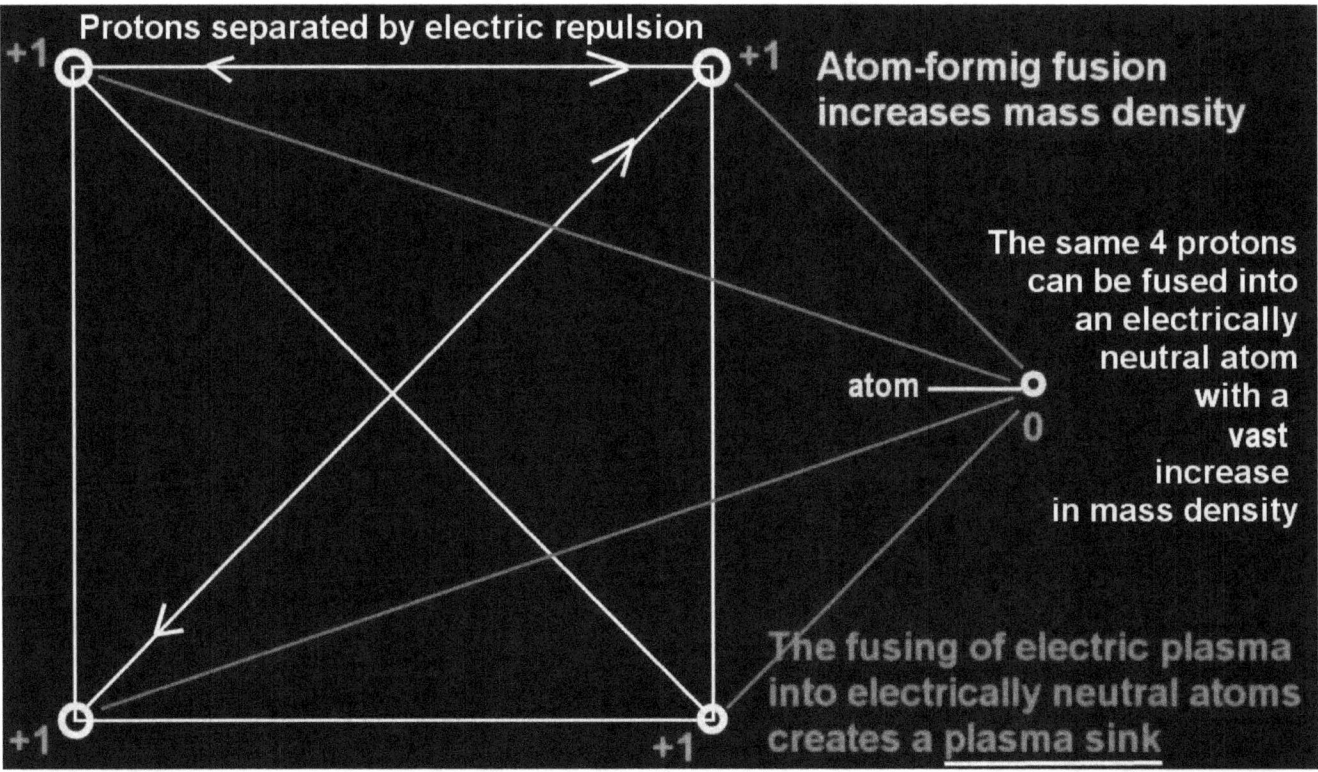

It is interesting to note that free plasma gets 'consumed' by the solar electric-fusion process, as plasma is converted into atoms. The consumption of plasma by the fusion process, creates a plasma sink that gets plasma moving. Plasma has no energy in itself, but becomes energetic when it is set in motion. And for plasma to move, we need a source and a sink that it flows into. The cold fusion process is this sink. But what is the source? We have monumental evidence on hand for the sink process, and almost no evidence that there is a source.

Evidently it is not enough for the Sun to have streams of plasma to flow to it, to energize it, because every physical flow process needs a source and a sink for it to function. The fusing of protons into electrically neutral atoms, renders the Sun a plasma sink.

If plasma would merely flow into the Sun

The Sun at a plasma node

David LaPoint - The Primer Fields

If plasma would merely flow into the Sun to energize it, it would simply pile up there, and nothing would happen. Plasma needs to flow to be intensely energetic, just like water needs to flow for hydro-electric generating systems to work.

For hydro-electric generating to work

For hydro-electric generating to work, water needs to flow from a high point, the source, to a low point.

The energy of the water

water turbine at the Grand Coulee Dam

The energy of the water, flowing from a high point, pulled by the force of gravity, delivers the energy that turns the turbine.

The energy that is generated on the Sun must have a similar potential. It must have a source and a sink, for an energetic flow to happen. On the Sun, the conditions are met. The nuclear fusion process energizes the dynamics that also create energy in the form of light and heat.

The synthesizing fusion on the surface of the Sun

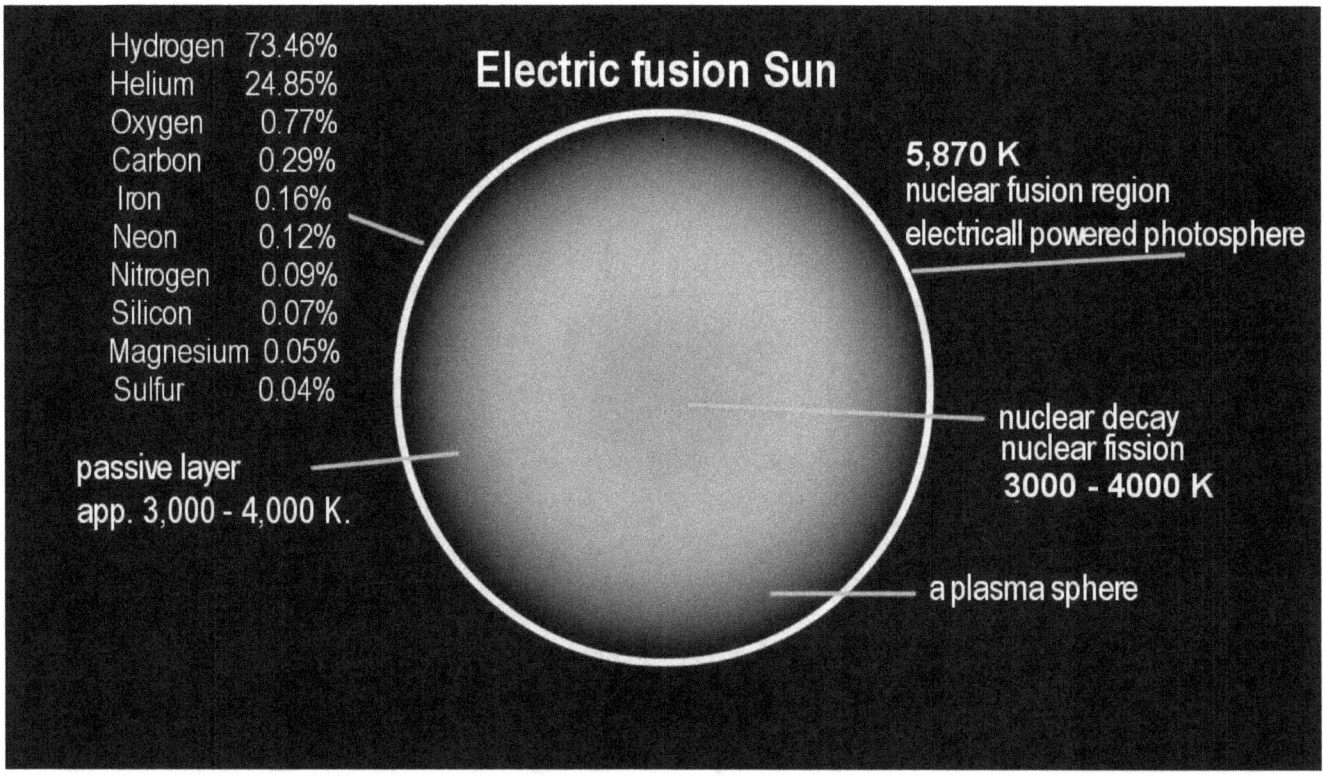

The synthesizing fusion on the surface of the Sun literally eats up plasma, packs it together, and spits it out as electrically neutral atoms that flow away with the wind. Without this process, of the Sun eating plasma, nothing would happen. Nothing would flow.

The process may be termed 'cold' fusion, as the process is not initiated by intense thermal agitation in the range of millions of degrees, but functions by cold electro-magnetic plasma acceleration and plasma pressure that also produce a modest amount of thermal and light energy along the way at a temperature of 5,505 degrees Celsius.

Plasma-fusion maybe the sink that activates the source

Electric plasma-fusion maybe the sink that activates the source.

The Sun and its fusion process is easily identified by its functioning as a consuming sink. In comparison the source is far-more vague and difficult to conceptionalize. Nevertheless, this too, is of critical importance to us living on this planet.

Even in hydro-electric generating system, the source of the water that delivers the energy is often miles distant and is rarely perceived as related to the sink system.

When the plasma-flow into a sink activates the source

When the plasma-flow into a sink activates the source, no matter how far away the source may be located, just as it would in a water-supply system, then a faint picture for a possible concept for a plasma source comes to light. For the water-supply system, the source may be a lake far distant in the mountains.

The American theoretical physicist David Bohm

For the plasma supply system the source may simply be the vast expanse of space, that, as it is defined by the American theoretical physicist David Bohm, is not really empty space at all. He introduced the concepts of Implicate Order and Explicate Order, which appears to be exotic theory, but according to evidence may be most fundamental to everything. David Bohm stated that "Space is not empty. It is full, a plenum as opposed to a vacuum, and is the ground for the existence of everything, including ourselves. The universe is not separate from this cosmic sea of energy." The explicate is then merely a specific expression of enfolded implications that leave on the surface but ripples of countless waves of energy that coming together take on a specific form. Since each wave carries an amount of energy, it has been theorized that a cubic inch of space may contain more energy than is found in all the galaxies of the universe.

The speed of light

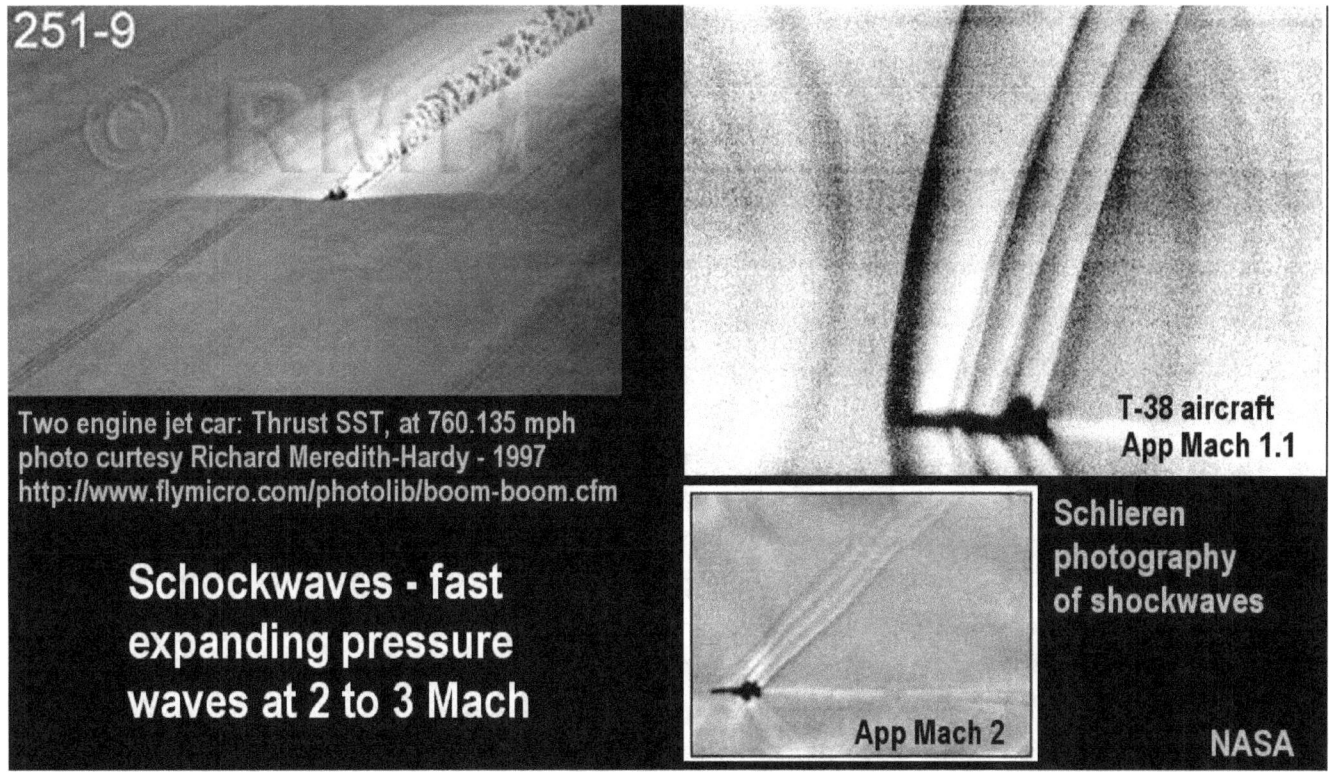

That this concept may not be far off the mark is evident in the speed limit of light as it propagates through space. The speed of sound in the air is limited by the fluid dynamics of the air. The speed of light may be similarly determined by a type of 'fluid' dynamics that governs the vast energy background in what is deemed empty space.

In the sea of latent energy that is cosmic space

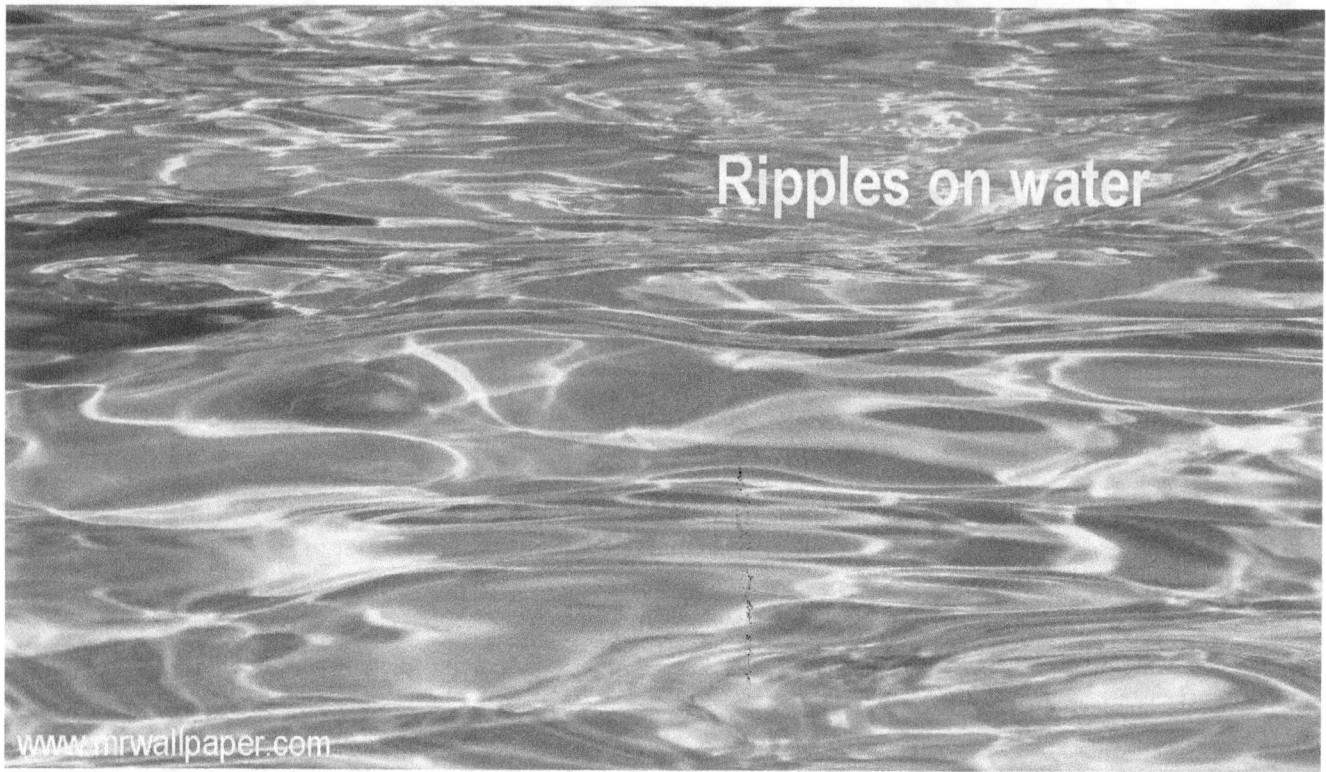

In the sea of latent energy that is cosmic space, the ripples that form explicitly on the 'surface' may form the quarks that in turn form the electrons and protons that make up the plasma in the universe.

Quarks cannot be divided

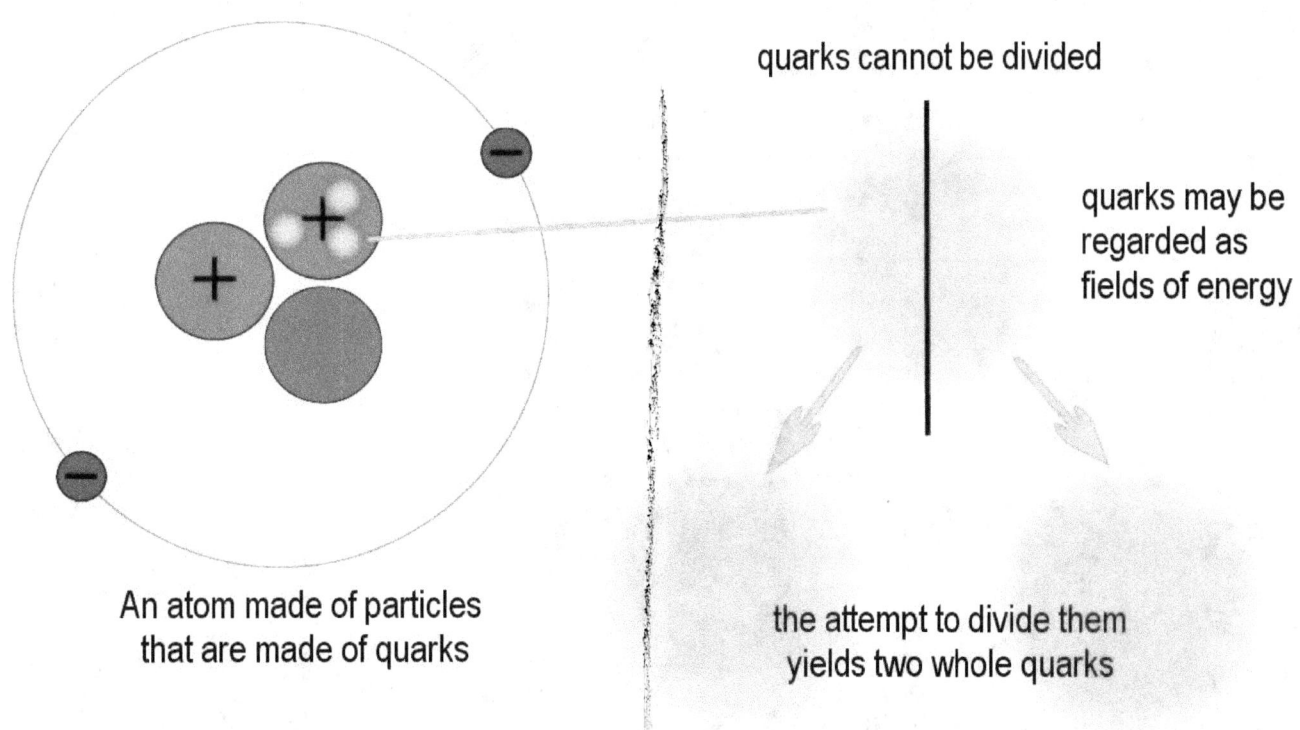

It is interesting to note that quarks cannot be divided. An attempt to break them apart, yields not two halves, but two whole quarks.

David Bohm,

As exotic as this may sound, the originator of the Implicate Order and Explicate Order, David Bohm, who lived between 1917 and 1992, may have solved this puzzle for us, and also the puzzle of the origin of plasma. David Bohm is considered to be one of the most significant theoretical physicists of the 20th century, whom Einstein had referred to, as his successor.

We see two very-long climate cycles expressed

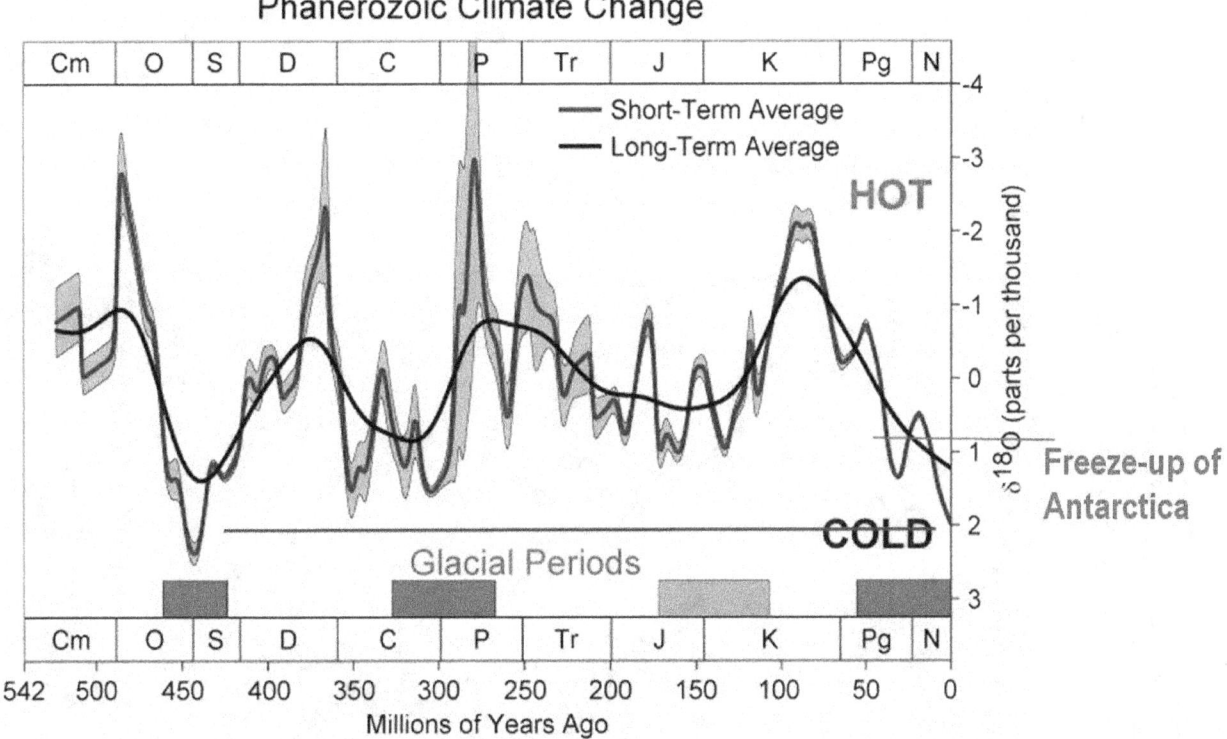

Exotic as the theory may seem, we have some amazing physical evidence for it, which is expressed in the way our climate on Earth has been modulated in long cycles over the last 500 million years. We see two very-long climate cycles expressed here, that are overlaid over each other. We see a 150-million-years cycle, and a 31-million-years cycle expressed. These are huge cycles with enormous effects. Note, where on this gigantic scale the freeze-up of Antarctica is located, and where the current stage of the world is located, that puts us at the lowest and coldest point in 440 million years.

The very long cycles can be seen as evidence

The Capodimonte Deep Field - of 35,000 galaxies

The very long cycles can be seen as evidence that the plasma that powers our Sun may have its origin primarily in intergalactic space. Researchers at the Los Alamos National Laboratory have come to the recognition that vast plasma streams extend through all cosmic space, combining all galaxies with networks of plasma streams.

Plasma streams that have galaxies at their node points

The plasma streams that have galaxies at their node points, with plasma being electric in nature, are inherently subject to electric resonance principles. For the Milky Way galaxy, which appears to be located at one of the node points between large galaxies, the resonance-characteristics in the plasma streams would become expressed in our galaxy as very long plasma-density cycles, with the length of the cycles corresponding to the length of the plasma streams.

The Andromeda galaxy

This means for us, that the closest galaxy to the Milky Way, named the Andromeda galaxy, which lays 2.5 million light years distant, appears to correspond to the 31 million-year resonance that we find modulated in the Milky Way, and the climate cycles on Earth.

The other connecting stream from the Milky Way

galaxy Messier 83

NASA, ESA, and the Hubble Heritage Team

The other connecting stream from the Milky Way would be correspondingly longer. It would likely lead to the gigantic M83 galaxy that lays 15.2 million light years distant, and cause the 150-million-year cycle.

The resonance waves become overlaid

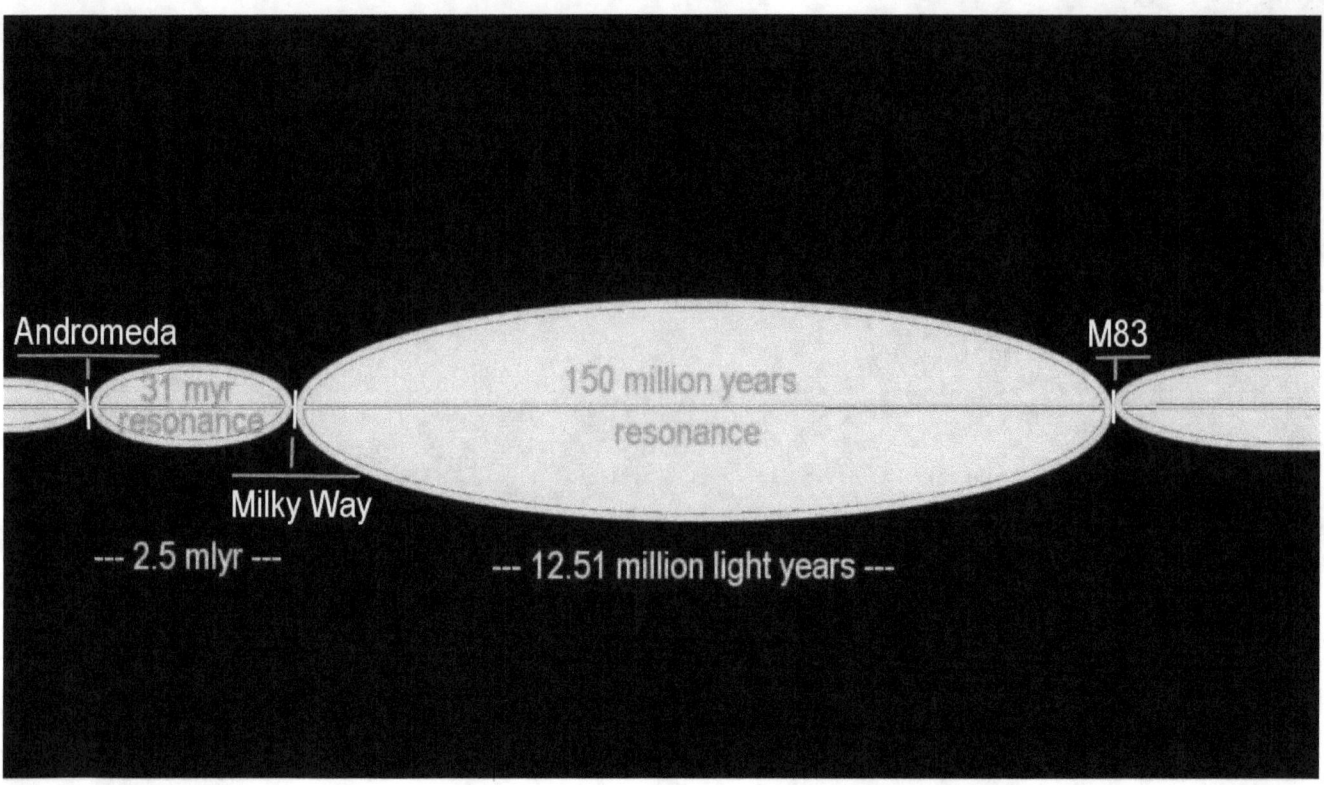

The resonance waves in both of these long-distance plasma streams coming together in the Milky Way, would give us the 31 and 150 million-years plasma density cycles respectively, that become overlaid and expressed as climate cycles on Earth.

The resulting evidence in historic climate cycles seems to tell us that the plasma source for our Sun, and for every sun in our galaxy, is located far-distant in intergalactic space, which the sink-effect of our sun, together with every star in our galaxy, draws from, which thereby keeps the entire plasma flow landscape in motion.

The fusion-sun sink process

However, the flow through the faucet depends on the pressure at the source. This means that the fusion-sun sink process depends for its functioning on the supply-line density, which is the plasma density in the galaxy.

The galactic plasma density is at a low point

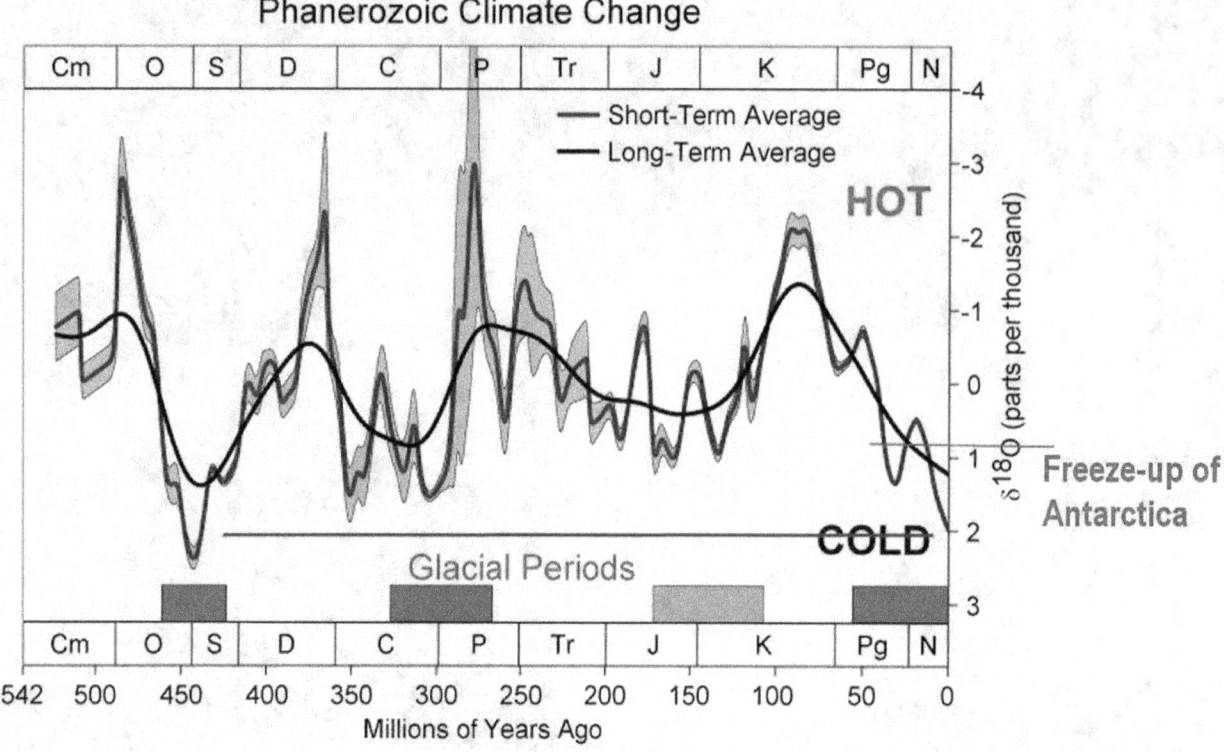

Right now, the galactic plasma density is at a low point. The 31 and 150 million-years plasma density cycles are both at or near their lowest point. This combined low has brought us into the density zone were ice ages erupt, according to the dynamics of the local resonance cycles that are specific for our solar system. Here too, we see several cycles overlapping near their low point, with the result that our Sun will go inactive for the lack of plasma density.

When the sink effect draws more than the supply line holds

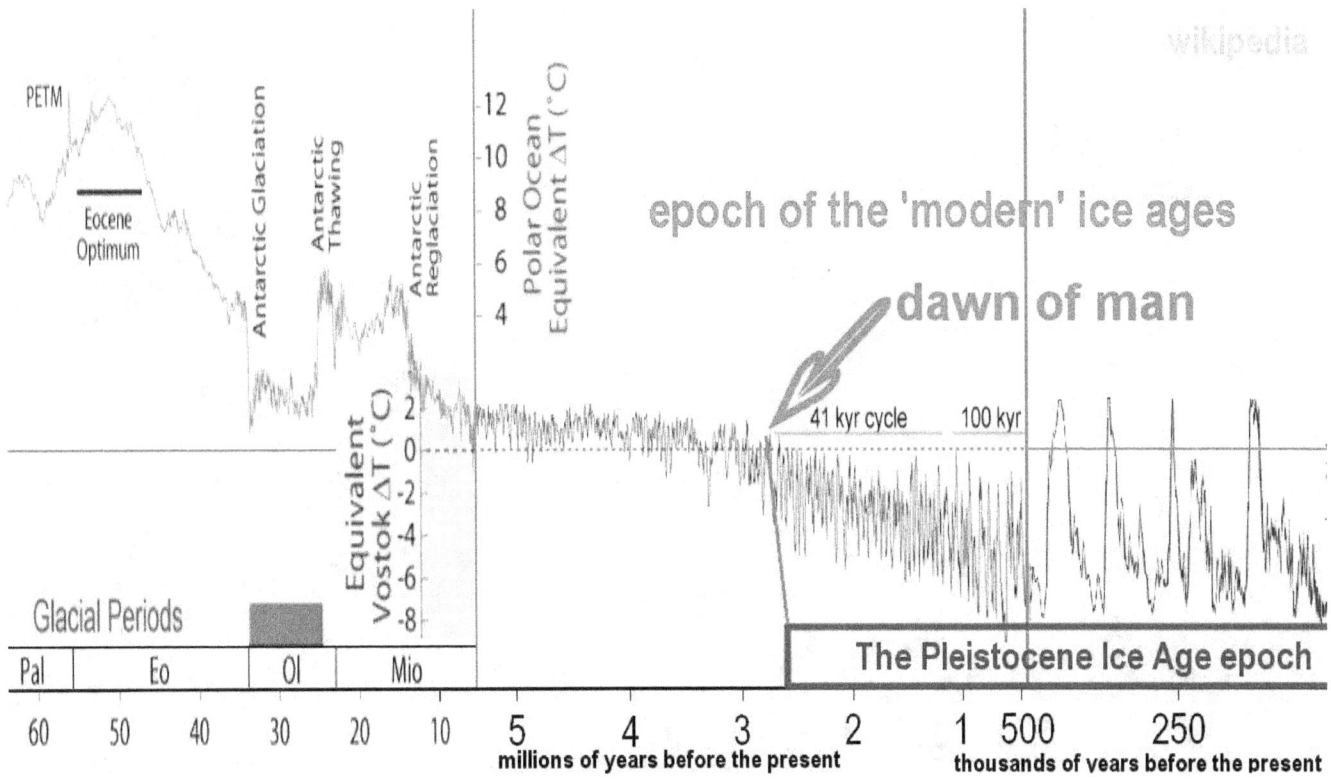

When the sink effect draws more than the supply line holds, the process becomes interrupted till the supply streams recover. Here the big Ice Age unfolds, with an inactive Sun, except for short periods, until the supply density becomes re-established. We are in this zone.

The universe is motivated by the plasma-sink process

The universe is motivated by the plasma-sink process. It affects everything.

The need for the sink feature

Since the need for the sink feature is not readily apparent, allow me to illustrate it once more, by comparing the plasma-flow system with a water supply system. Water supply systems typically have a faucet installed. When one opens the faucet, water flows. When this happens, the water in the system flows across the entire length of the supply line, all the way to the source that may be a lake high in the mountains. In such a system, the open faucet becomes a type of sink that enables the water to flow to where it is needed. With the faucet open, water flows throughout the entire length of the pipeline system. When the faucet is closed, nothing flows anywhere.

In the process of creating the uranium atom

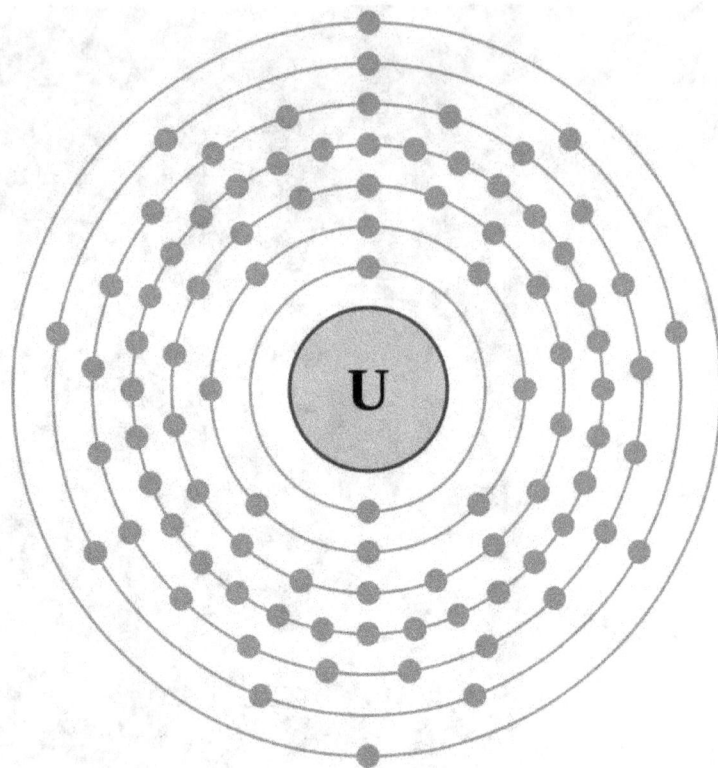

The Uranuim Atom

92 protons
146 neutrons
92 electrons
aranged in layers of
2, 8, 18, 32, 21, 9, 2

Now, let's look at an extreme example. In the process of creating the uranium atom, 238 protons were fused together to form the central nucleus for it. Before the protons were fused, they were a part of the plasma landscape and were repelling each other. After the atom is formed, the 238 protons suddenly are no longer a part of the electric landscape where they would add to the plasma pressure. The forming of the atom left behind a 'vacuum' as it were, in the plasma landscape. Of course, the vacuum is quickly filled with inflowing plasma. In this manner, a significant rate of flow is created in the movement of plasma, that becomes reflected throughout the entire system of plasma streams, reaching far back into interstellar space.

Interstellar plasma streams are set in motion by this consuming plasma fusion process that becomes a dynamic sink. This plasma sink is needed. Without a sink, no flow happens.

When plasma flows in interstellar space

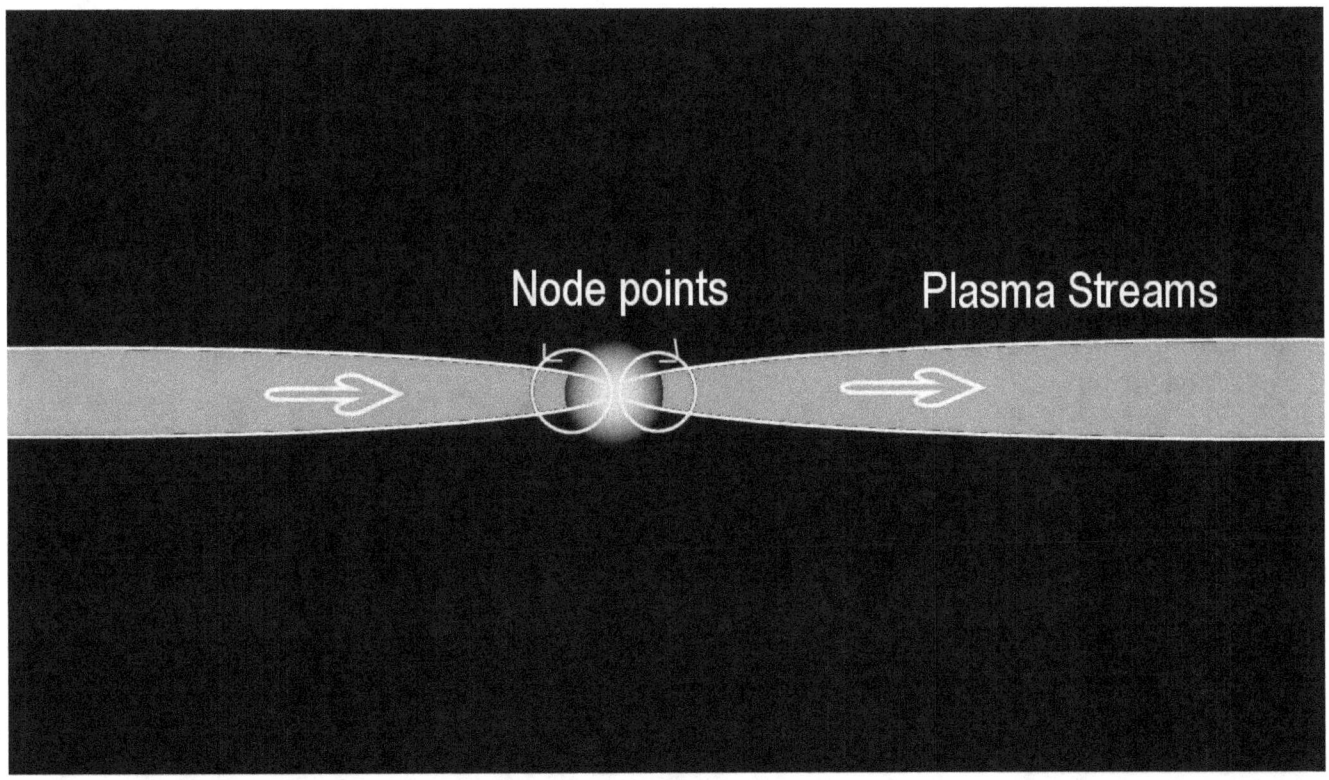

Now when plasma flows in interstellar space, the magnetic fields produced by the electric flow, draw the plasma streams closer together, by what is called the Lorentz force. Of course, when the plasma streams are drawn together, the electromagnetic pinching of them into an ever-smaller cross section increases the plasma density. The increased plasma density, in turn, increases the magnetic fields, which in turn pinches the current ever tighter. This self-feeding process forms the geometry of a bowl-type structure at the highest-density point. The magnetic fields become intensely tangled up at a point of high concentration, whereby the bowl-shaped end opens up and enables concentrated plasma to escape. The node points would logically occur near a plasma sink, which motivates the plasma flow in the first place.

Assume that the plasma sink is our Sun

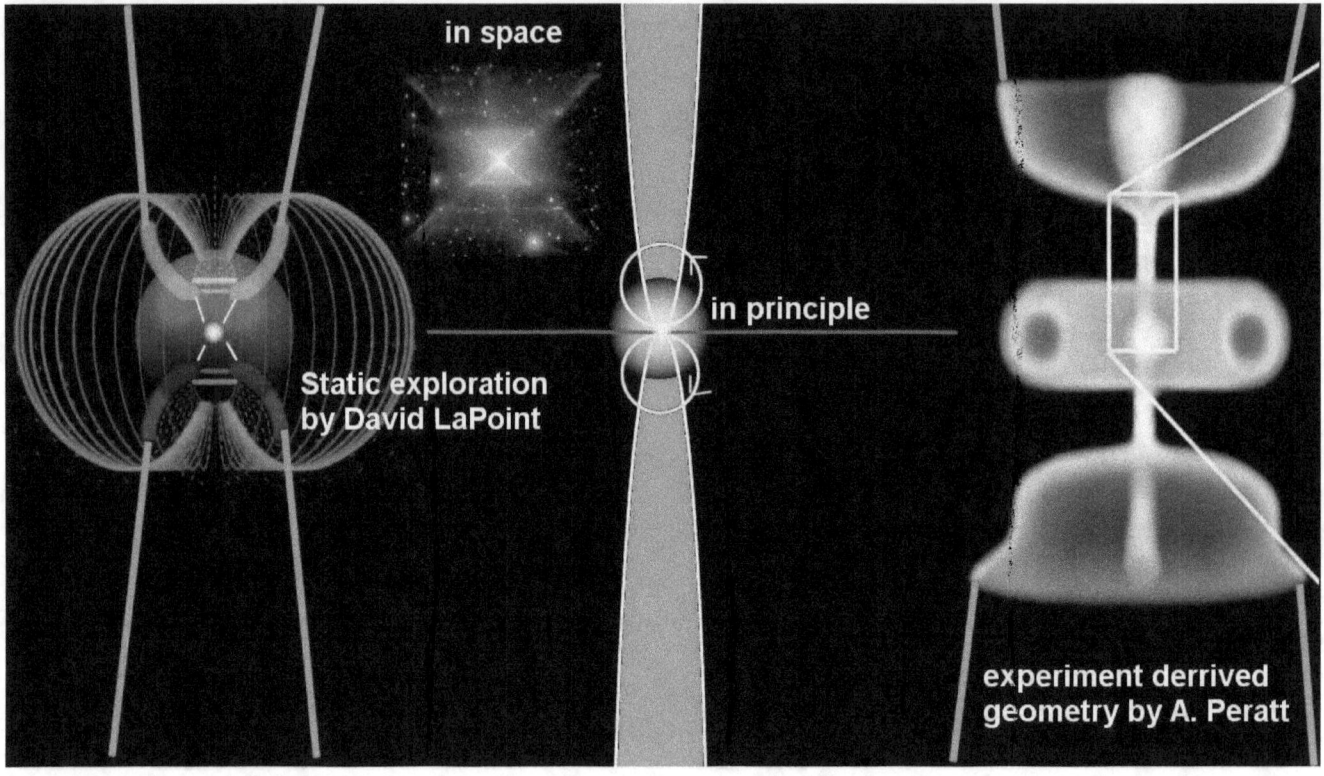

Let's assume that the plasma sink is our Sun, as the center of the system. In this case the bowl shaped magnetic fields would be focused onto our Sun.

In the Red Square Nebula

Normally, interstellar plasma streams are not visible, because the plasma particles are too small to be visible, but in a few extremely high-powered examples, such as in the Red Square Nebula, some parts of the bowl-shaped geometry do become visible in the areas of the highest plasma concentration and atomic concentration, where the dance of the electrons in plasma interacts with the synthesized atoms.

Contrary to general perception, nebulas are not the remains of exploded stars, but are high-power electromagnetic structures.

In the case of Red Square Nebula, a number of unique features of the high-power electromagnetic bowl-type structure become visible.

The plasma researcher David LaPoint

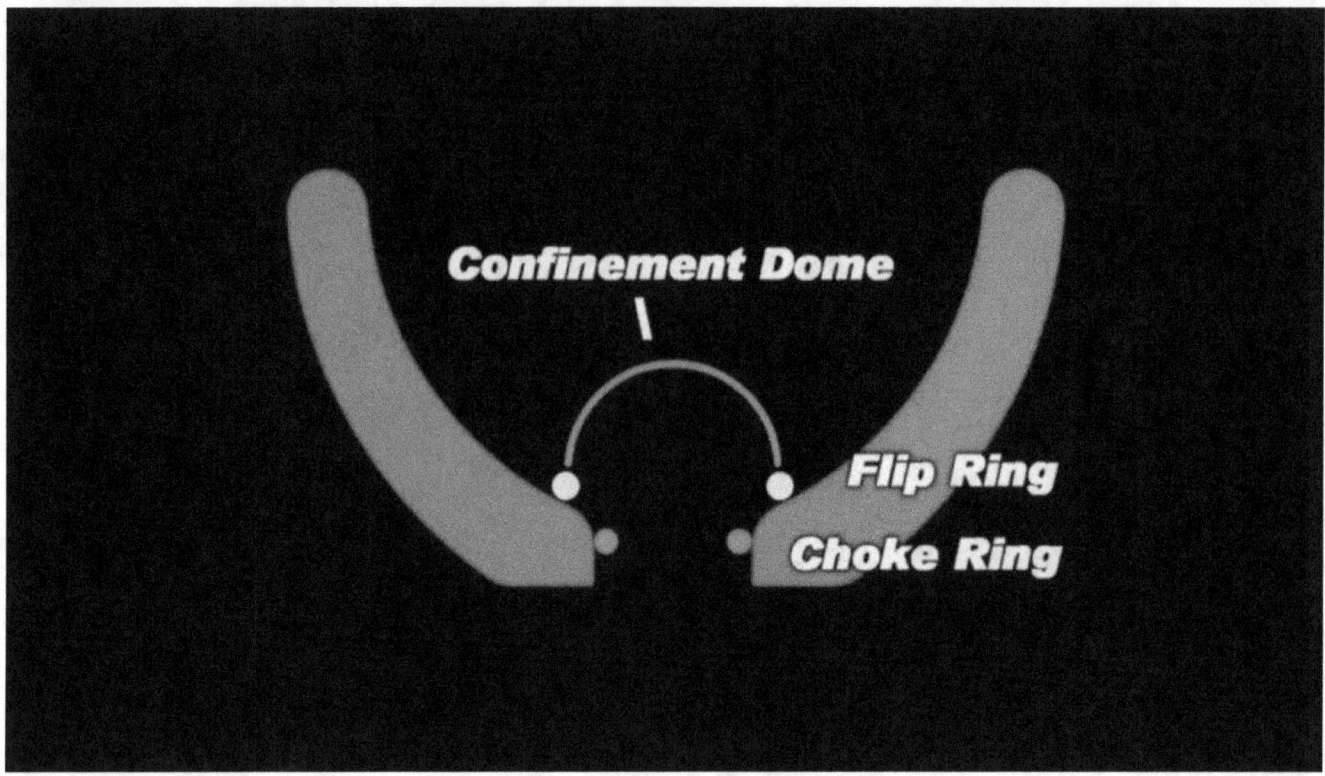

The plasma researcher David LaPoint has extensively explored the nature of the magnetic fields of bowl-shaped magnetic structures, which he has termed The Primer Fields. He identified three unique magnetic features. He has identified what he calls a confinement dome, a flip ring, and a choke ring.

The flip ring is a magnetic ring

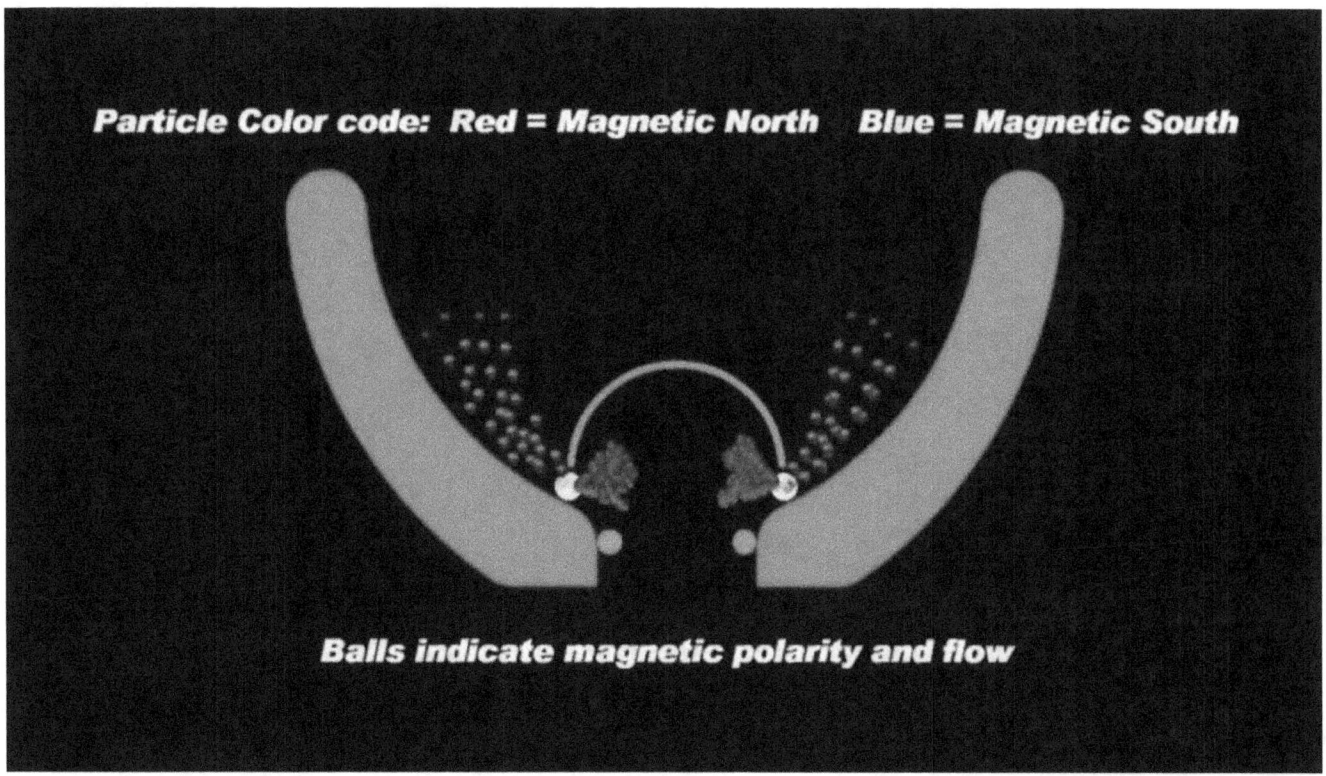

The flip ring is a magnetic ring that flips the incoming plasma stream upwards under the confinement dome where plasma becomes intensely concentrated. The concentrated plasma, becomes then focused downward by the choke ring that focuses the concentrated plasma onto a sun that acts as a sink for the plasma flow. All three features are visible in the Red Square Nebula.

The features that we see in the Red Square Nebula

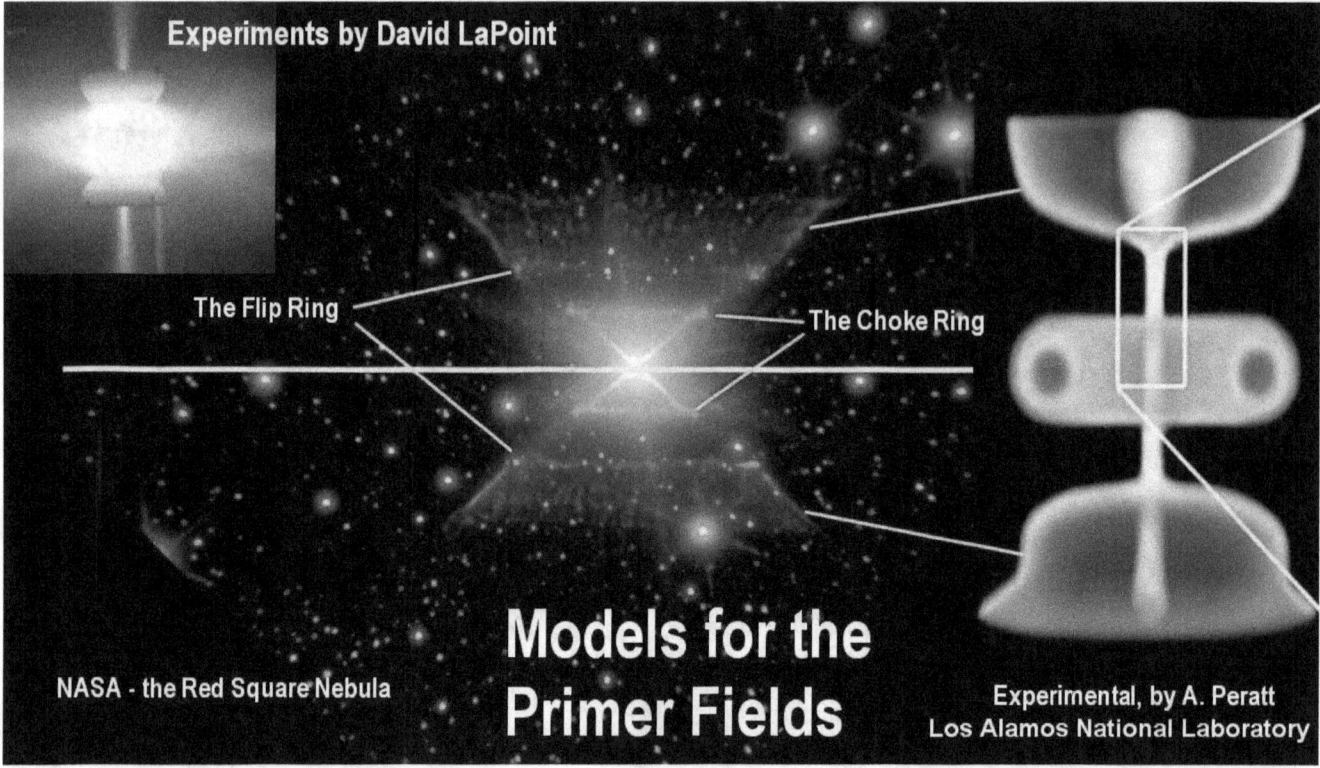

The features that we see in the Red Square Nebula correspond with what David LaPoint has been able to replicate in the laboratory. The features also match in principle the resulting plasma shapes observed in high-energy plasma flow experiments conducted at the Los Alamos National Laboratory, by Antony Paratt, director of experiments.

The flow of electricity is critical

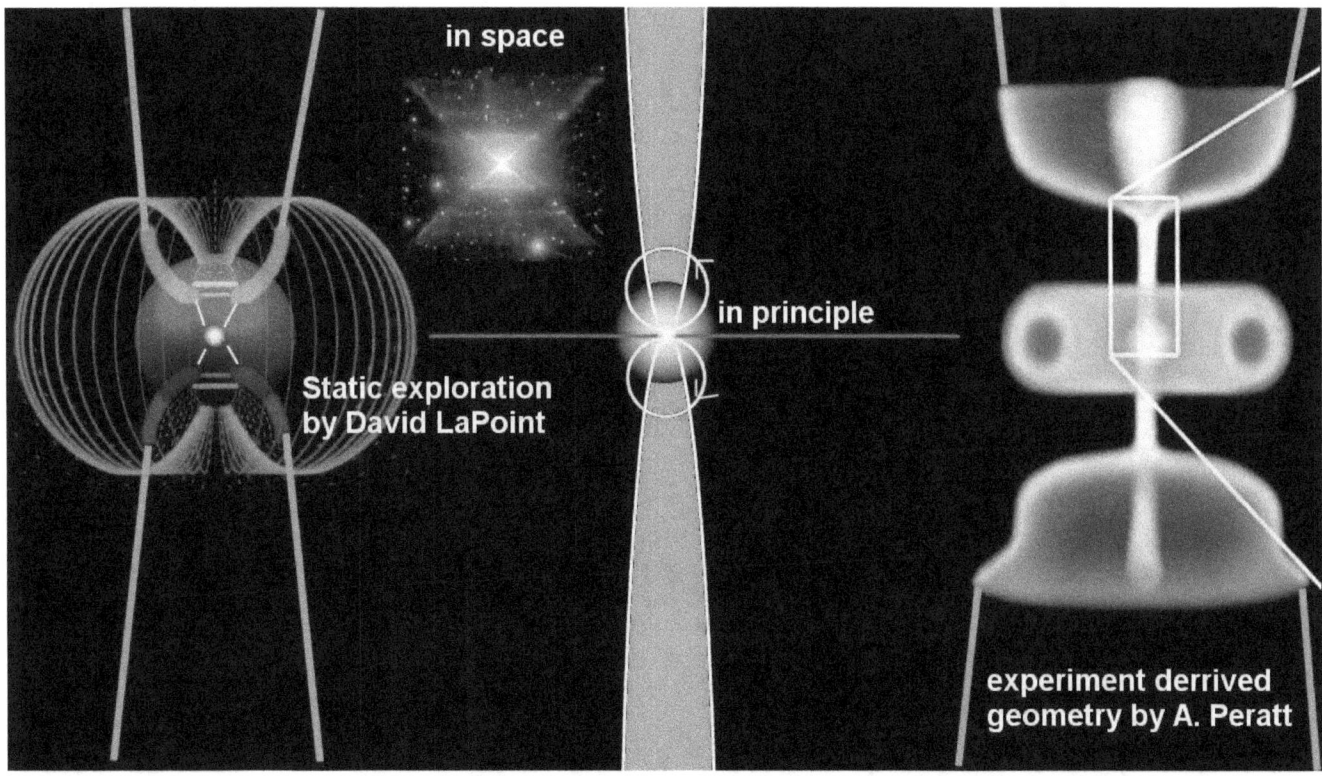

In electrodynamic systems, the flow of electricity is critical. It is exclusively the flow of electricity that creates magnetic fields. No other cause for magnetic effects exists.

By the resulting magnetic fields

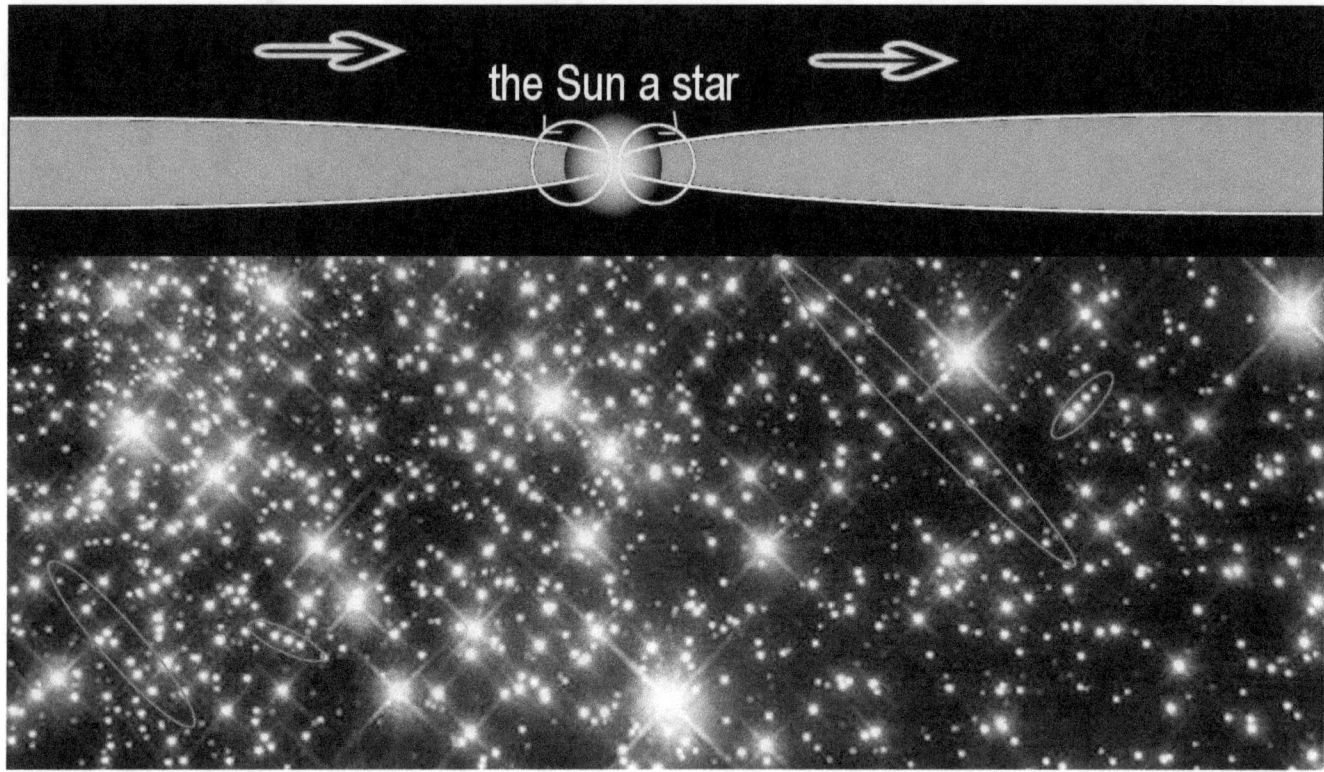

By the resulting magnetic fields, the electric plasma streams become aligned, they become pinched together, and become increasingly concentrated by the magnetic fields in such a manner that a high density plasma sphere forms around a sink object, like our Sun is.

Plasma surrounding our Sun becomes so dense

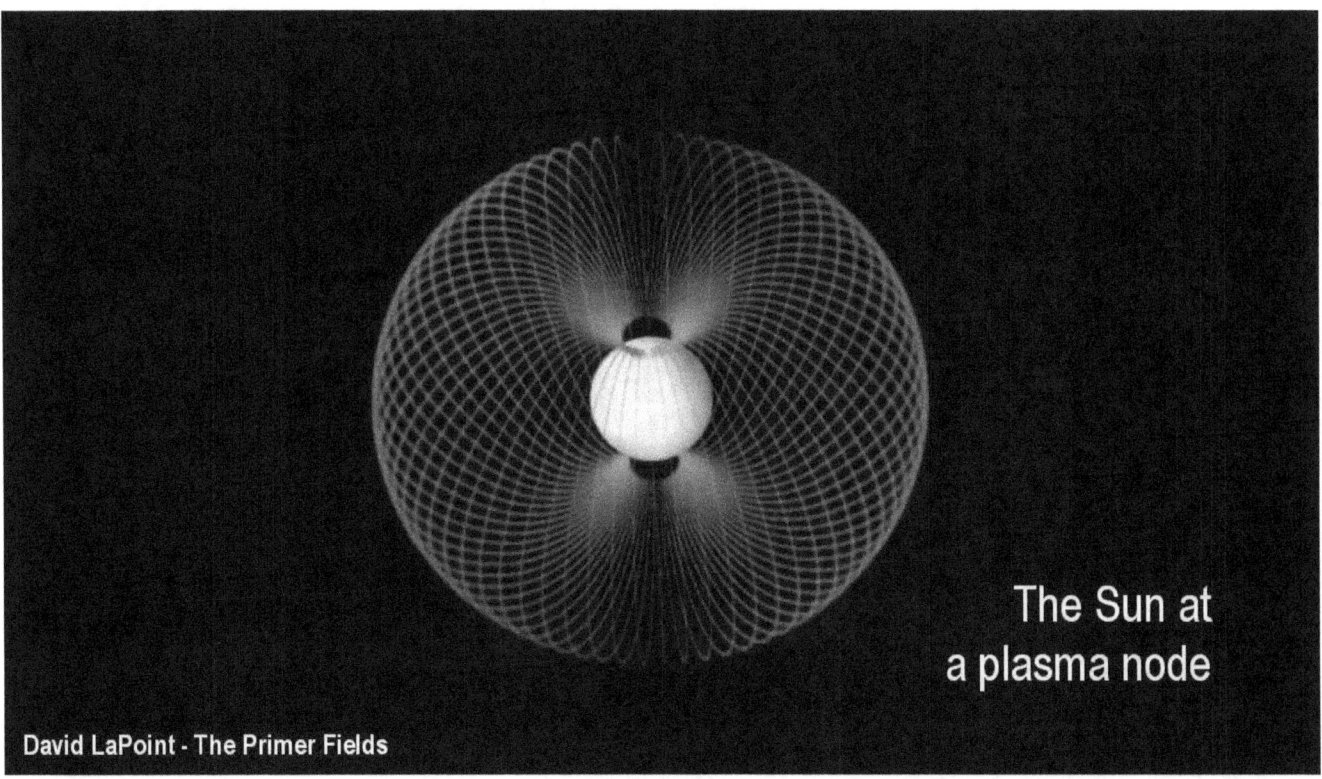

By this process the plasma surrounding our Sun becomes so dense that electric fusion reactions can, and do, occur at the Sun's surface in intensely concentrated plasma, magnetically confined by the primer fields.

This is what causes our Sun to become a sink

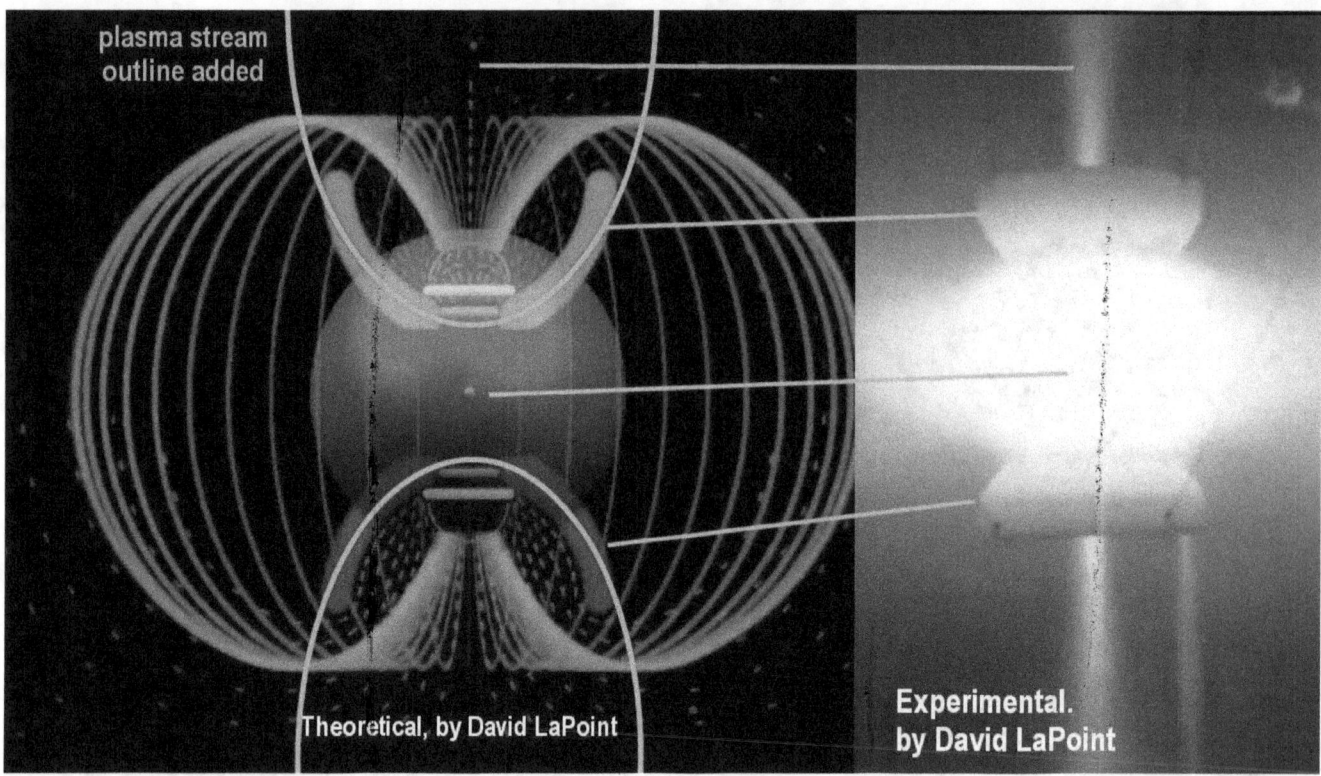

This is precisely what causes our Sun to become a sink for interstellar plasma, which it must be, in order to motivate the flow of plasma into it?

The Sun is seen as a vast sea of granular cells

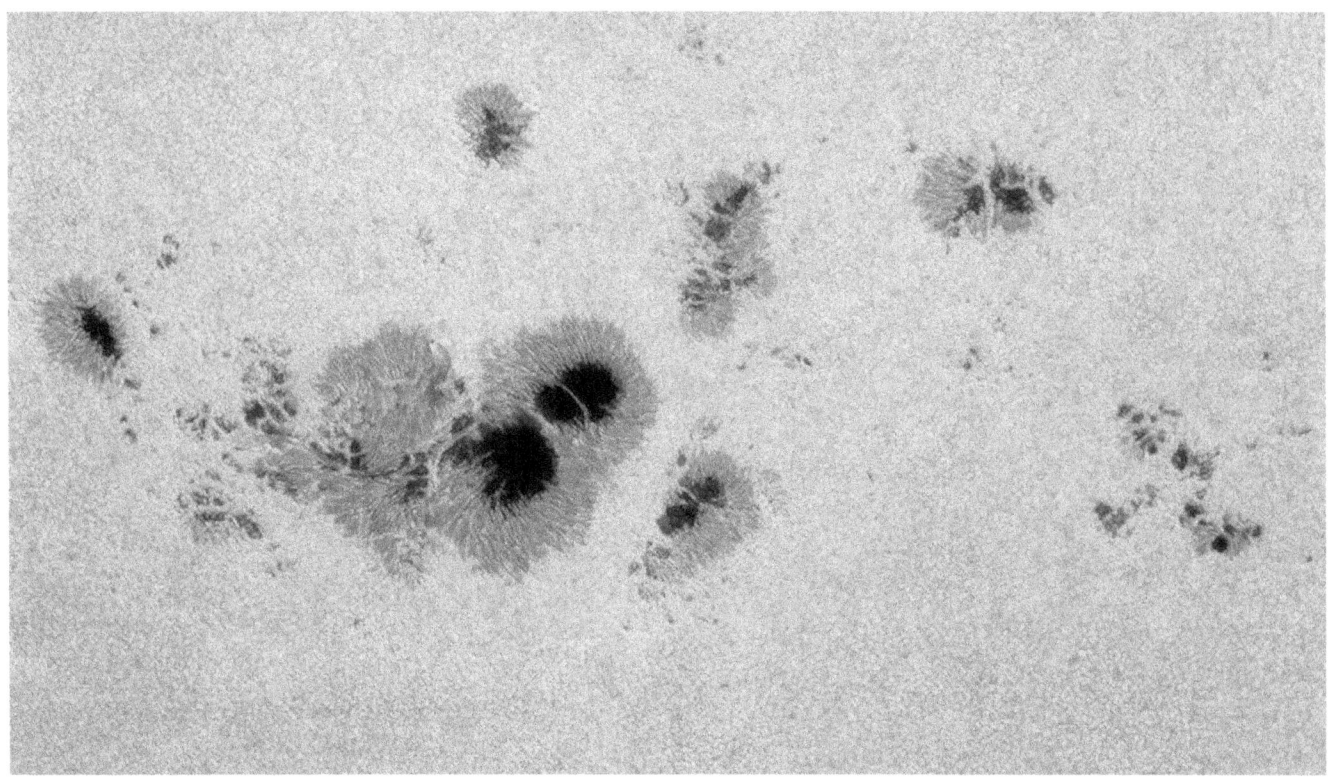

For a closer look, we need to look at the Sun's surface.

When the Sun is observed with an optical telescope, the Sun is seen as a vast sea of granular cells, which sometimes break down, leaving sunspots in the wake.

The graduals are cells of Primer Fields

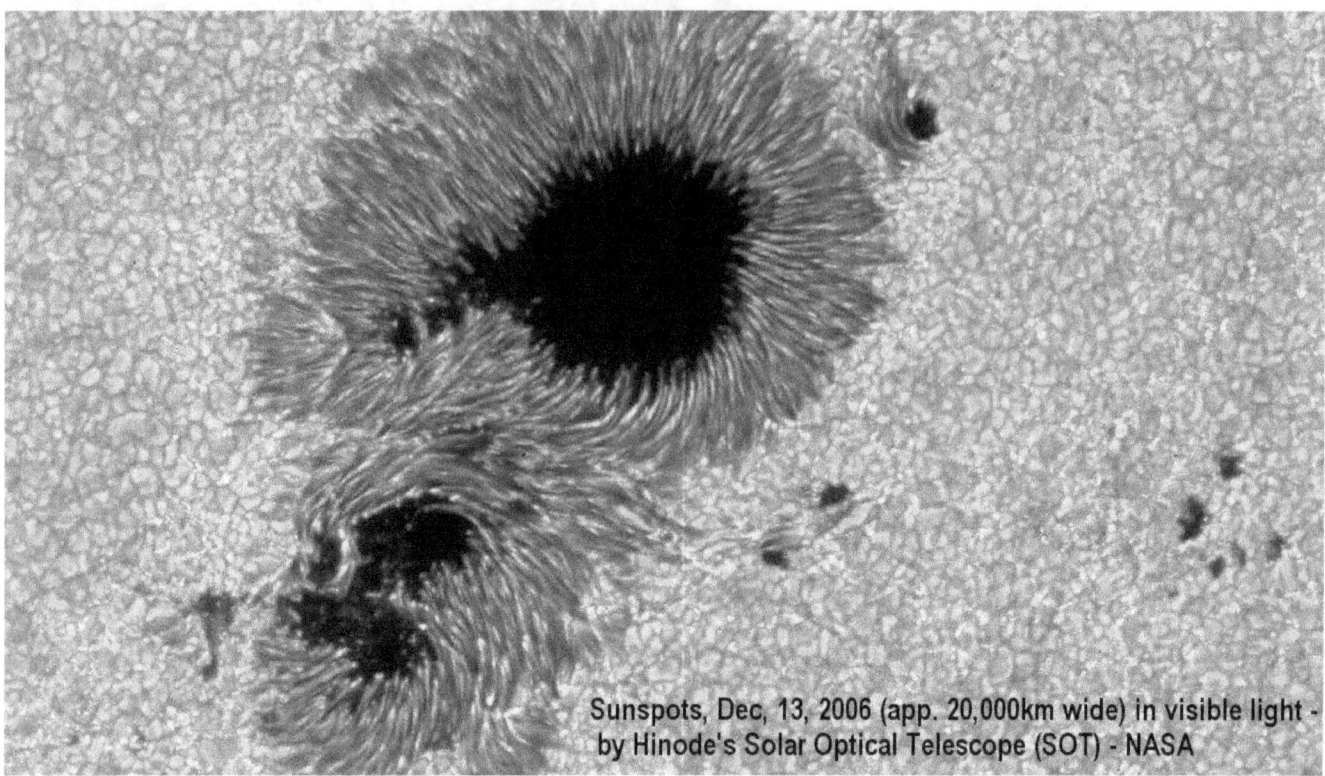

Sunspots, Dec, 13, 2006 (app. 20,000km wide) in visible light - by Hinode's Solar Optical Telescope (SOT) - NASA

Since the Sun is powered on its surface by electric plasma interaction, the graduals are themselves cells of Primer Fields in operation, on a relatively 'small' scale. The cells are typically a thousand kilometers across.

The Sun is a vast sea of cellular primer field structures

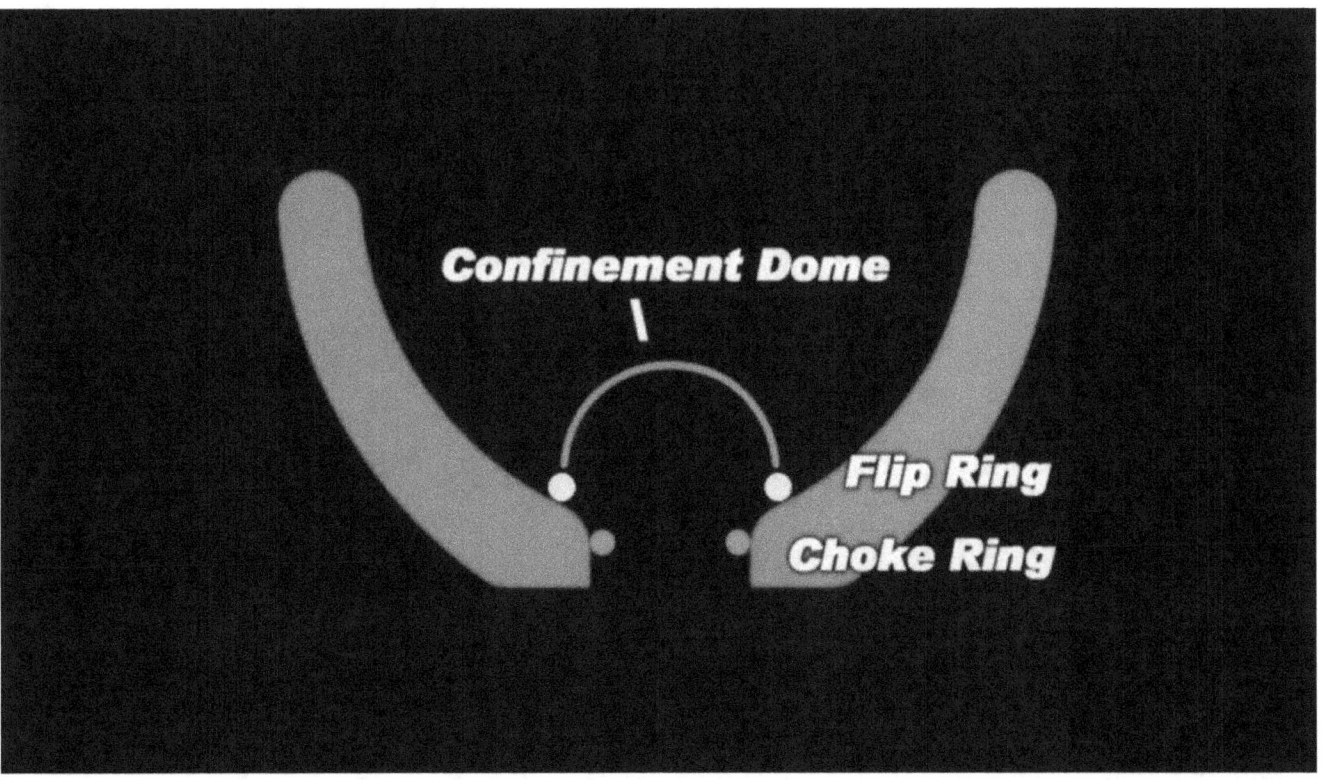

This means that the surface of the Sun is a vast sea of cellular primer field structures, all operating side by side, with each having its own flip ring and confinement dome.

Here again, the inflowing plasma

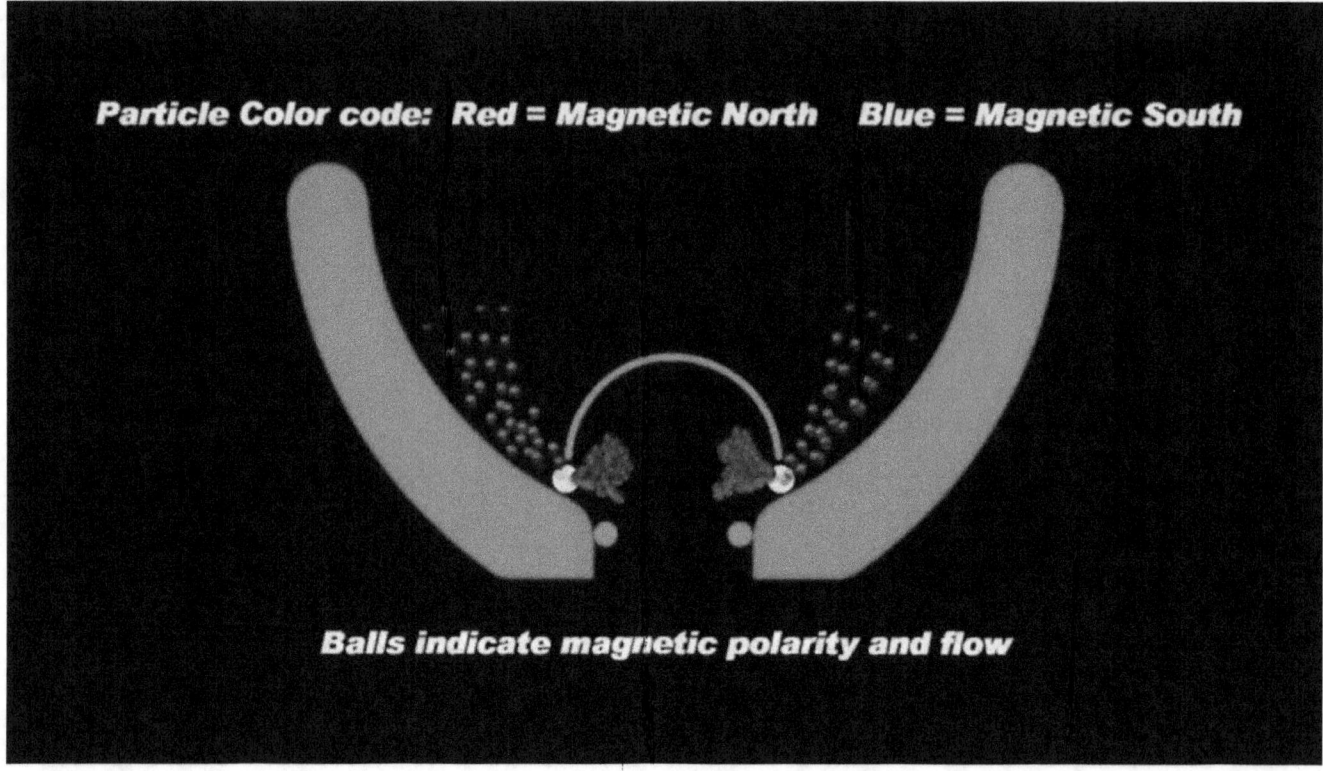

And here again, the inflowing plasma, that is at this stage highly concentrated around the Sun, becomes even more extremely concentrated under each cell's confinement dome.

Nuclear fusion causes the Sun to act as a plasma sink

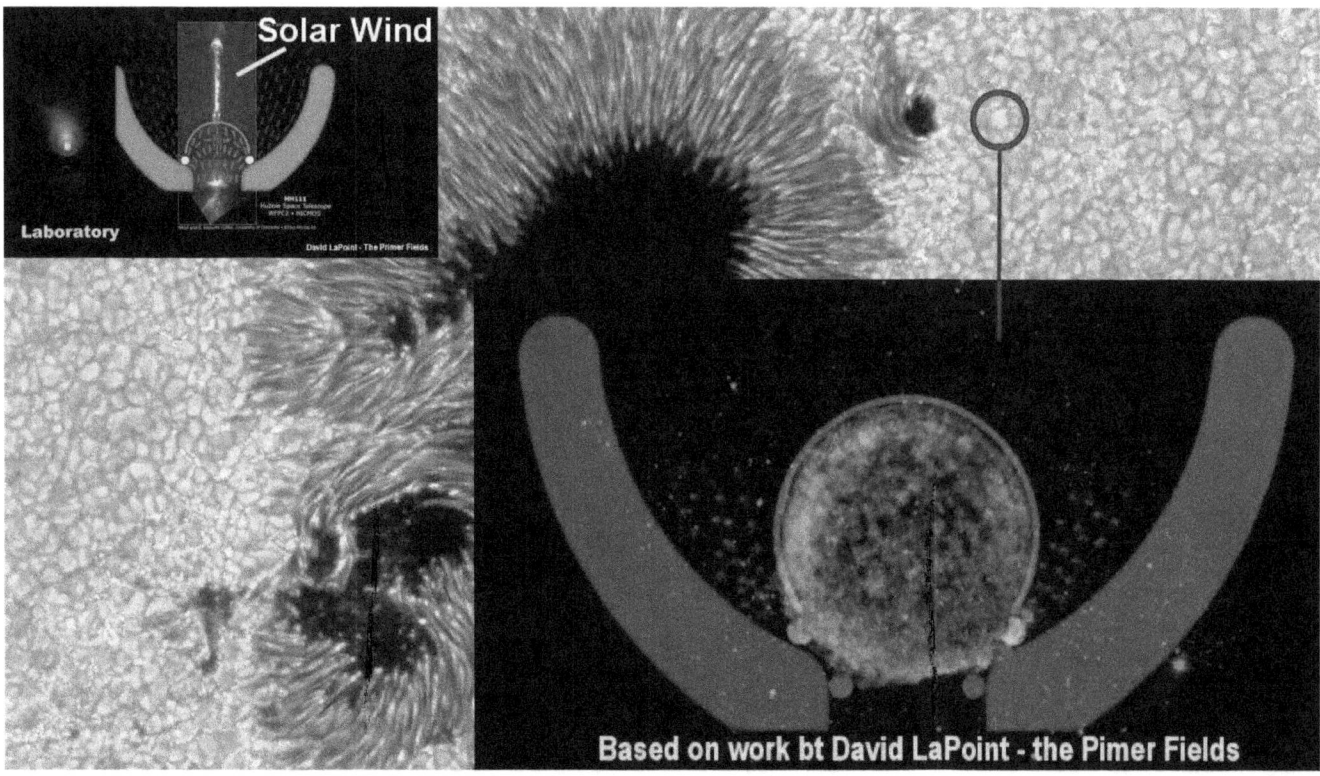

In this case the plasma becomes so immensely concentrated, and accelerated, under the confinement dome, that nuclear fusion occurs. As I said before, the resulting nuclear fusion process causes the Sun to act as a plasma sink.

Without the fusion process on the surface of the Sun, which synthesizes all natural atoms, the Sun could not be powered with external energy. The reactions simply would not occur. Nothing would flow. However, when atoms are formed, the previously electrically active plasma particles suddenly become so tightly packed and perfectly balanced that they become electrically neutral. As a consequence they cease to exist in the electric landscape.

Synthesized atoms flow away with the solar wind

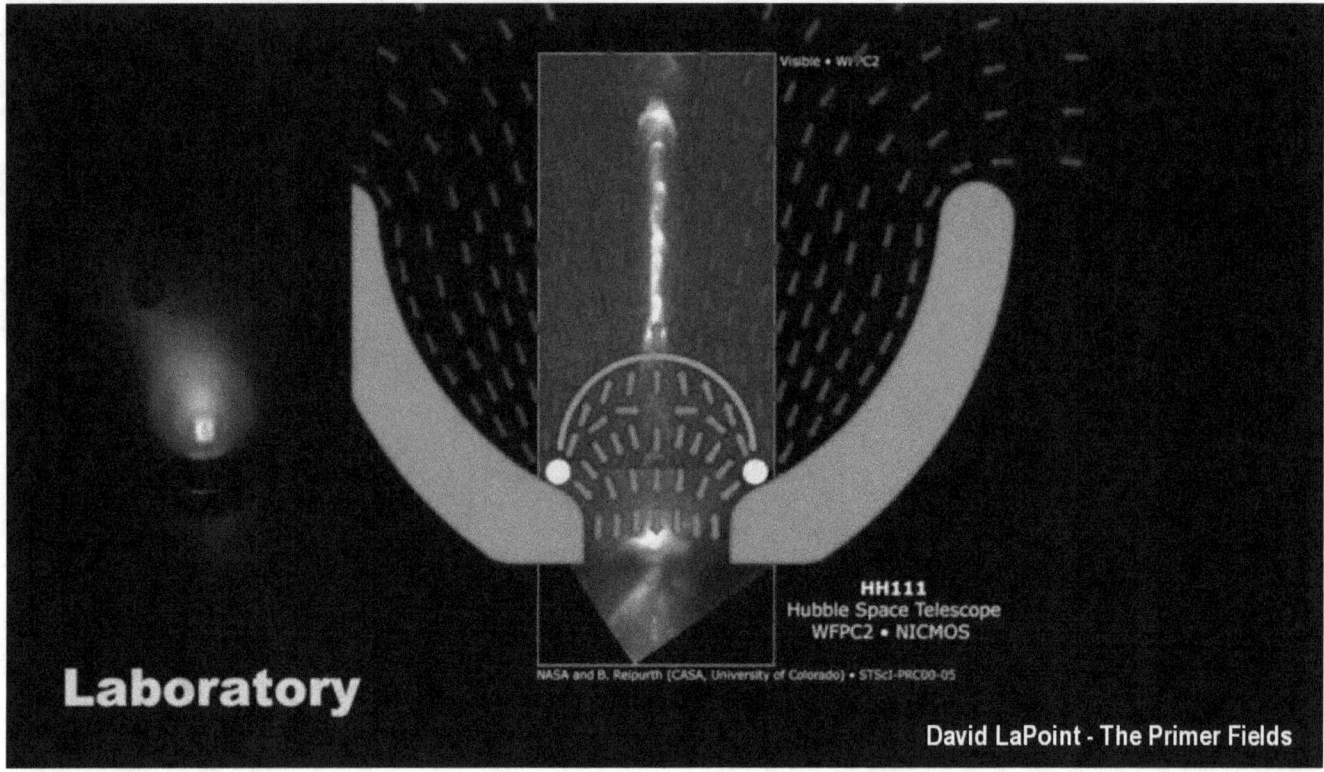

By being no longer tied to anything, the synthesized atoms flow away with the solar wind. By this process of electric atomic synthesis, a powerful sink effect is created, in which, in addition, all the natural elements that exist, that the planets are made of, were and are created by the process.

The synthesis extends far beyond the helium fusion stage

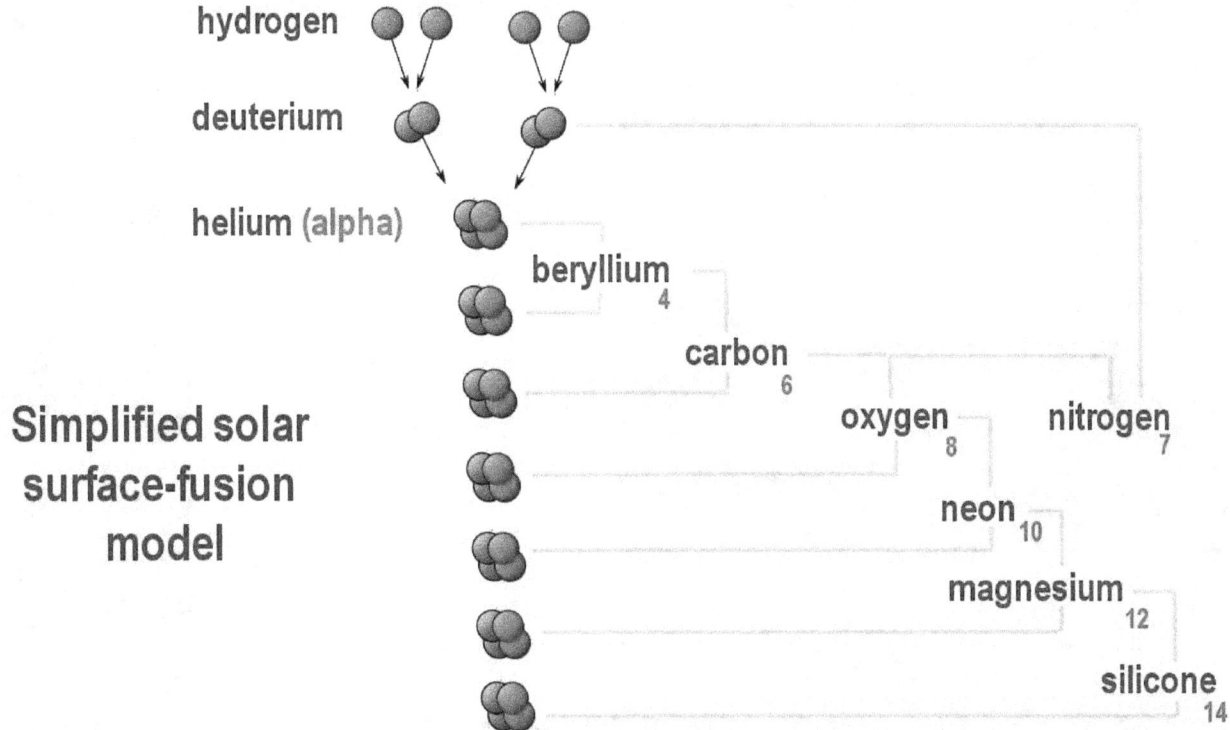

Instead of the p-p chain of fusion that is theorized for the internally heated Sun, which ends at the helium-4 stage, no such limit is inherent in the plasma powered surface fusion process. The synthesis extends far beyond the helium fusion stage.

In the remarkably close agreement

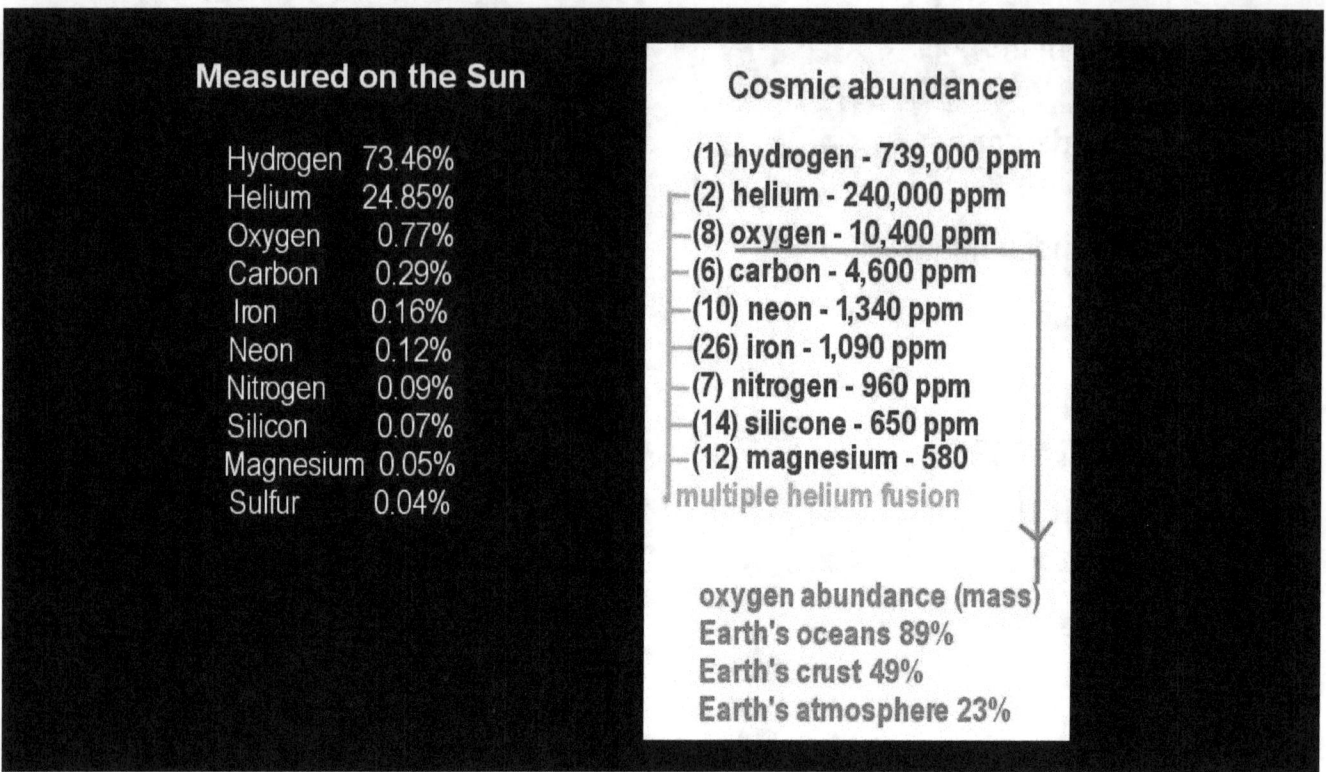

The evidence that a wide range of surface fusion is happening, is found in the remarkably close agreement of the measured element ratios on the surface of the Sun, shown on black, with the known cosmic abundance ratio for the same elements in the solar system, shown on white.

The heavier elements, past the helium stage

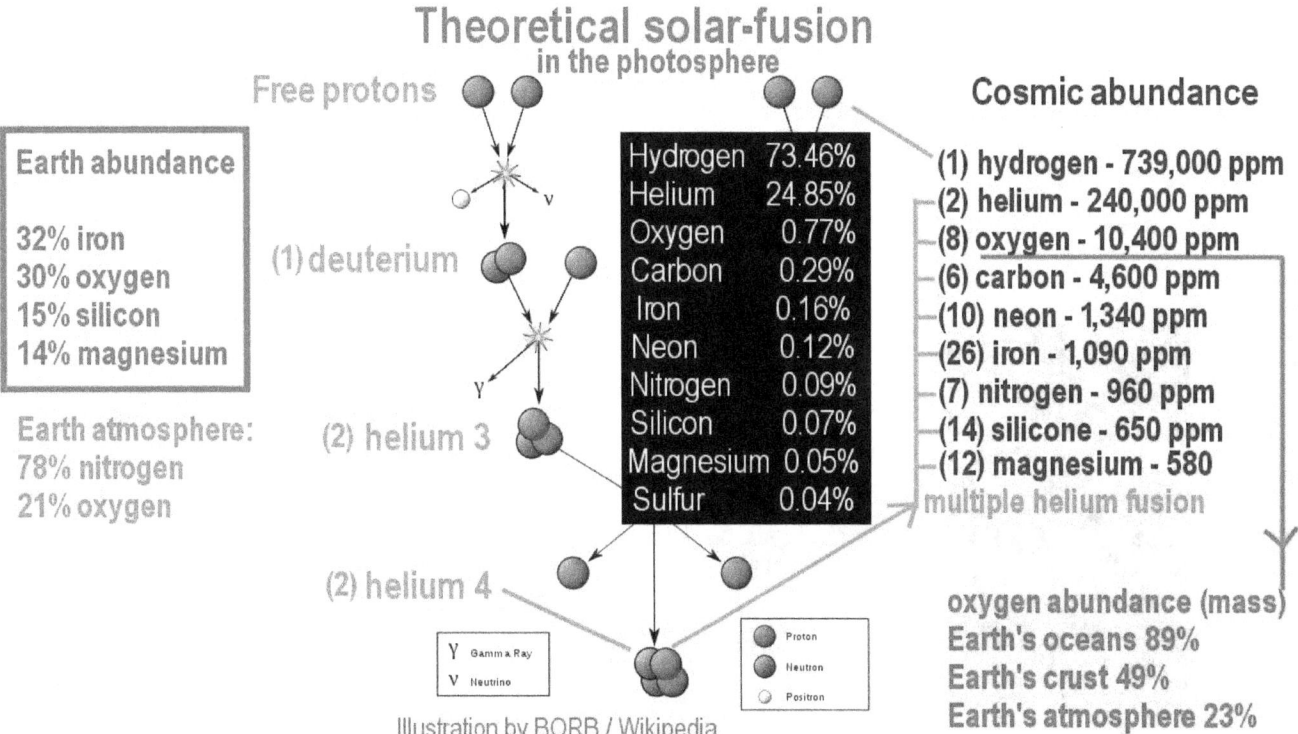

The heavier elements, past the helium stage, should therefore not be present in the atmosphere of the Sun at all. None should be there. Still, they have all been measured there. Their presence, that shouldn't be, invalidates the internal-fusion theory still further.

The presence of 'heavy' elements in the solar atmosphere

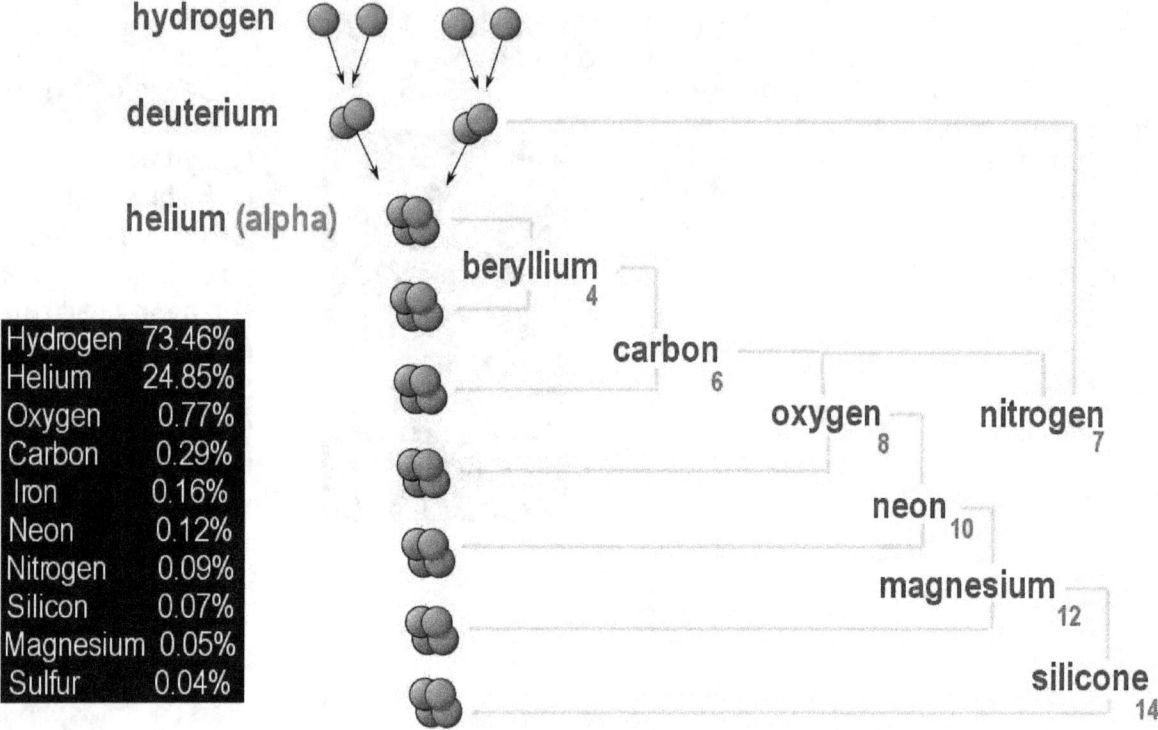

The presence of 'heavy' elements in the solar atmosphere proves the electric surface-fusion concept, which alone enables atomic synthesis to be possible on the Sun, in a process in which atoms are fused directly from plasma. This includes all the known natural elements in the periodic table.

If all of these heavy elements were synthesized in the core of the Sun, these elements would have accumulated there and would have become the core, a very heavy core, which the Sun does not have, and cannot have.

This type paradox is not found on the external fusion platform.

Evident by the existence of noctilucent clouds

Noctilucent clouds over Kuresoo bog, Viljandimaa, Estonia, app. 75-85 Km high - wikipedia

That strong fusion synthesis is happening on the surface of the Sun, is visibly evident on Earth. It is evident by the existence of noctilucent clouds high in the stratosphere. We see rivers of water vapor present there, frozen into ice crystals, that were evidently not lifted up from the surface 80 kilometers high into the stratosphere, but came instead directly from the Sun, carried by the solar winds.

Cold fusion drives the universe

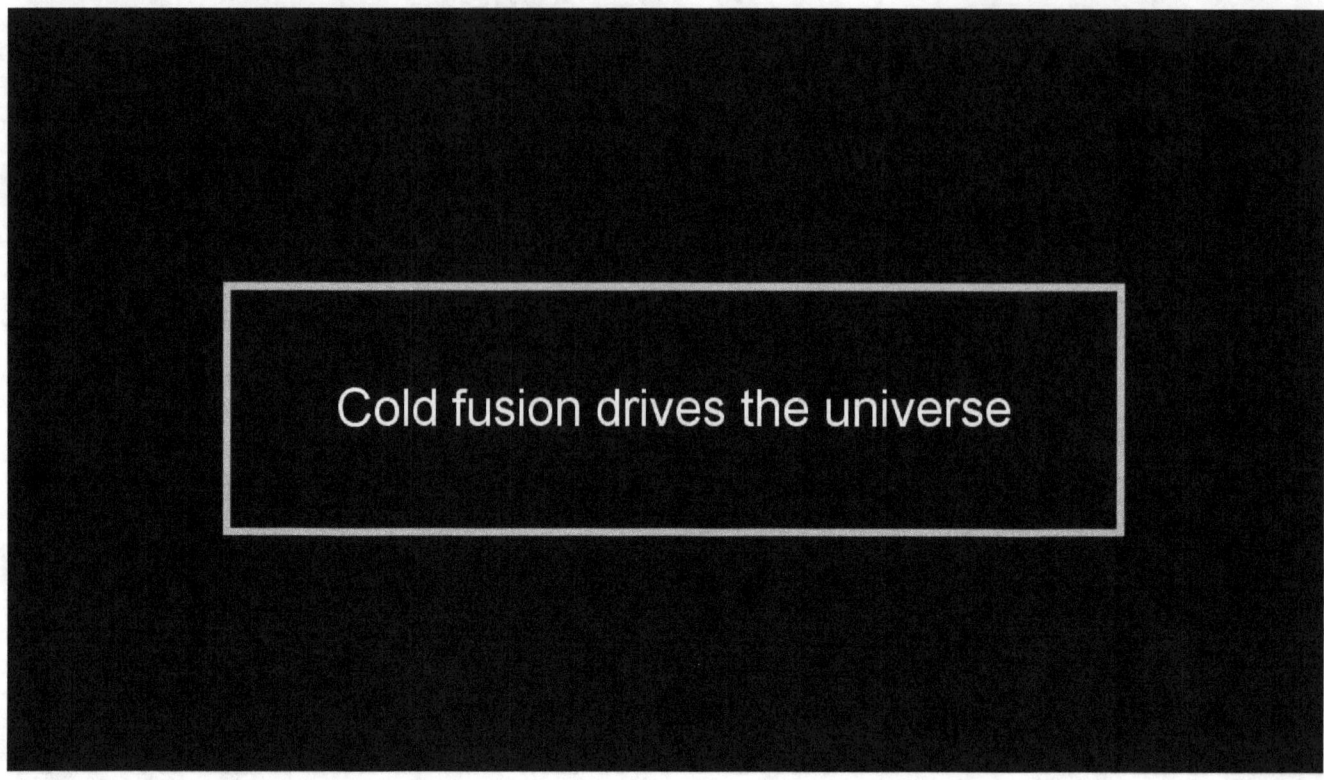

Cold fusion drives the universe

Electric fusion is cold fusion

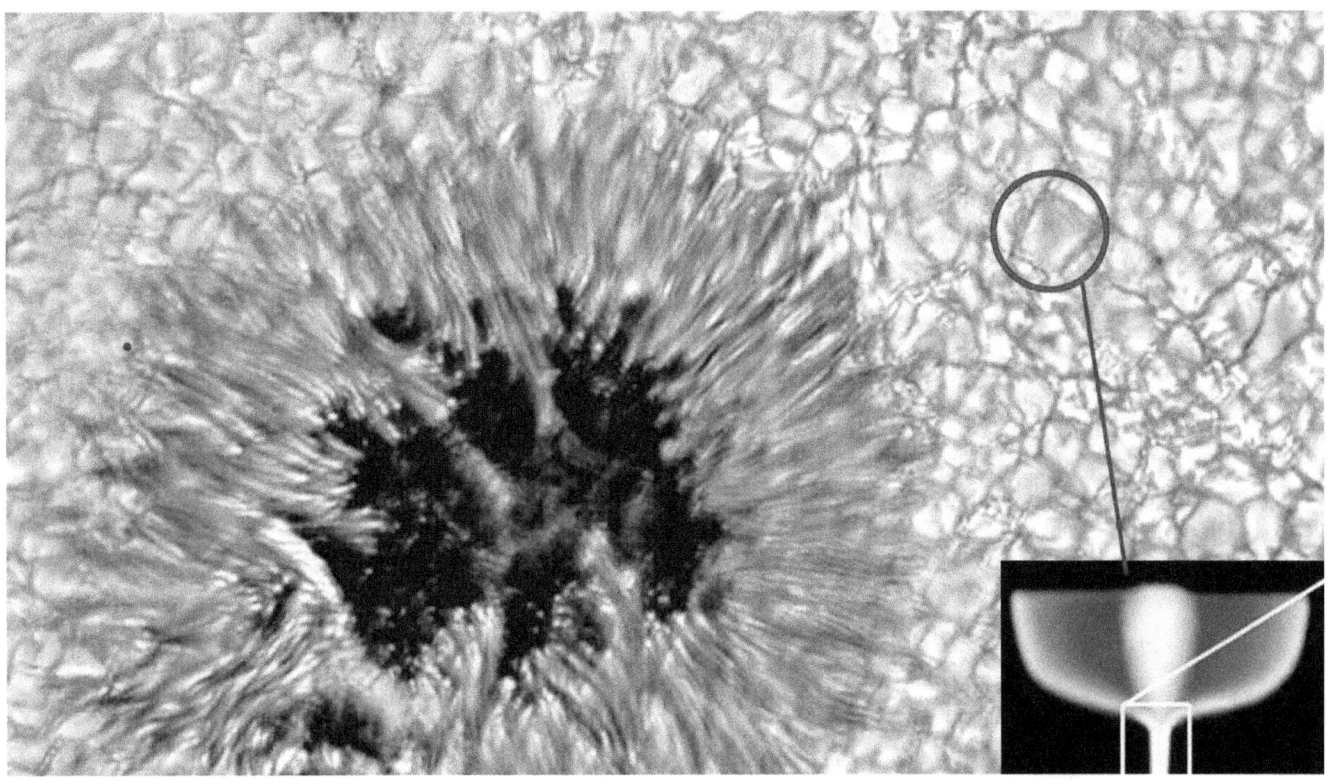

Electric fusion on the surface of the Sun is cold fusion. It is enabled by magnetic plasma compression and is maintained by the sink effect of the fusion itself. The heating of the photosphere is a subsequent phenomenon. It is comparatively minuscule phenomenon, at 5,505 degrees Celsius.

All the nuclear fusion experiments on Earth are based on the illusion that the fusion process inside the Sun, which the experiments aim to replicate, require those 15 million degrees that are theorized to exist in the Sun for the internal heating that is deemed to so violently agitate the atoms that they become forced to fuse. For this reason, all fusion experiments on Earth, are hot fusion experiments, even though hot fusion doesn't actually happen on the Sun, and likely nowhere in the universe either.

Cold nuclear fusion is the fusion that has created all the elements that the planets are made of. The atoms in the Universe are not the result of the mythical, primordial Big Bang, but are the product of cold-powered electric fusion that is motivated exclusively by the electric force and is maintained by the fusion-process itself, as it creates an electric sink where plasma electricity is being bottled up and neutralized into packages of electrically neutral atoms.

Atomic elements are extremely rare in the universe

Group → Period ↓	1	2	3	4	5	6	7	8	9	10	11	12	13	14	15	16	17	18
1	1 H																	2 He
2	3 Li	4 Be											5 B	6 C	7 N	8 O	9 F	10 Ne
3	11 Na	12 Mg											13 Al	14 Si	15 P	16 S	17 Cl	18 Ar
4	19 K	20 Ca	21 Sc	22 Ti	23 V	24 Cr	25 Mn	26 Fe	27 Co	28 Ni	29 Cu	30 Zn	31 Ga	32 Ge	33 As	34 Se	35 Br	36 Kr
5	37 Rb	38 Sr	39 Y	40 Zr	41 Nb	42 Mo	43 Tc	44 Ru	45 Rh	46 Pd	47 Ag	48 Cd	49 In	50 Sn	51 Sb	52 Te	53 I	54 Xe
6	55 Cs	56 Ba		72 Hf	73 Ta	74 W	75 Re	76 Os	77 Ir	78 Pt	79 Au	80 Hg	81 Tl	82 Pb	83 Bi	84 Po	85 At	86 Rn
7	87 Fr	88 Ra		104 Rf	105 Db	106 Sg	107 Bh	108 Hs	109 Mt	110 Ds	111 Rg	112 Cn	113 Uut	114 Fl	115 Uup	116 Lv	117 Uus	118 Uuo

Lanthanides		57 La	58 Ce	59 Pr	60 Nd	61 Pm	62 Sm	63 Eu	64 Gd	65 Tb	66 Dy	67 Ho	68 Er	69 Tm	70 Yb	71 Lu
Actinides		89 Ac	90 Th	91 Pa	92 U	93 Np	94 Pu	95 Am	96 Cm	97 Bk	98 Cf	99 Es	100 Fm	101 Md	102 No	103 Lr

It is not unreasonable therefore, to assume, that all the atomic elements in the solar system, which the planets are made of, which together make up fourteen-one-hundredth of a percent of the visible mass of the solar system, where all synthesized on the surface of the Sun in a nuclear fusion process that creates atomic structures. The remaining 99.86% of the mass of the solar system is plasma located in the Sun, and this ratio does not even include the mass of the plasma in the Primer Fields and in the solar winds, and so on.

The extremely high plasma-mass ratio tells us that atomic elements are extremely rare in the universe, even though they are the heart of the world in which we live.

Every element of our world was synthesized on the Sun over time. Most of this synthesis occurred during the epoch in which the Sun itself was forged in the high-density plasma-landscape near the center of the galaxy. This means that the synthesizing system that produces the vast abundance of different atomic elements that make up our world, must necessarily be enormously efficient. The great efficiency is rooted in the dynamics of electrically motivated cold fusion that creates its own sink.

The cold-fusion process in the Sun is simple

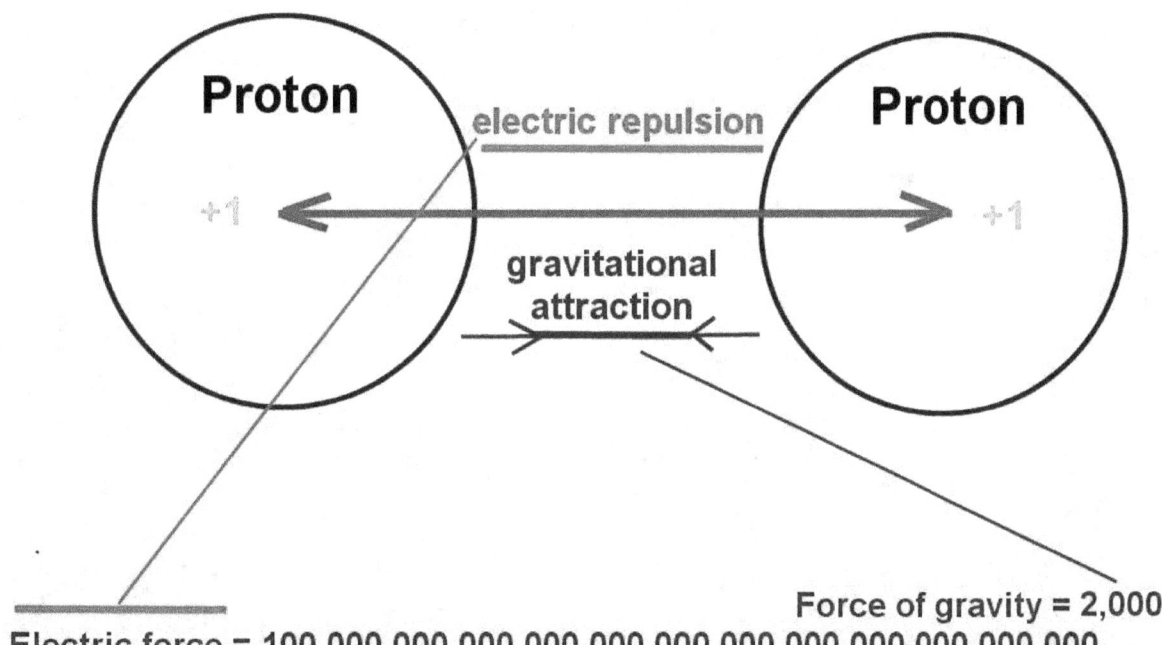

The cold-fusion process in the Sun is simple. All protons repel each other strongly by the electric force that is one of the basic forces of the universe. Each proton also packs a substantial mass that is roughly 2,000 times greater than the mass of an electron.

When energy is invested

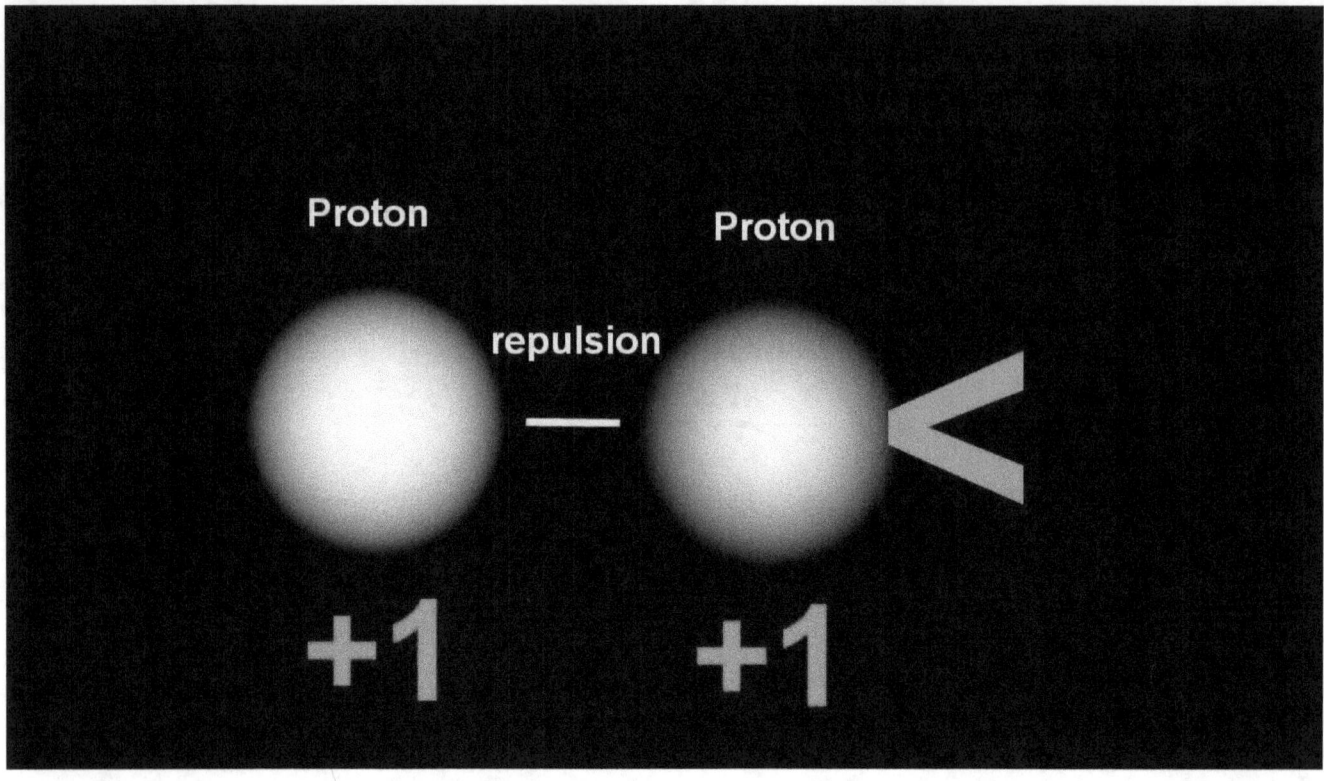

When energy is invested into putting the mass of a proton into motion, and this kinetic energy becomes great enough to overcome the electric repulsion, the protons will simply snap together. They will fuse.

As the protons fuse

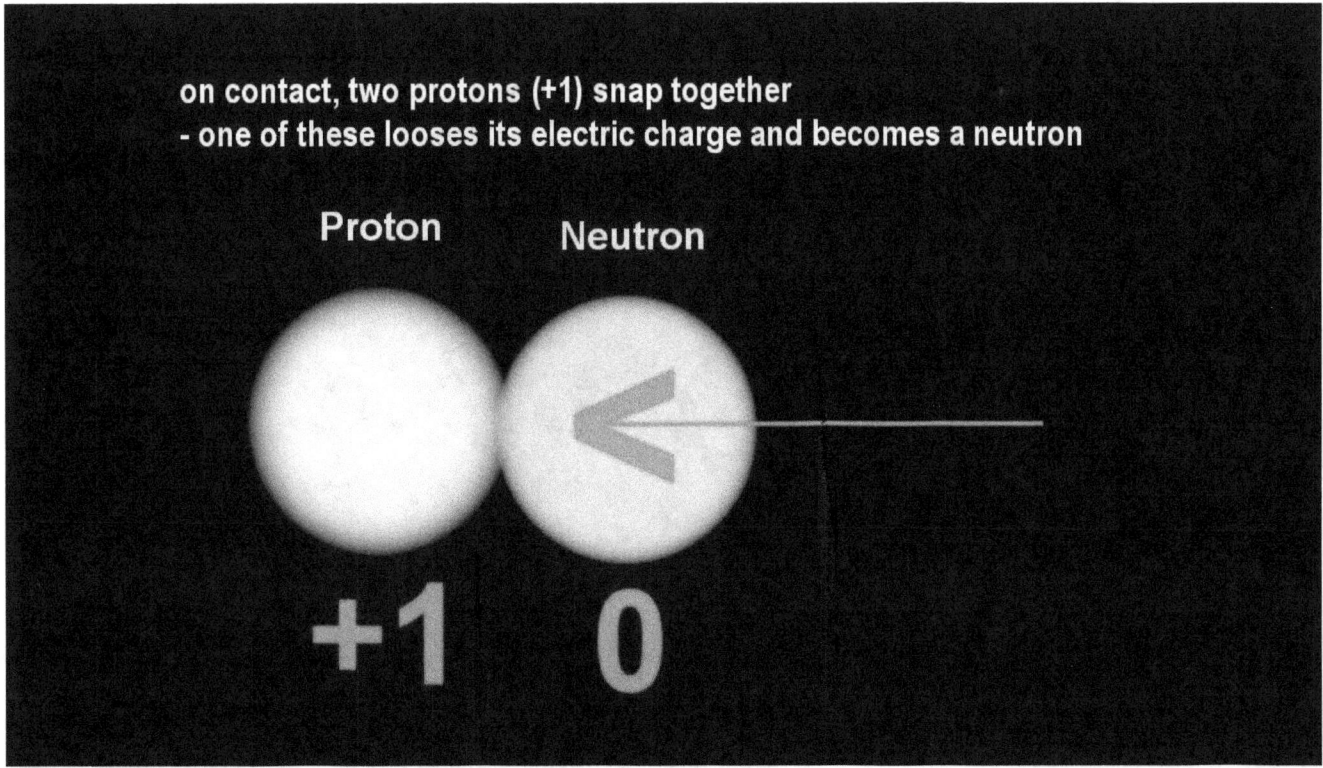

As the protons fuse, some of the kinetic energy becomes 'loaded' into one of the fusing protons. The proton becomes transformed thereby. It becomes transformed by the process, into a neutron. The resulting neutron carries no electric charge. As the result of this fusion process, the electric field of one of the joining protons simply ceases to exist.

The resulting field reduction

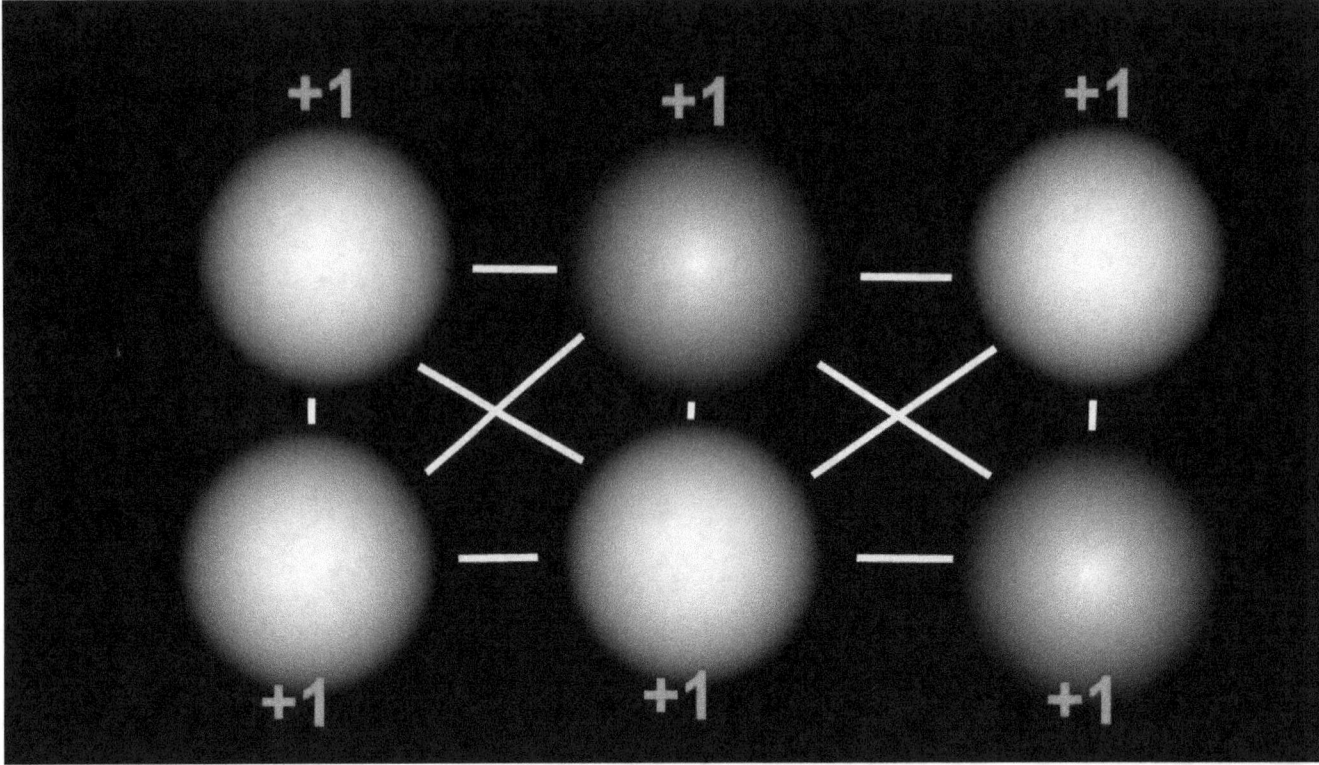

The resulting field reduction has an immensely dramatic effect on the plasma stream where the protons repel one another, as in the case shown here.

The dynamics create an imploding effect

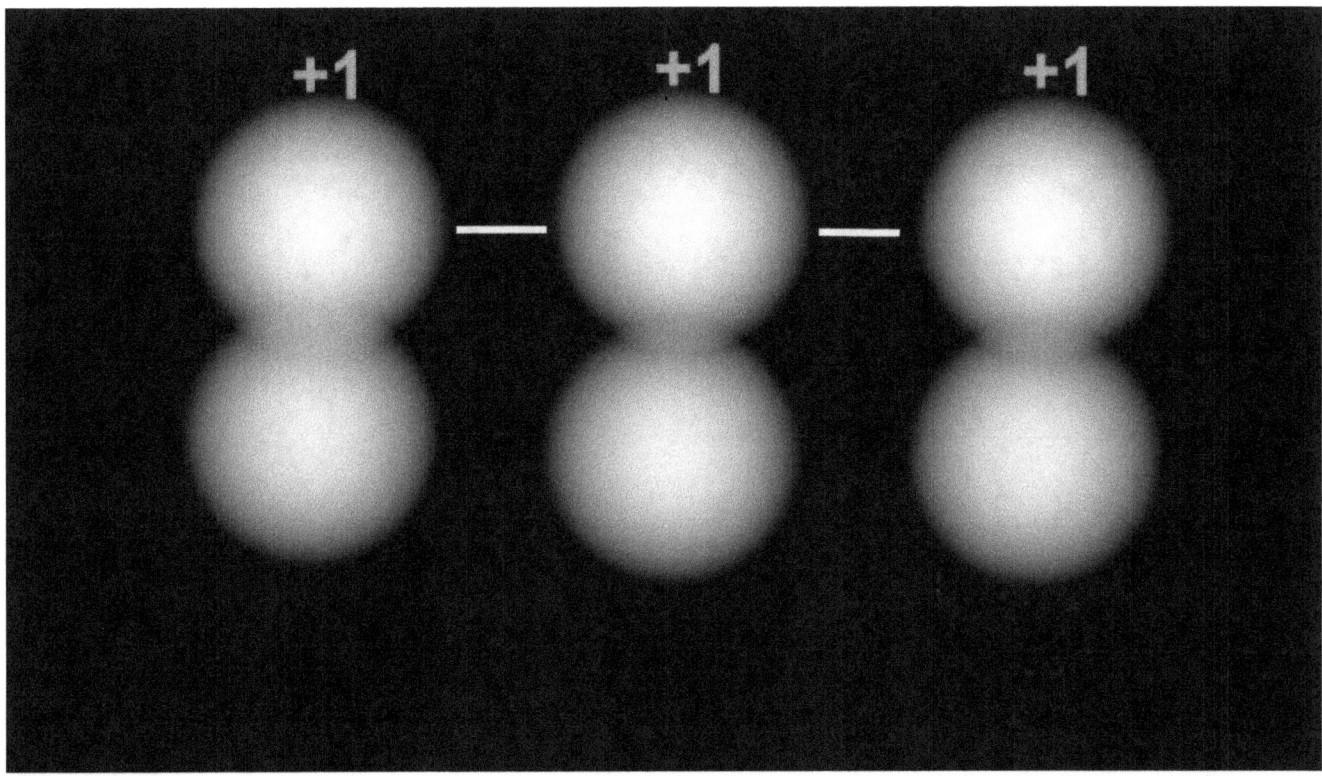

In this manner, the overall repelling effect that affects the plasma density, is much more-dramatically reduced. The dynamics create an imploding effect.

The resulting 'shock' of the vanishing electric fields and the lost repulsion effect, creates a partial 'void' in the plasma pressure, which the adjoining protons are propelled to fill. Thus, the fusion effect creates, by the electric force, some highly energized movement in the plasma stream towards the Sun. The plasma movement, of course, further aids the plasma compression, and thereby aids the fusion process.

The resulting voids accelerates the remaining plasma

The resulting voids accelerates the remaining plasma, just as if one had opened a water-faucet more fully. This process is self-escalating.

The more protons are joined, the more neutrons are created, and the greater the void in the repelling electric field becomes, which in turn accelerates the process in leaps and bounds. This means that the faucet is effectively opened evermore, the stronger the process flows. Nor does it end here.

As more protons are being used up

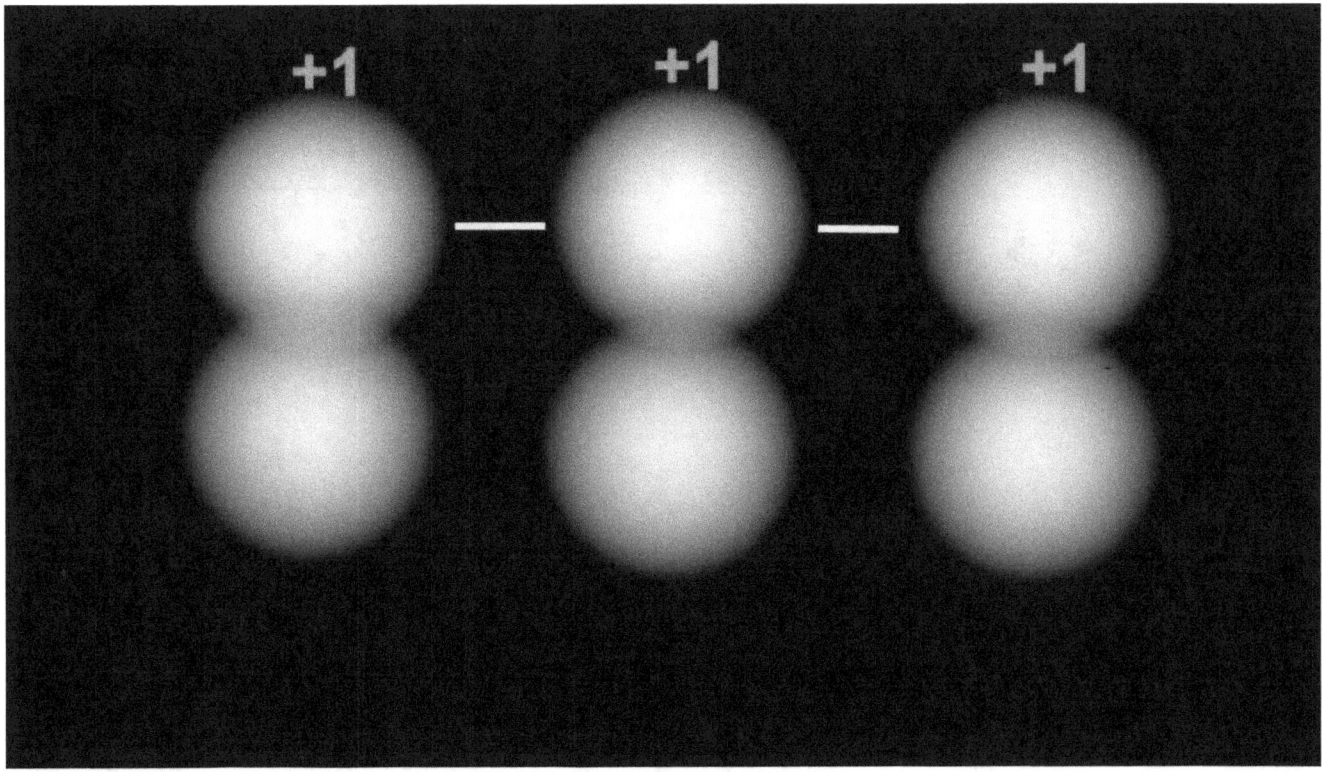

As more protons are being used up, by becoming neutrons that have no electric effect, the ratio between the remaining electrons and protons the in the plasma stream, suddenly doubles, and in some cases triples.

As the fused proton and neutron clusters are joined

As the fused proton and neutron clusters are joined to one-another themselves, forming ever larger clusters, the overall electron density surrounding the larger clusters is thereby increased evermore, to the point that the electrons, too, become highly energetic, and become bound up by the now increasing electric energy strength of the accumulating nuclei. When this happens, the result is even more dramatic.

The entire electric scene becomes a great void

When the electric charge of the electrons balance the charge of the remaining protons, the entire electric scene becomes a great void. All protons and electrons in the entire atomic construction loose their overall electric field potential as if they had vanished from the universe.

The void is so great that substantial plasma streams are drawn into it, like into a bottomless pit. In this manner, the electric nuclear-fusion synthesis creates in effect a huge bottomless pit that eats up electric fields in plasma. This bottomless pit sucks up plasma, by which atomic elements emerge.

In comparison, the waterline carries the plasma streams

In comparison, the waterline carries the plasma streams. The faucet is the fusion process that the plasma flows into. Out of the faucet comes a steam of assembled atoms that are electrically neutral and dissipate into the landscape. In the process the faucet gets exceedingly hot. The energy that drives the system comes from the active fusion process, enabled by the plasma-sink principle.

The same principle applies

The same principle applies here that we utilize on Earth to drive hydroelectric generating plants, powered by the pressure differential between the reservoir level, and the discharge level.

Energy is carried away by electric transmission lines

From hydro-electric plants, the generated energy is carried away by electric transmission lines.

The process heat is radiated as light

From the Sun, the process heat is radiated as light and heat, enabling life to be.

Solar fusion IS cold fusion

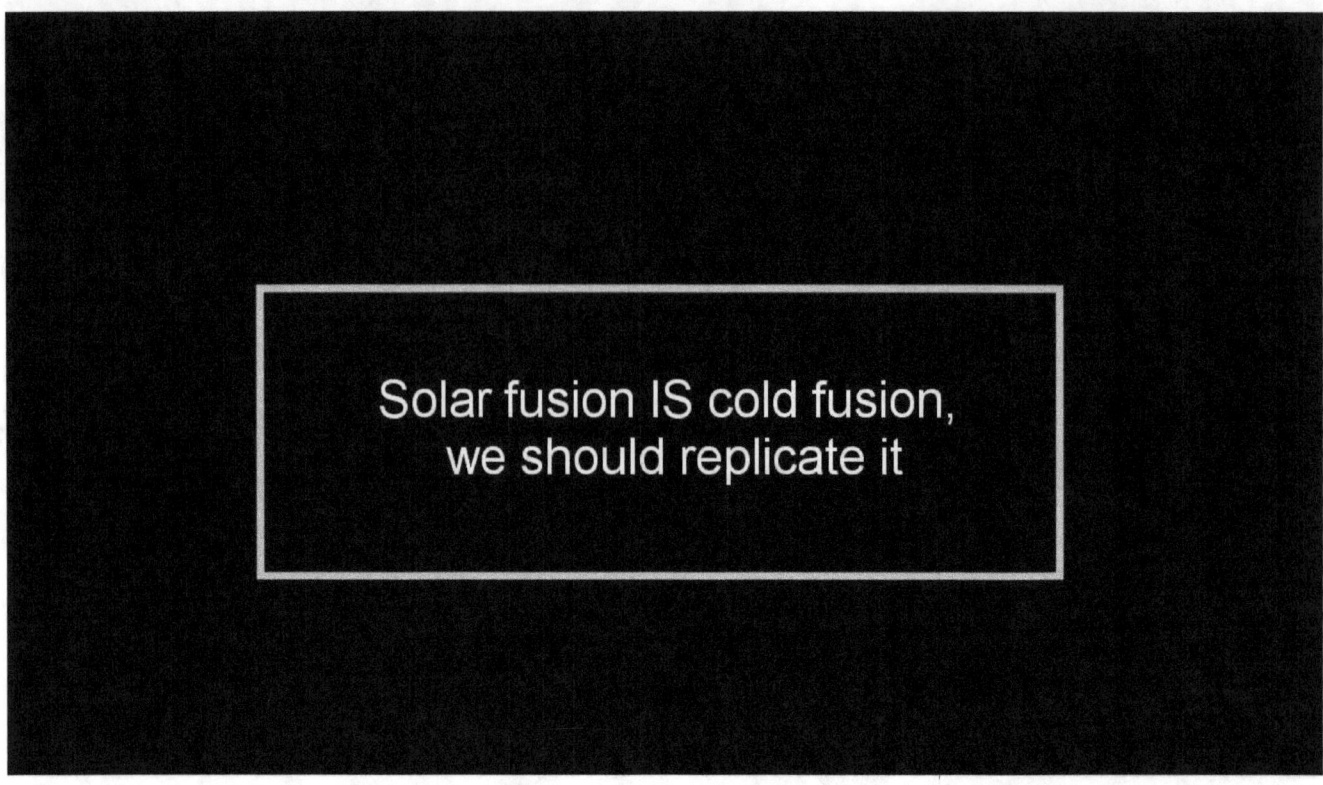

Solar fusion IS cold fusion, we should replicate it

Cold fusion, as we see the principle applied on the Sun, is the most efficient energy producer in the universe. It is anti-entropic in nature, with a resource that is self-renewing.

Cold fusion is efficient as a solar process

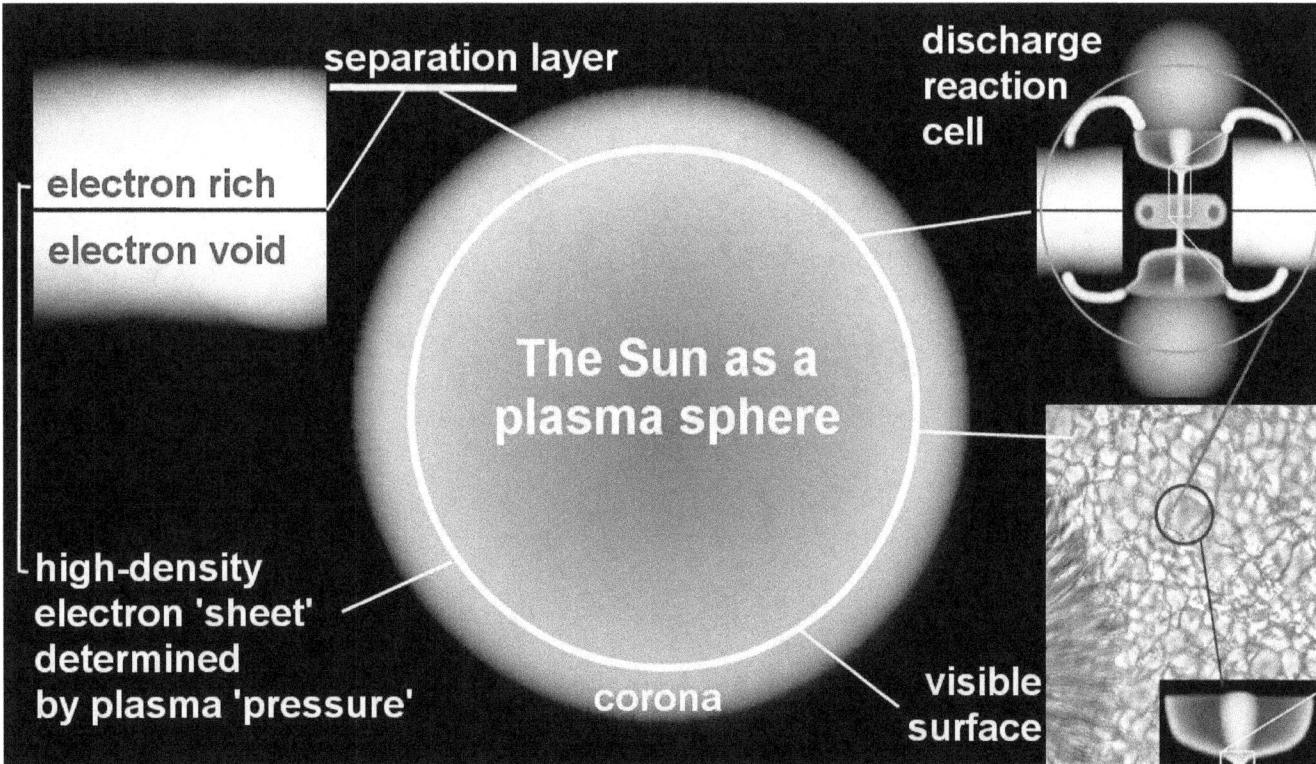

Cold fusion is efficient as a solar process, because its active component is self-accelerating. Thereby, large amounts of energy and atomic elements are generated with relative 'ease,' on the solar surface. Nothing is forced there. Everything happens naturally. The 'sink' effect is so efficient in accelerating plasma that the millions of degrees in thermal energy that the fusions experiments on Earth require, to force atoms to fuse together, are not required on the Sun. The Sun operates with a 'cold' fusion process in which thermal and light energy is produced as a secondary effect by the mechanics of the process.

Artificial nuclear-fusion power processes fail

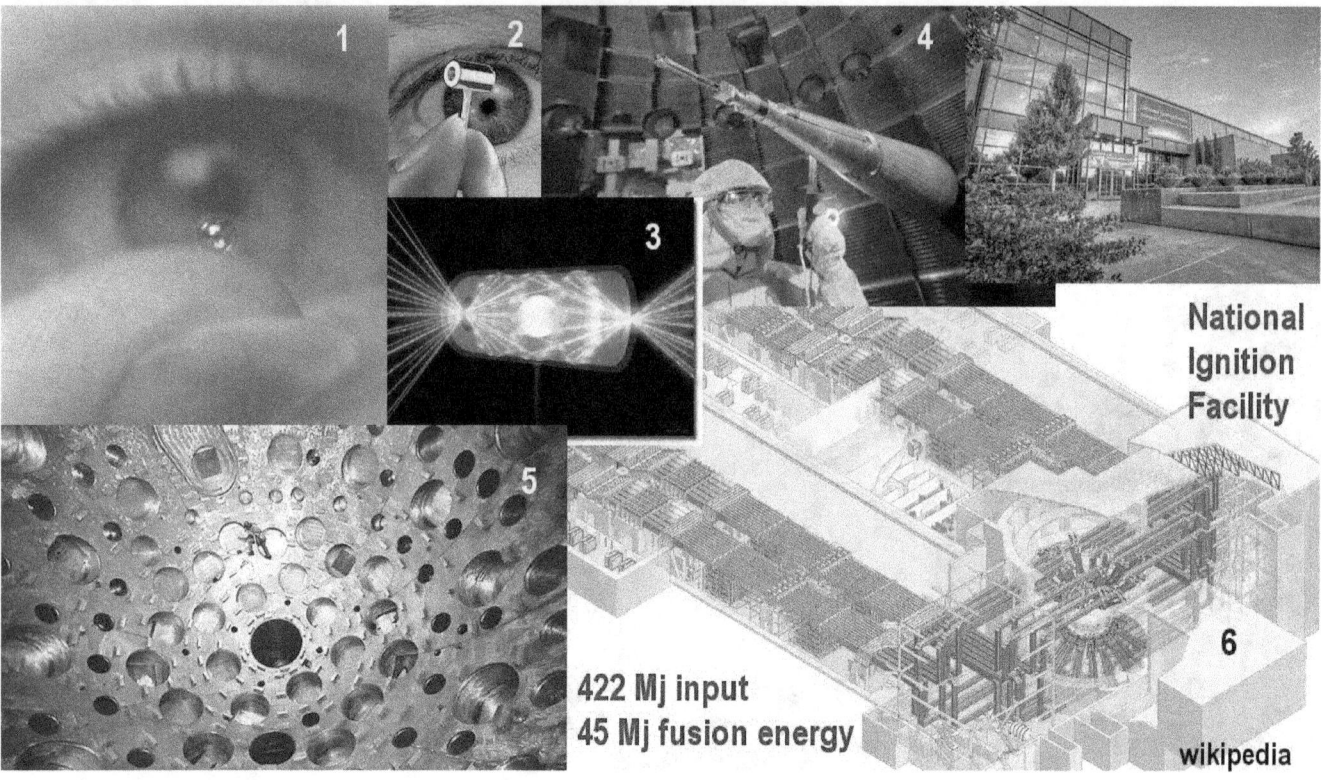

National Ignition Facility

422 Mj input
45 Mj fusion energy

All of the current, artificial nuclear-fusion power processes fail, because they fail to utilize the critical 'sink' principle that motivates cold fusion. They employ enormously large external forces in an effort to break the electric repulsion barrier that develops between approaching atomic nuclei, once the electron shells are penetrated.

The kind of agitation of atoms that breaks this barrier, requires enormous energies as input to make the thing possible. And even with all that, efficient fusion has not yet been achieved, not even with such giant efforts as are made at the National Ignition Facility, which aims for inertial fusion by means of light-energy compression. The effort has failed so far, in spite of the most ideal fusion fuel being used for the experiment.

The easiest-to-fuse fusion fuel

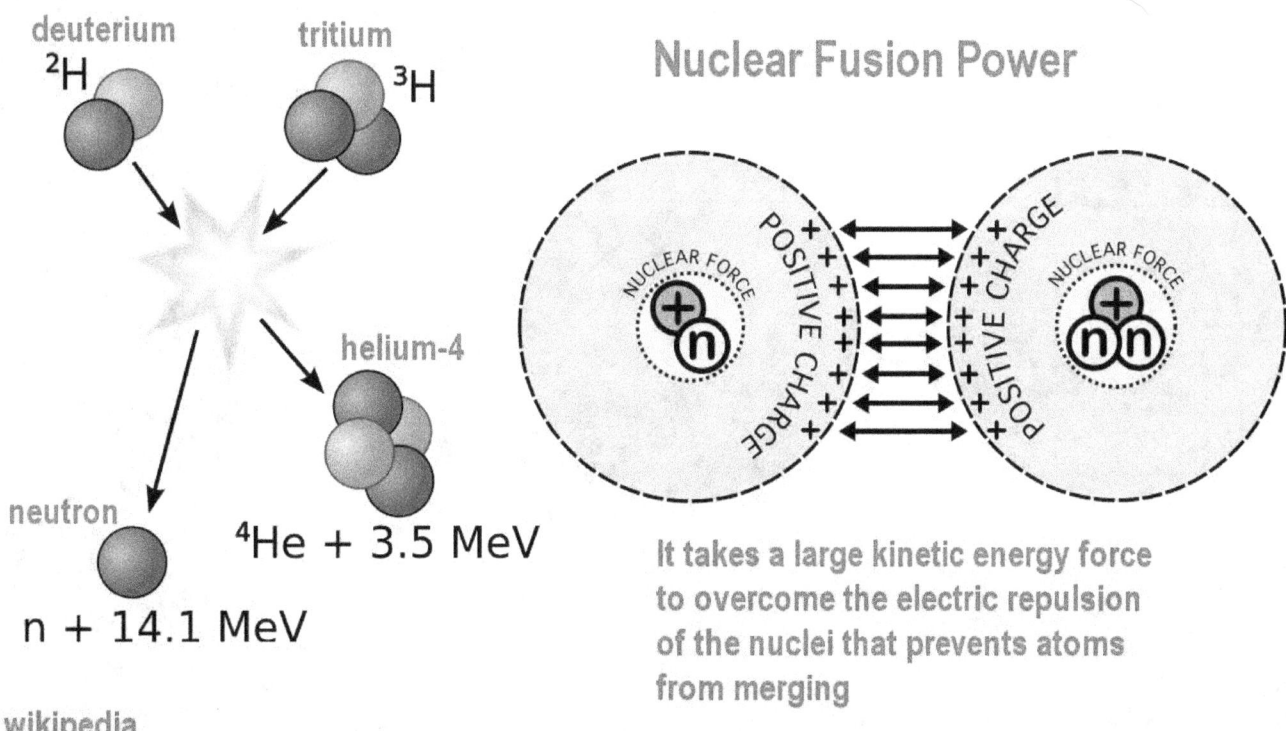

wikipedia

The easiest-to-fuse fusion fuel is made up of two types of over-built atomic isotopes, deuterium and tritium, that are violently forced to connect with each other. When they do connect and fuse, one of the overbuilt isotopes releases a redundant neutron that is not required for the end product. When this happens, its previously invested binding energy becomes liberated. This liberated energy is intended to be used to power electricity generating systems. This, in short, is the dream scenario.

Attempted power production processes all invariably fail

The attempted power production processes all invariably fail, because the intended fusion process does not aim for the natural 'sink' effect, but forces results with the injection of brute agitation for which enormous energies are expended to generate the agitation. Consequently, it takes radically more energy as input, to operate such a fusion process, than the result gives back.

The inherent inefficiency that results from substituting the natural 'sink' principle, with brute force, renders nuclear-fusion power production on Earth, a hopeless proposition. This basic failure in principle stands in addition to all the other problems associated with the projects that render the hoped-for success an empty dream.

Natural solar processes are not yet utilized for power production

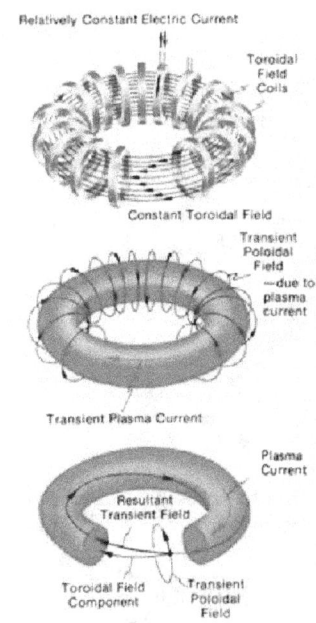

wikipedia

When it is being said that terrestrial nuclear-power projects aim to replicate the process that powers the Sun, a fundamental error is incurred, because what is being replicated in experiments, does not occur, naturally, either on the Sun, nor anywhere else, as a process. The actually solar processes are infinitely more efficient than what the experiments aim for. The natural solar processes that are in operation on the Sun, are not yet utilized for power production anywhere on Earth, experimentally or otherwise, for the simple reason that they are deemed not to exist.

This does not mean that the real solar-energy processes

This does not mean that the real solar-energy processes will never be replicated on Earth? Not necessarily.

Plenty of evidence exists

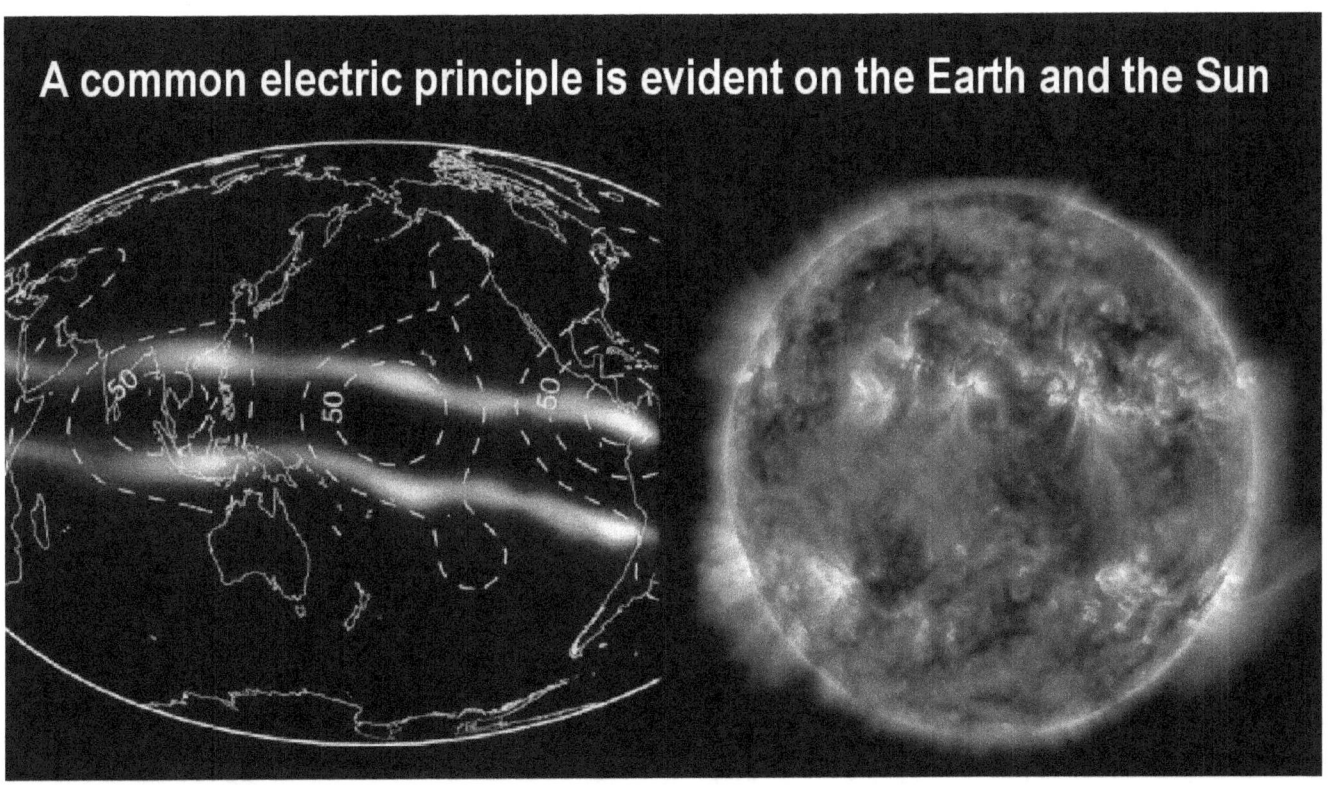

Plenty of evidence exists that cosmic electric plasma surrounds the Earth, in a pattern that is also visible on the Sun.

A plasma-flow pattern is 'visible' in the ionosphere

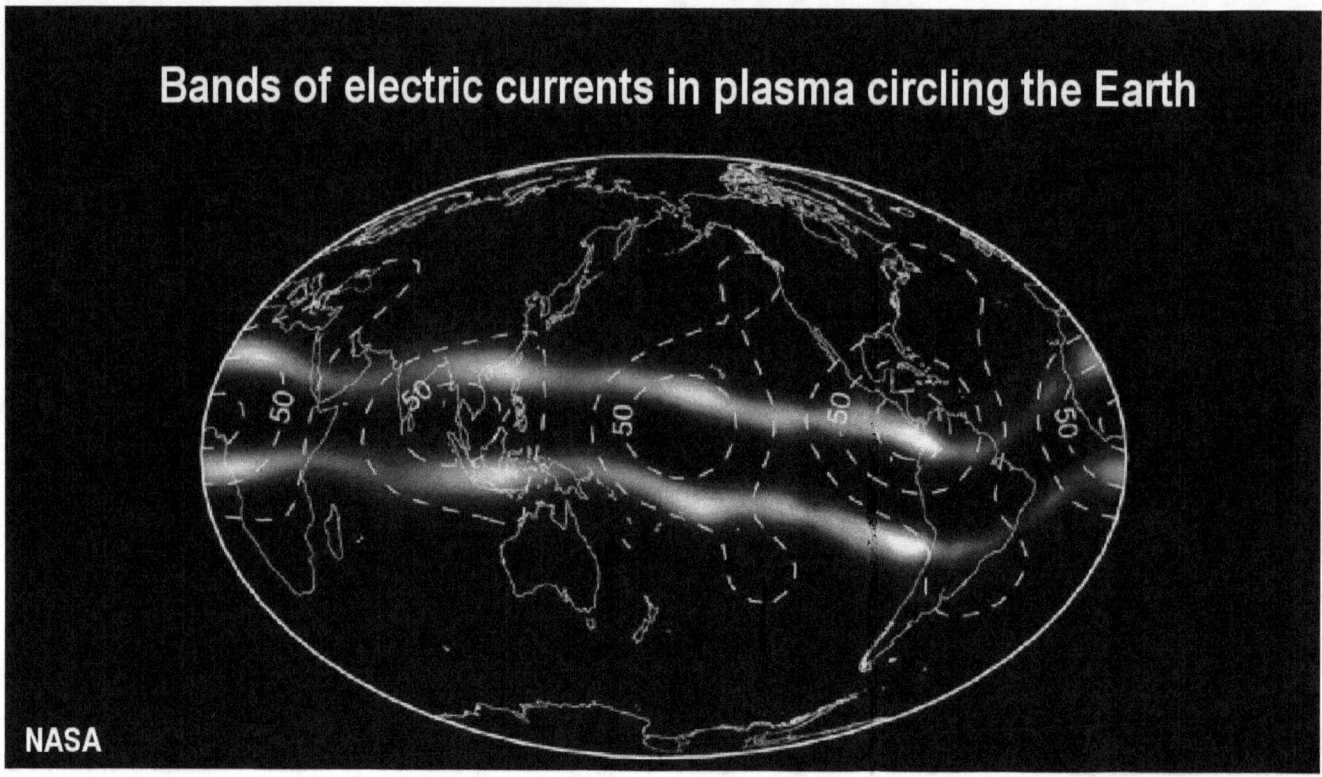

On Earth, a plasma-flow pattern is 'visible' in the ionosphere. The enormous electric potential that is visible there, appears to supply the driving energy for a long list of natural processes on Earth.

The sky is not the limit

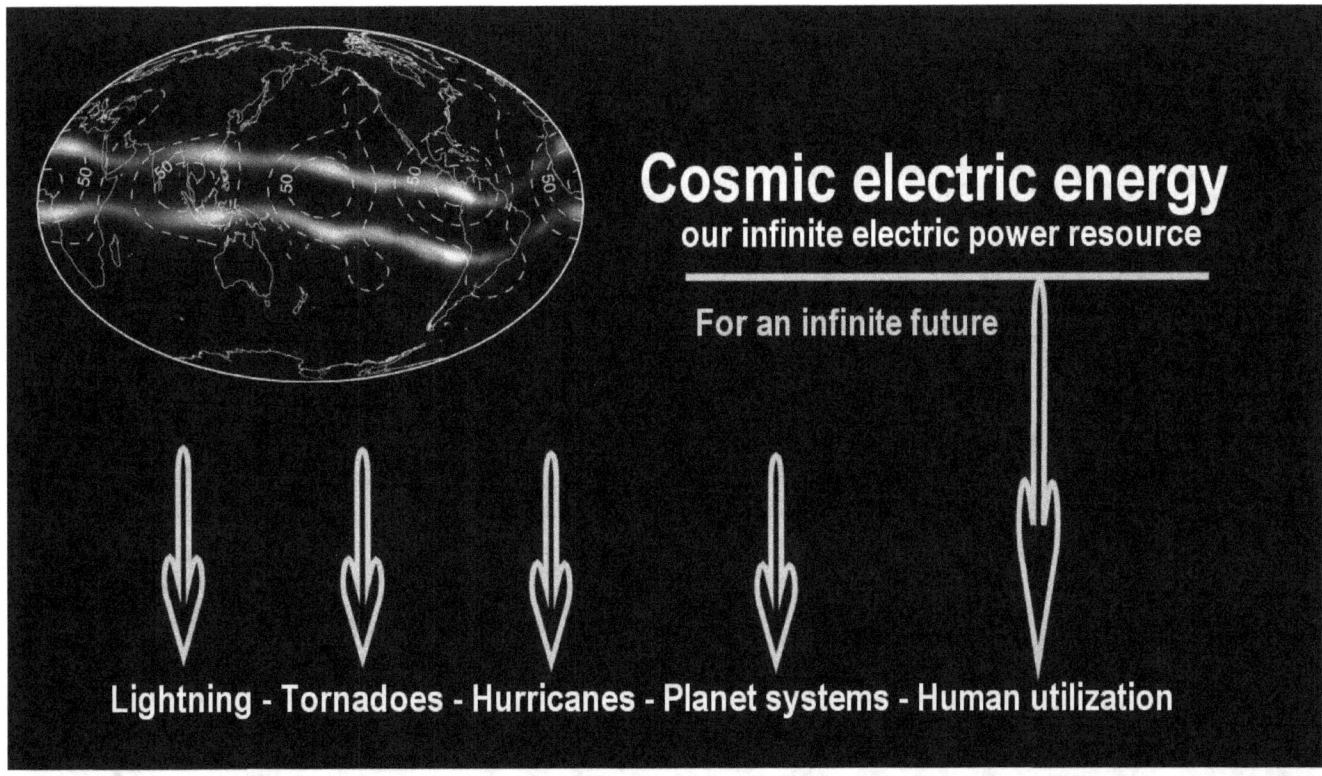

Among the electric systems powered from the ionosphere are hurricanes, lightning, and tornadoes, whenever thermal events create a sufficiently conductive connection to the ionosphere.

We can step up to the plate and participate in the utilization of the cosmic power system, and beyond that, by replicating on Earth the cold fusion principle that the Sun has utilized for its past billions of years. When we get to this point, nuclear fusion, as a synthesizing fusion, will power the economies of humanity. Then the old saying will have a new meaning attached, that "the sky is not the limit." The limit is in the mind. The Sun is not an energy producer, but is an energy converter. The universe doesn't need to produce energy. It is energy. It employs its energy as a creative force.

Part 3: Historic evidence of large cosmic plasma structures

Our Electric Cold Fusion Sun (Part 3) Historic evidence

Historic paradoxes

**Historic paradoxes
resolved in plasma physics**

* The heating of the Sun's corona, paradox
* The solar-wind acceleration paradox
* The Giza Pyramid's alignment paradox
* The Stone Henge paradox

Historic paradoxes, resolved in plasma physics.

* The heating paradox of the Sun's corona

* The solar-wind acceleration paradox

* The Giza Pyramid's alignment paradox

* The Stone Henge Paradox

The coronal heating paradox

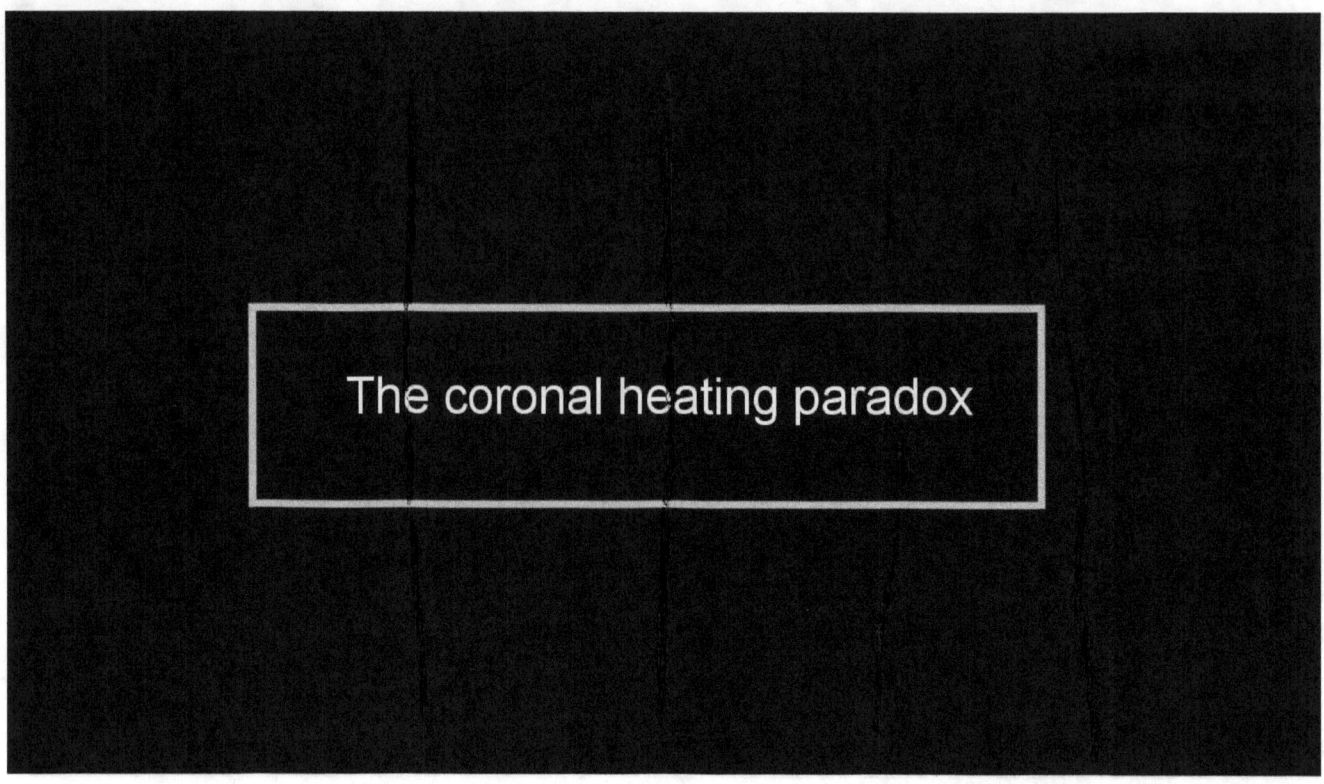

The coronal heating paradox

The principle is simple that causes the 'high temperatures' that occur in the Sun's corona.

As if the corona had been heated

The effect that has been measured is the same as if the corona had been heated to 2 or 3 million degrees? The so-called heating results from electric plasma interaction with atomic elements in the corona.

The same effect cause the solar-winds to be visible

http://www.zam.fme.vutbr.cz/~druck/Eclipse/ - an example of the amazing solar eclipse photography of Milloslav Druckmueller

The same effect cause the solar-winds to be visible. During a total solar eclipse, the solar winds and the corona surrounding the Sun can be seen by the naked eye.

In the standard internal-fusion theory

In the standard internal-fusion theory, the super-heated corona poses a huge problem. It should not be possible by this theory for plasma, that is deemed to flow away from the Sun, to be up to 450 times hotter than the Sun itself is.

Exotic theories have been spun around the paradox created by the internal-fusion doctrine, to save the doctrine.

On the electric-sun platform

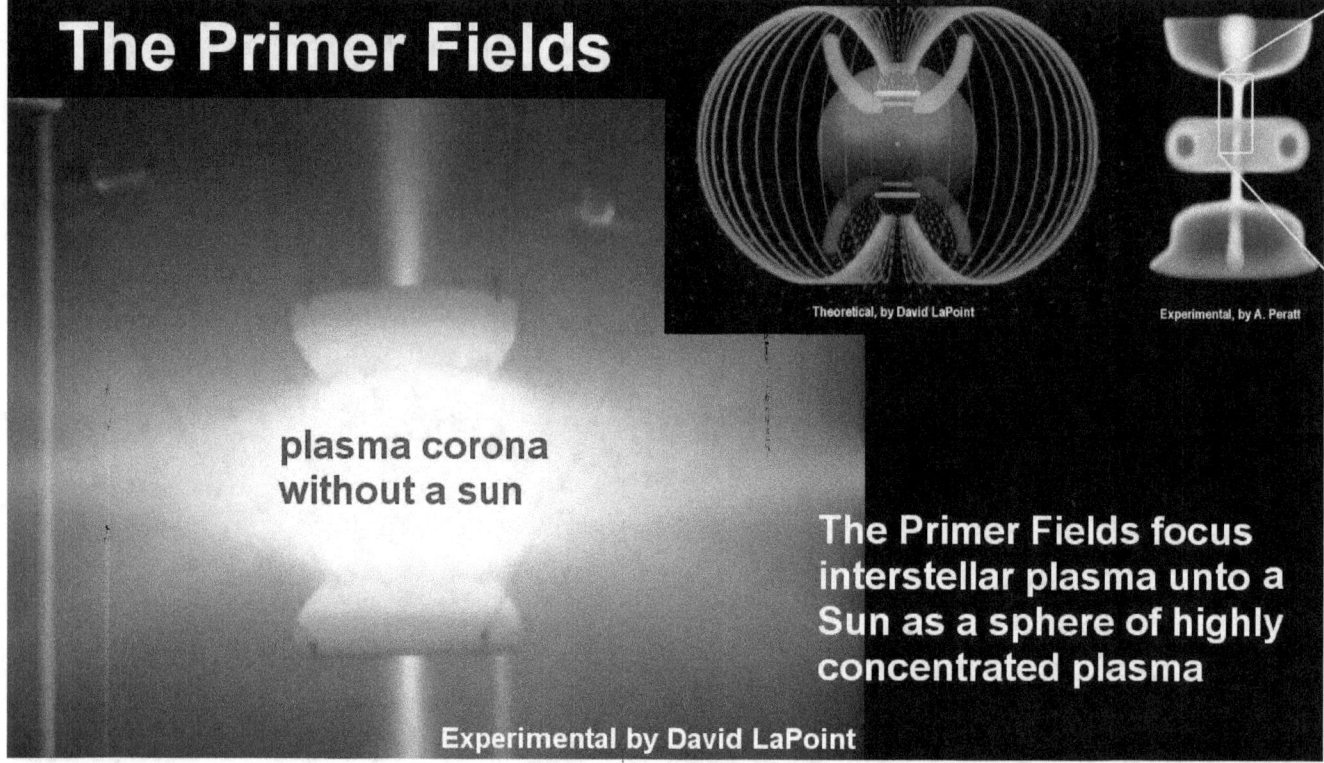

On the electric-sun platform, however, such a paradox does not exist. The super-heated corona is not a paradox there, because in the electric universe the super-heated plasma in the corona does not originate with the Sun at all. It is a part of the external supply stream for the Sun that enables the electric fusion process on the solar surface.

Interstellar plasma streams

Interstellar plasma streams are focused around the Sun by the primary Primer Fields.

Verified in static experiments by David LaPoint

The concept has been verified in static experiments by David LaPoint, and in high-energy dynamic experiments by Anthony Peratt at the Los Alamos National Laboratory.

The superheated corona can be seen as

The superheated corona can be seen as the external layer of plasma, focused onto the Sun by the Primer Fields. The extreme heating that is evident, does not come from the Sun, but results from the agitation of atoms in dense concentration of incoming plasma that is magnetically confined, spatially compressed, and kinetically activated, which together create an intense thermal environment. The corona is not the solar wind.

The accelerating solar-wind paradox

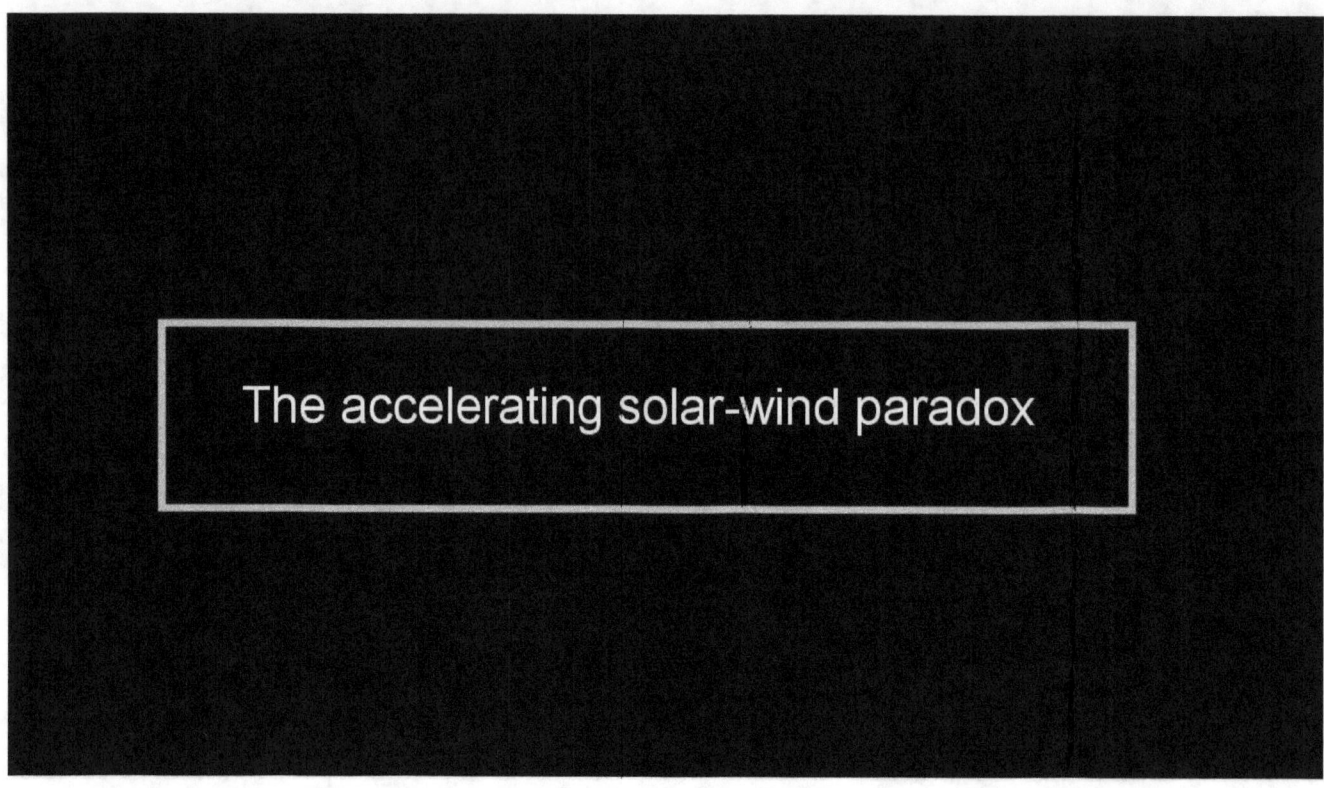

The accelerating solar-wind paradox

Under the internal-fusion theory for the Sun

http://www.zam.fme.vutbr.cz/~druck/Eclipse/ - an example of the amazing solar eclipse photography of Milloslav Druckmueller

Under the internal-fusion theory for the Sun, the solar wind that we see, shouldn't be happening at all, and should definitely not be accelerating away from the Sun against the force of gravity. But it all happens. That's a paradox, isn't it?

Under the standard solar theory the acceleration of the solar wind is a paradox. The very existence of the solar wind under this theory is a paradox.

It should not be possible for anything flowing away from the Sun, against the Sun's enormous force of gravity, to accelerate, much less to reach speeds up to 800 kilometers per second. Still, this impossible, happens.

In plasma solar physics accelerating solar wind not a paradox

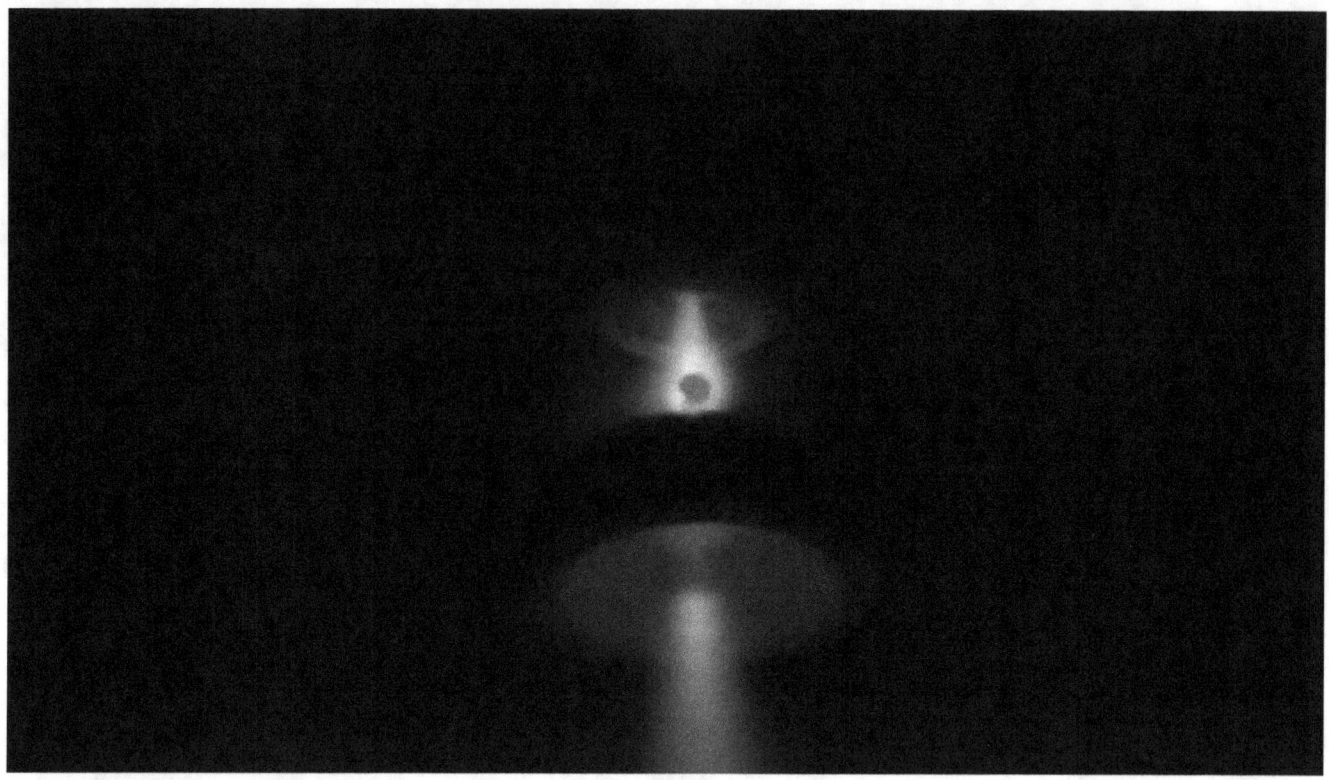

In plasma solar physics, however, the phenomenon of accelerating solar wind not a paradox. It is expected. It would be surprising if it didn't happen..

The solar wind originates in the confinement domes

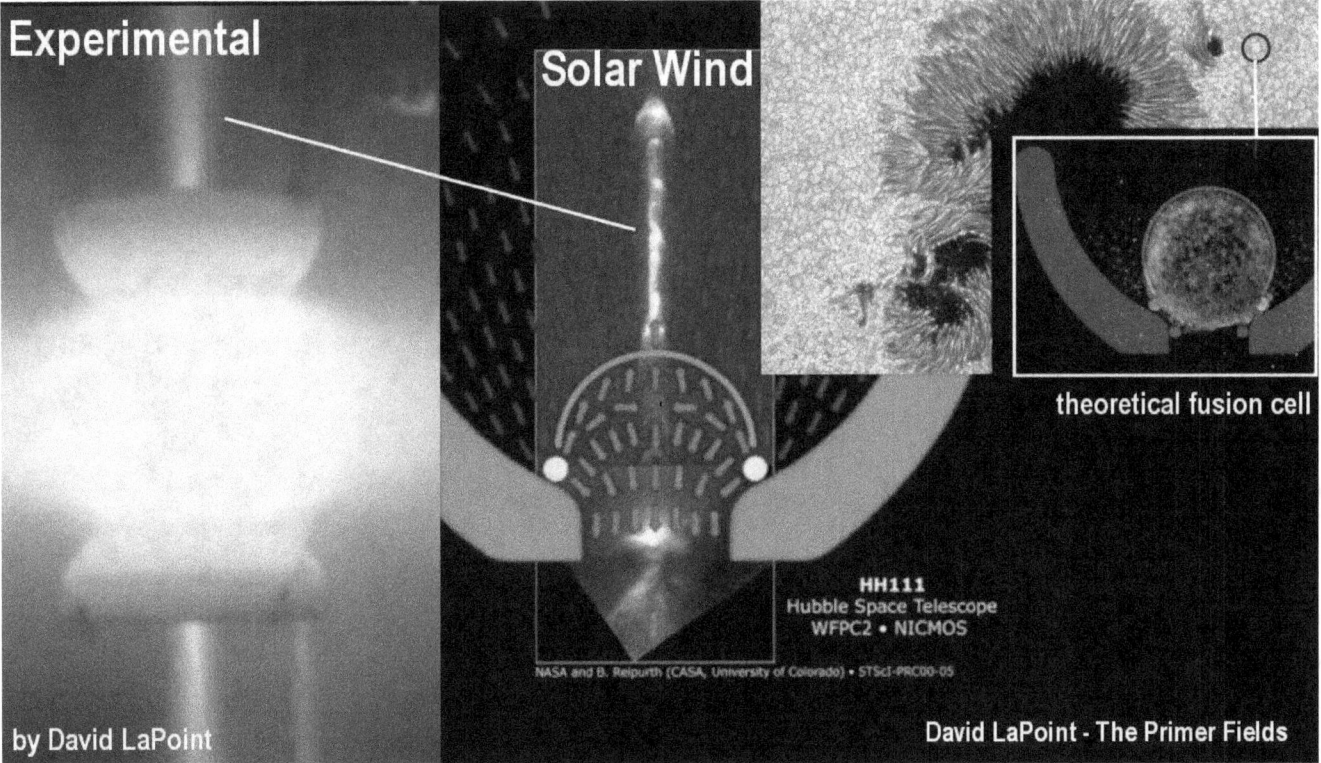

The solar wind originates in the confinement domes of the fusion-reaction cells on the surface of the Sun. When the magnetic barrier is breached, the escaping wind-particles form a plasma jet.

The plasma jet tunnels through the corona

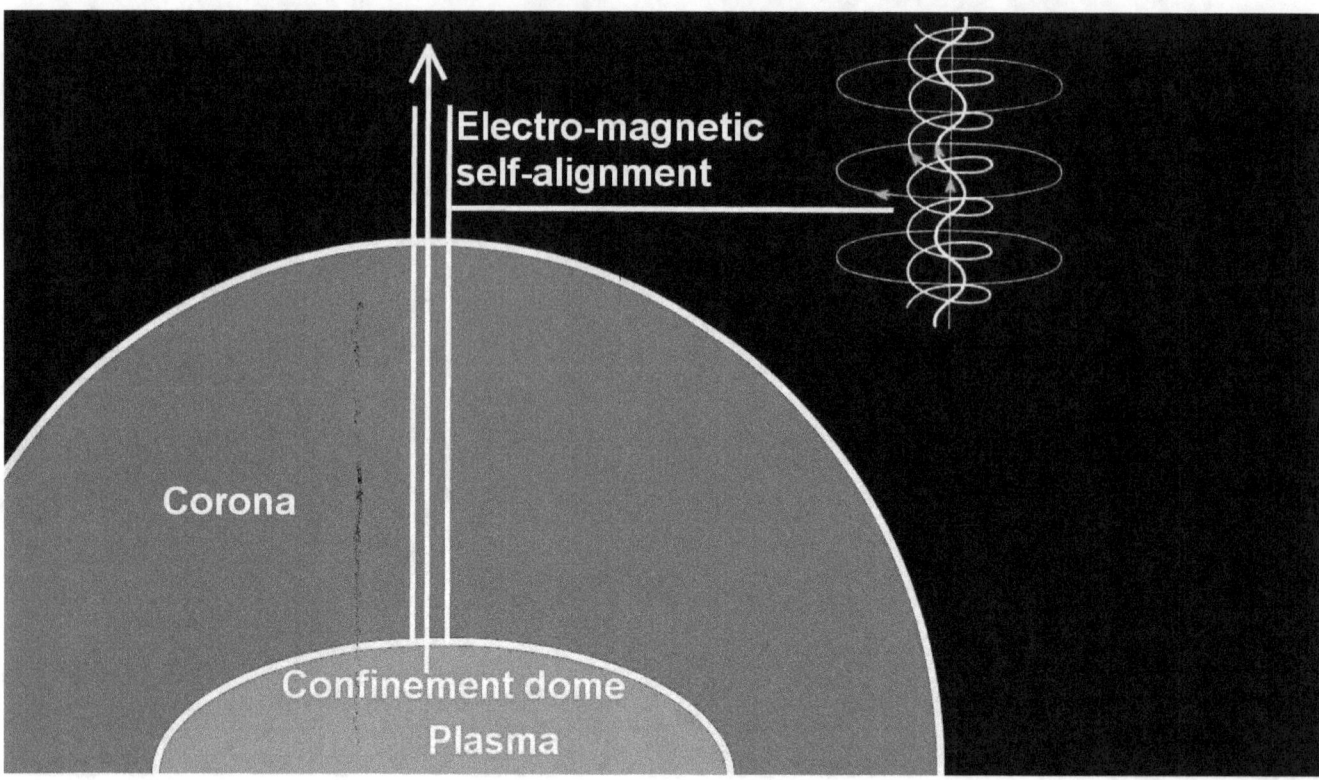

The plasma jet tunnels through the corona in electromagnetically self-confined streams, similar to Birkeland currents. Within the streams, the plasma particles push each other apart with the immense force of electric repulsion, and accelerate each other as they expand explosively for some distances, to typically 800 kilometers per second.

When plasma becomes highly concentrated

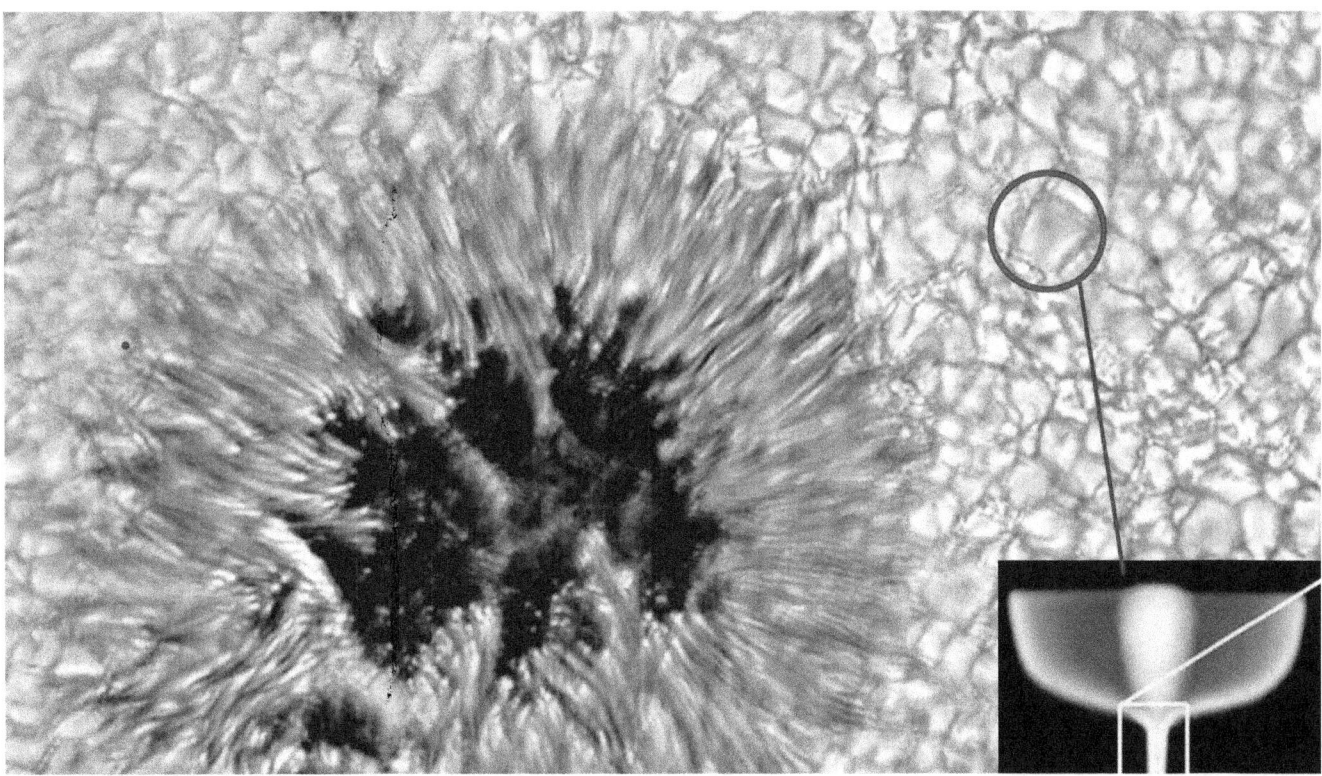

When plasma becomes highly concentrated under the magnetic confinement domes of the reaction cells on the surface of the Sun, and the plasma pressure under the dome exceeds the confinement field strength, some of the highly compressed plasma leaks through the confinement field and is suddenly free to move, to expand, to explode.

When plasma is compressed

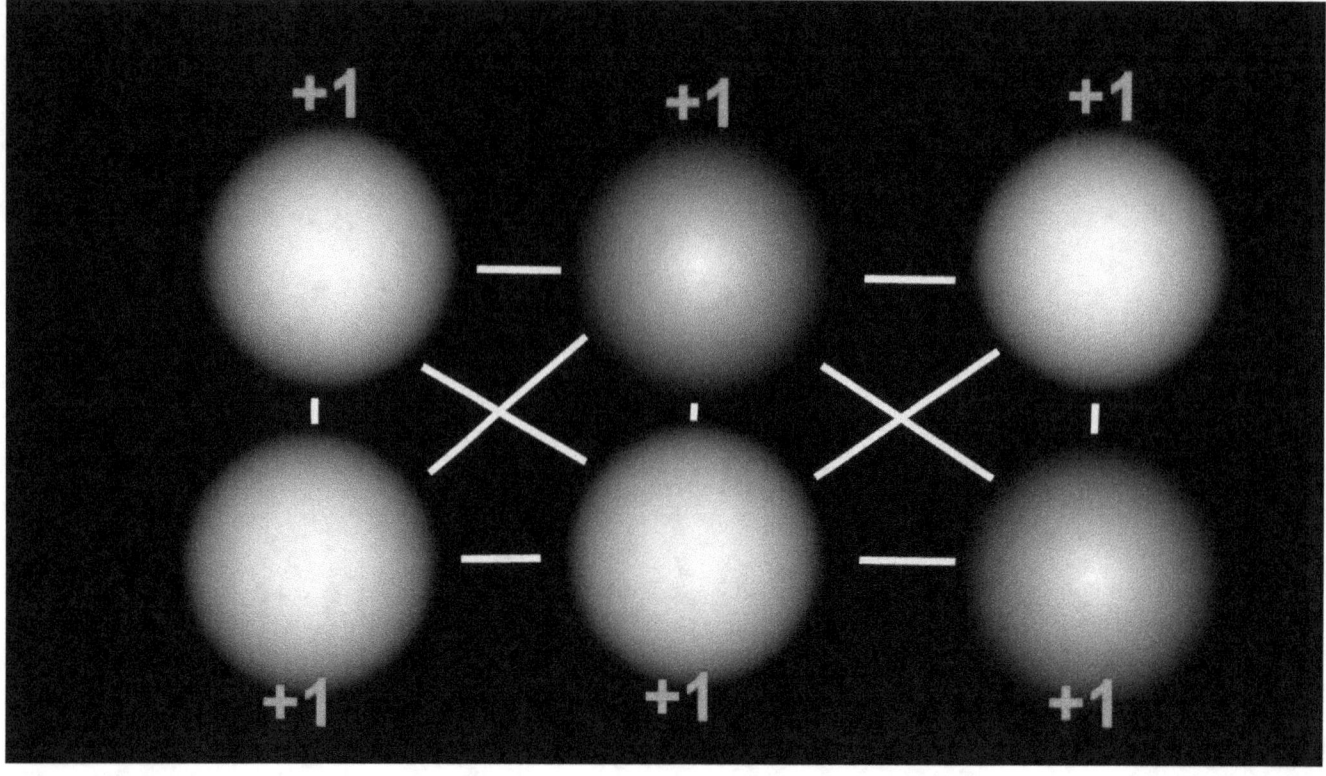

When plasma is compressed, it is compressed against the repelling electric force of the protons to one another. When the pressure is released, the repelling force drives the entire complex apart, explosively.

Protons would push each other apart

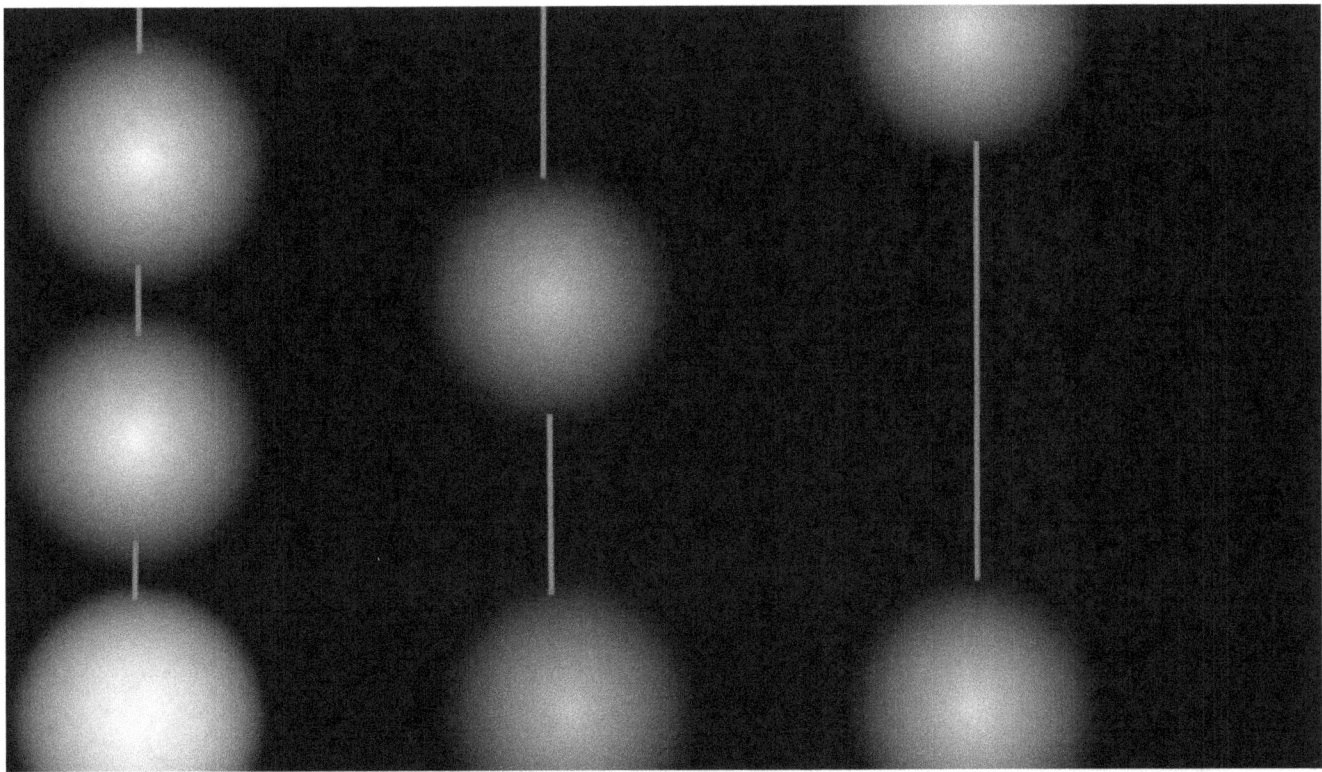

This means that the green protons would push each other apart with the strongest long-reaching force in the universe, the electric force. As a result, the spaces between them become larger, as in the blue example, and then larger again, as in the purple example.

Plasma streams generate a magnetic field aground them

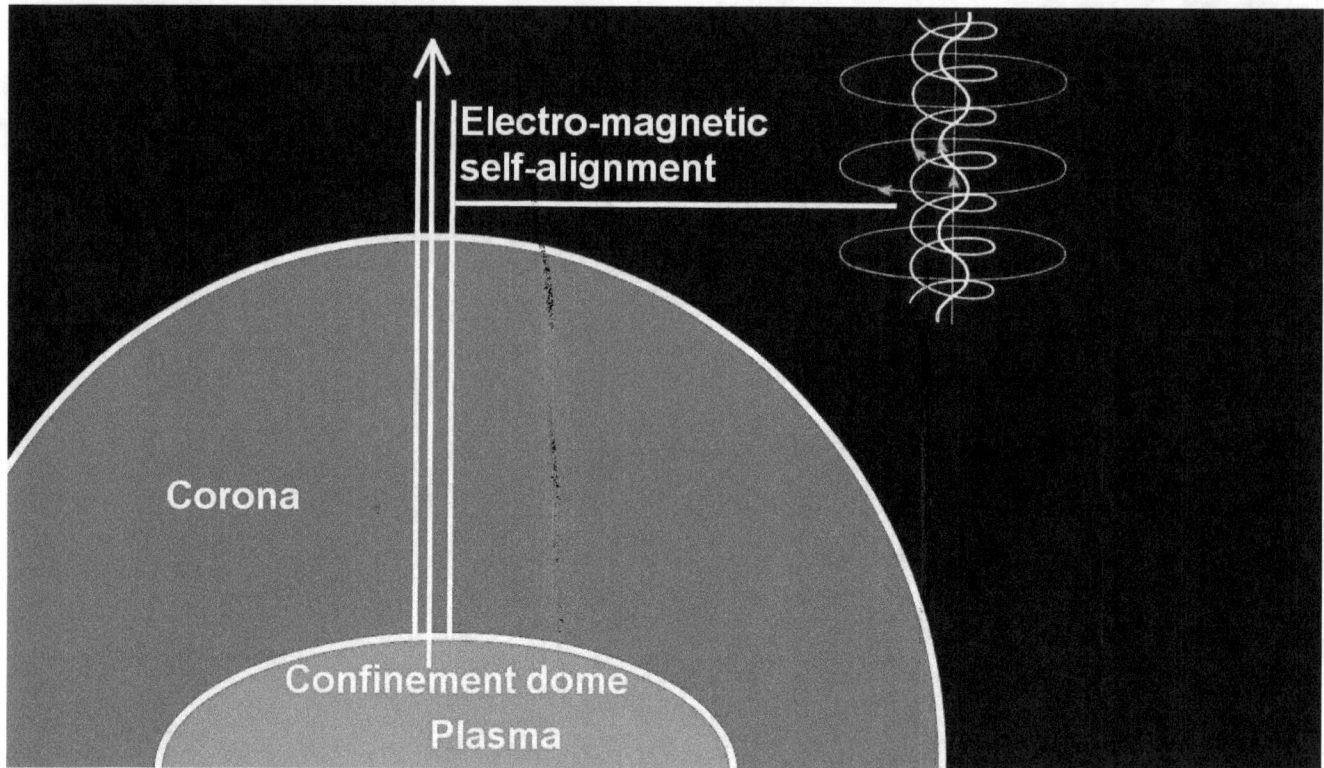

Fast moving plasma streams generate a magnetic field aground them that keeps the flowing stream narrowly concentrated., somewhat like a lightning bolt. The magnetic field tunnels through the corona, while the plasma particles push each other apart, further and faster. The process is similar a bullet fired through the barrel of a gun, propelled by exploding gas. The difference is that plasma is propelled by the vastly stronger repelling electric force.

Under the Big Bang model where only gravity is recognized

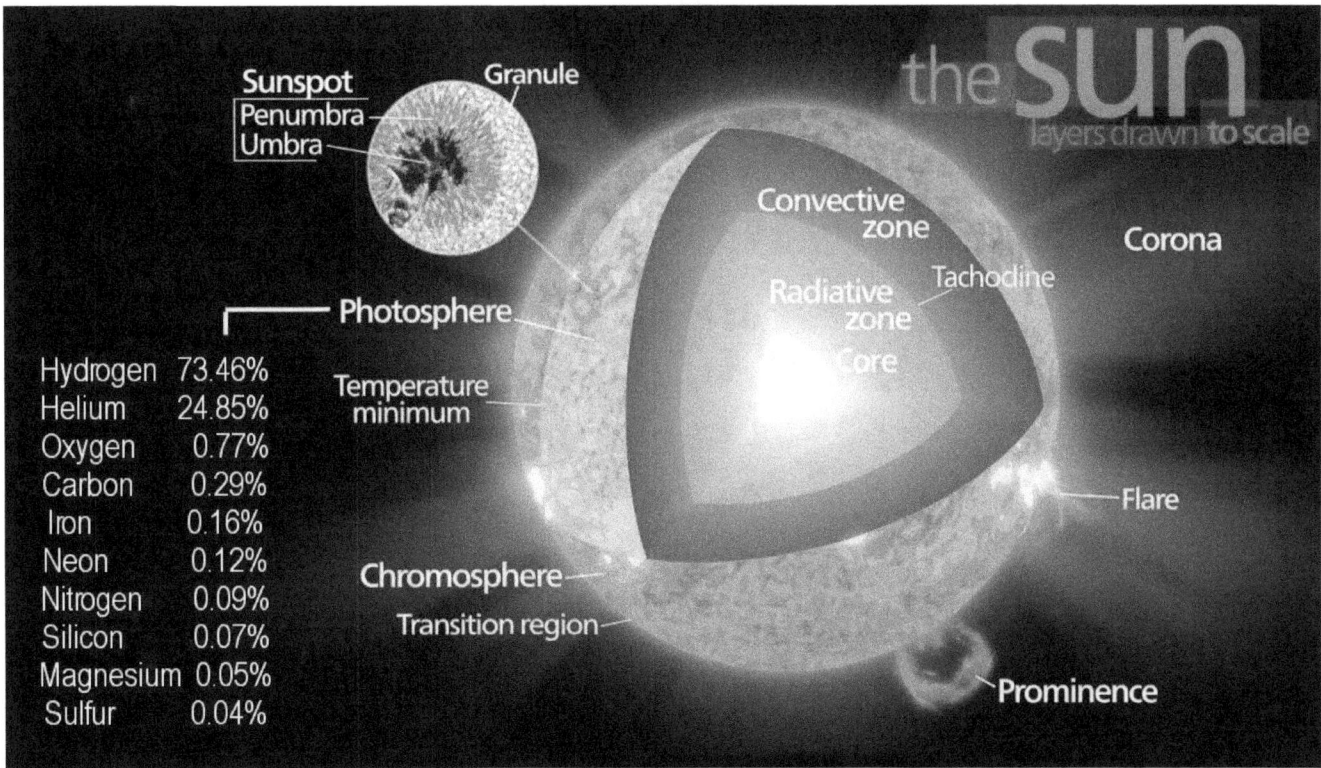

Under the Big Bang model where only gravity is recognized as a universally causative force, the solar wind shouldn't even be possible, as plasma streams are not recognized to exist. In the empty box of science perception - where plasma and its associated electric force and effects in the universe are deemed not to exist as an effective cosmic organizing impetus - the resulting constraint has imprisoned science to the task to rationalize an empty view of the world contrary to the obvious visual evidence. It is possible to break out from this empty box.

Plato, the great science genius

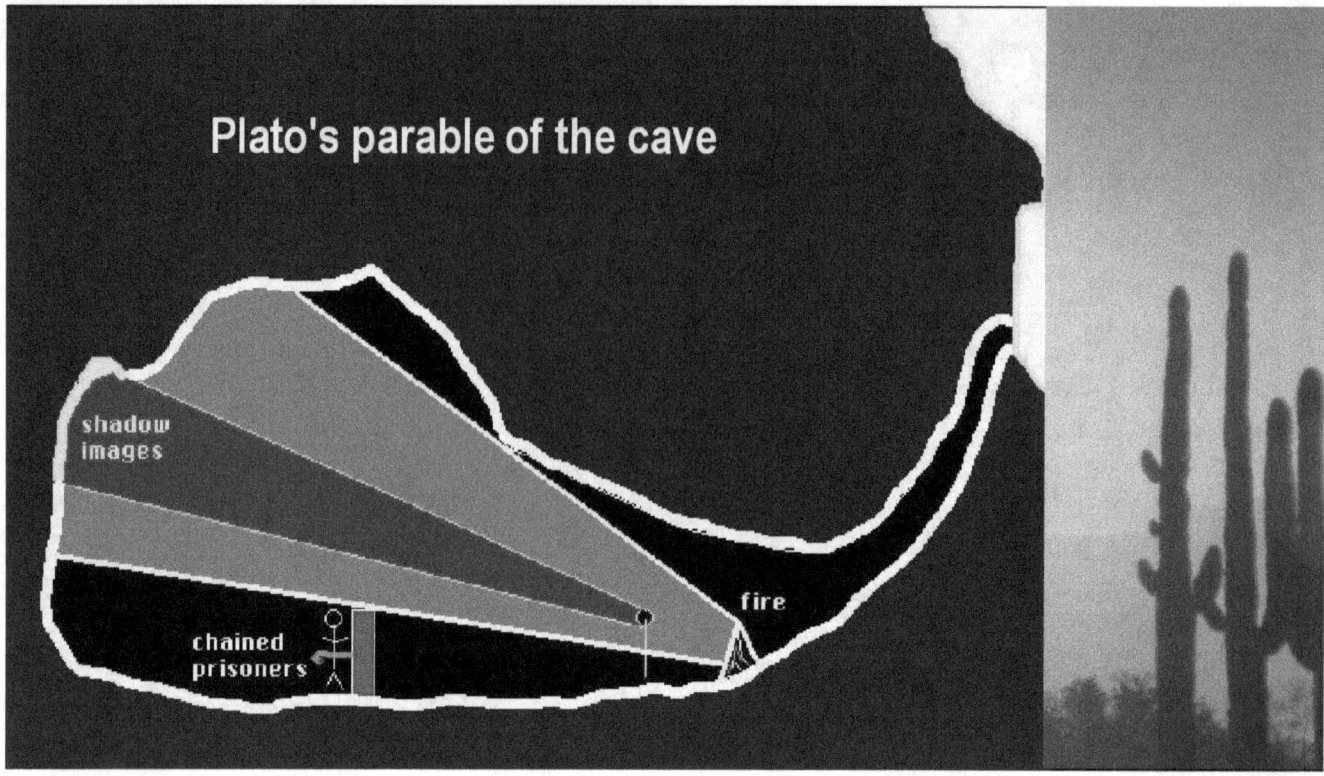

Plato, the great science genius of more than two thousand years ago, had illustrated the dynamics involved of breaking out from imprisoned perceptions with his allegory of the cave in his book, The Republic. He illustrates the case of a prisoner who had been long conditioned by a small sense of reality where nothing is actually real. One day he pokes his head over the barrier that he had lived behind. He notes that his world had been a world of illusions. As he ventures further, he notes that he lives in a cave, and that the cave has an exit.

As the prisoner ventures past the exit

As the prisoner ventures past the exit, he discovers that there exists a vast bright world outside the cave of his imprisoned past. He finds the brightness painfully blinding at first, but in time he celebrates the light and his new-found freedom.

Historic plasma evidence

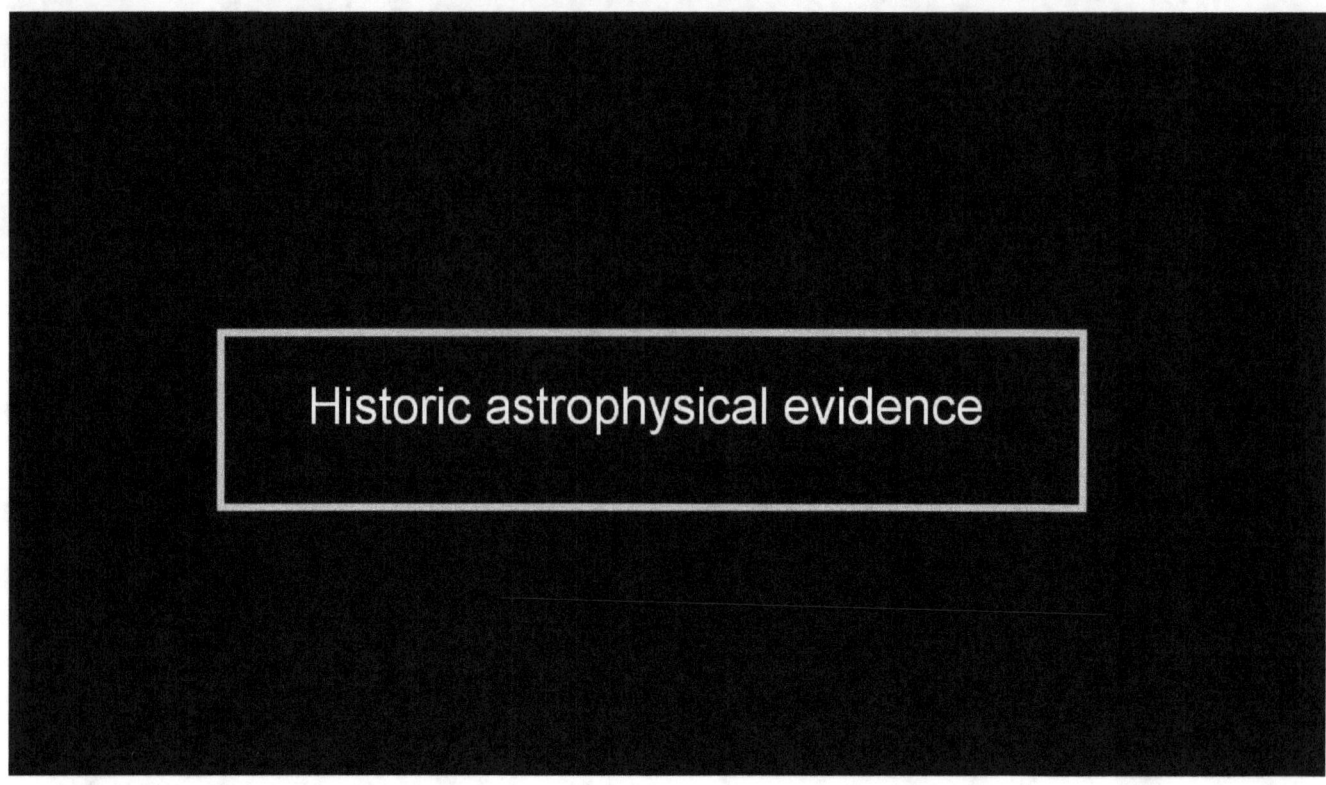

Historic plasma evidence in the Giza Pyramids' alignment.

Some amazing aspects of evidence

Some amazing aspects of evidence take us far away from the currently accepted theory of the internally heated, nuclear-fusion powered Sun, including some that apparently have no connection with electro-astrophysics, like the Giza pyramids.

Some extremely larger items of evidence

We have some extremely larger items of evidence before us, that relate to the Sun being externally powered.

Some related evidence exists that suggests

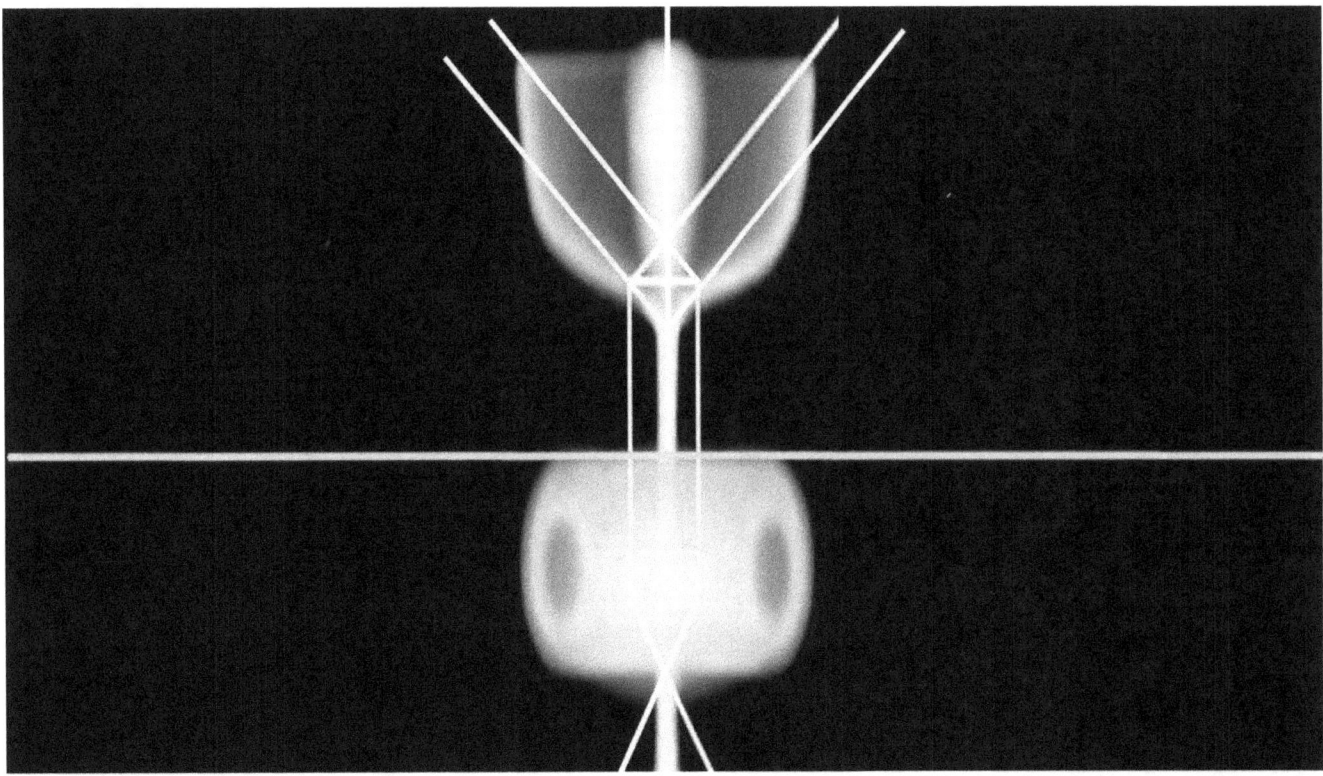

While the physical features of this evidence can no longer be seen, some related evidence exists that suggests that some truly gigantic features of large-scale plasma-flow phenomena had once been visible. In today's electrically weak time, the once visible features can nevertheless be experimentally replicated, in laboratory experiments.

Historic evidence found in the Giza pyramids

One aspect of historic evidence, of large-scale plasma-flow phenomena, is found in the Giza pyramids themselves. The most perplexing feature of the pyramids is the amazing accuracy of their alignment with the line of the Earth's meridian that no one can actually see.

The baseline of the pyramids is perfectly aligned

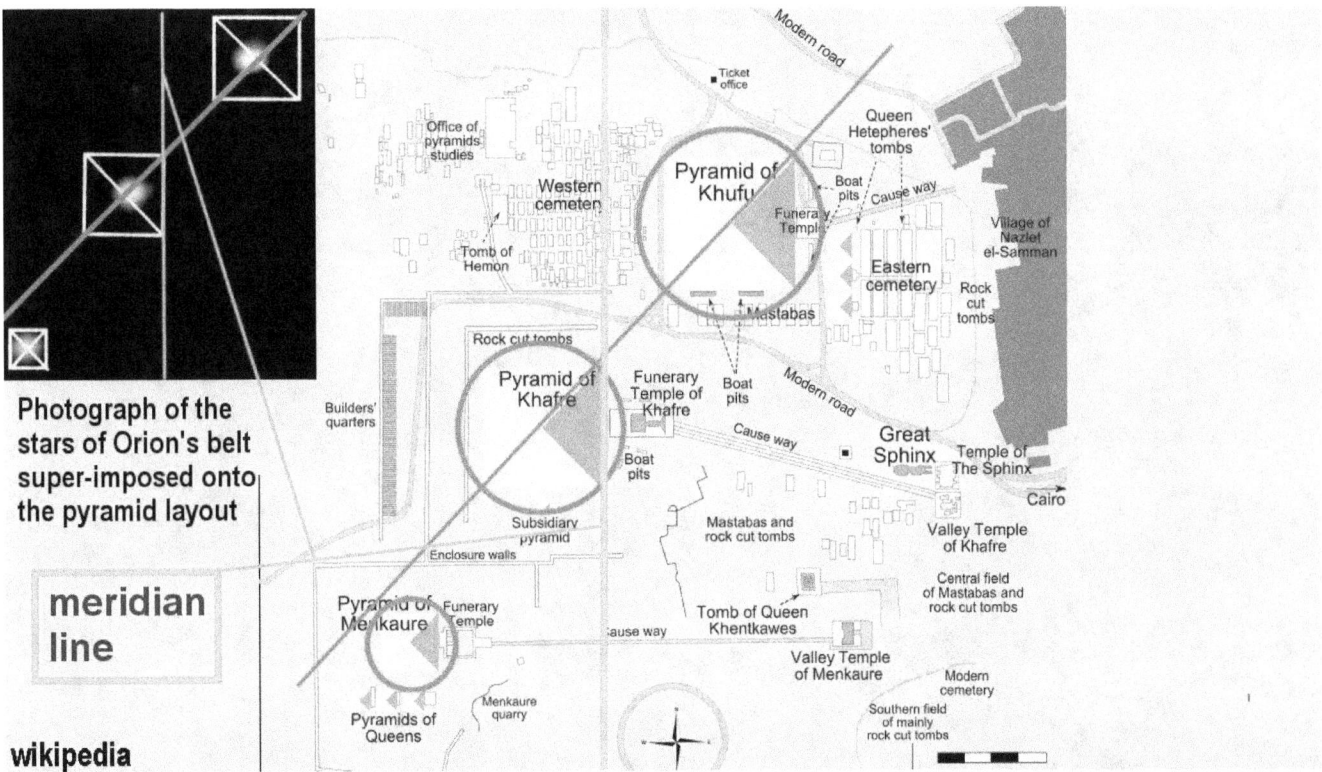

The baseline of the pyramids is perfectly aligned with this this physically invisible geographic orientation. These gigantic structures have been built with an accuracy of orientation of a mere four minutes of arc, by which the Giza pyramids have become the most accurately celestially oriented structures in the world. How has this amazing accuracy been achieved without a visible reference line for it. The evident answer is, that some type of reference line had likely been clearly 'visible' at the time when the pyramids were built.

The orientation of the sphinx

wikipedia - from a photo by Usuario:Barcex

Researchers discovered that the orientation of the sphinx is such, that it would have faced its image in the stars, the constellation Leo, roughly 12,800 years ago. This alignment suggests that the pyramids were built at this early time.

This distant timeframe coincides

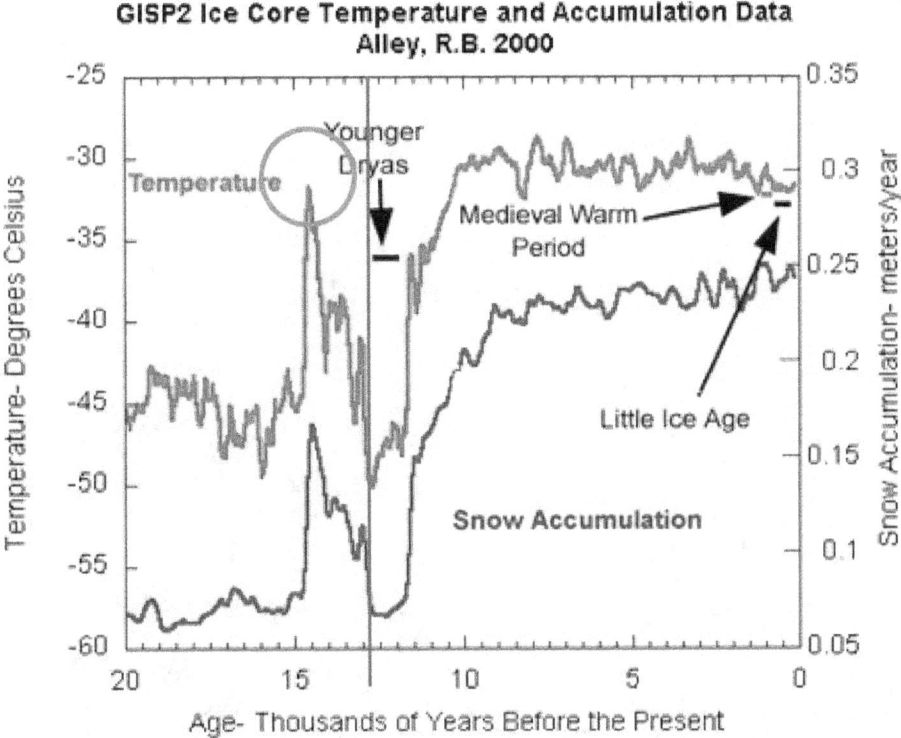

This distant timeframe coincides exactly with the start of the gigantic re-warming of the Earth that marks the end of the last Ice Age and the start of the present interglacial period.

The experiment-derived, magnetically-shaped dynamic flow geometry

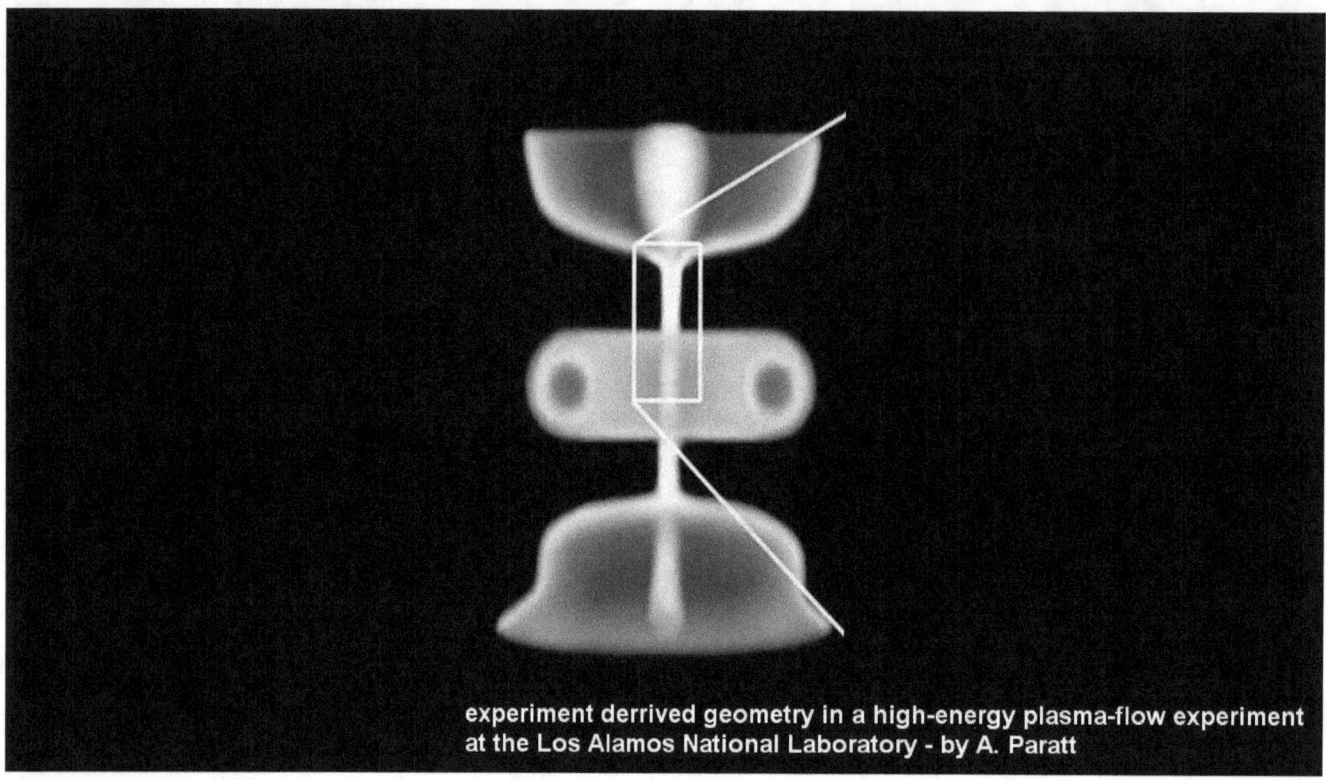

experiment derrived geometry in a high-energy plasma-flow experiment at the Los Alamos National Laboratory - by A. Paratt

If one utilizes the experiment-derived, magnetically-shaped dynamic flow geometry, which is shown here, as an example for what may have been visible in the night sky in ancient times, - and this would likely have been the case when the intensely powerful re-warming of the Earth began, - then a bright reference line would have been established in the sky that would have enabled the builders to accurately orient their pyramids with that clearly visible reference line in the sky that would have indicated the direction of the meridian line. The line might have had a religious significance for the builders.

The visible plasma geometry in the night sky

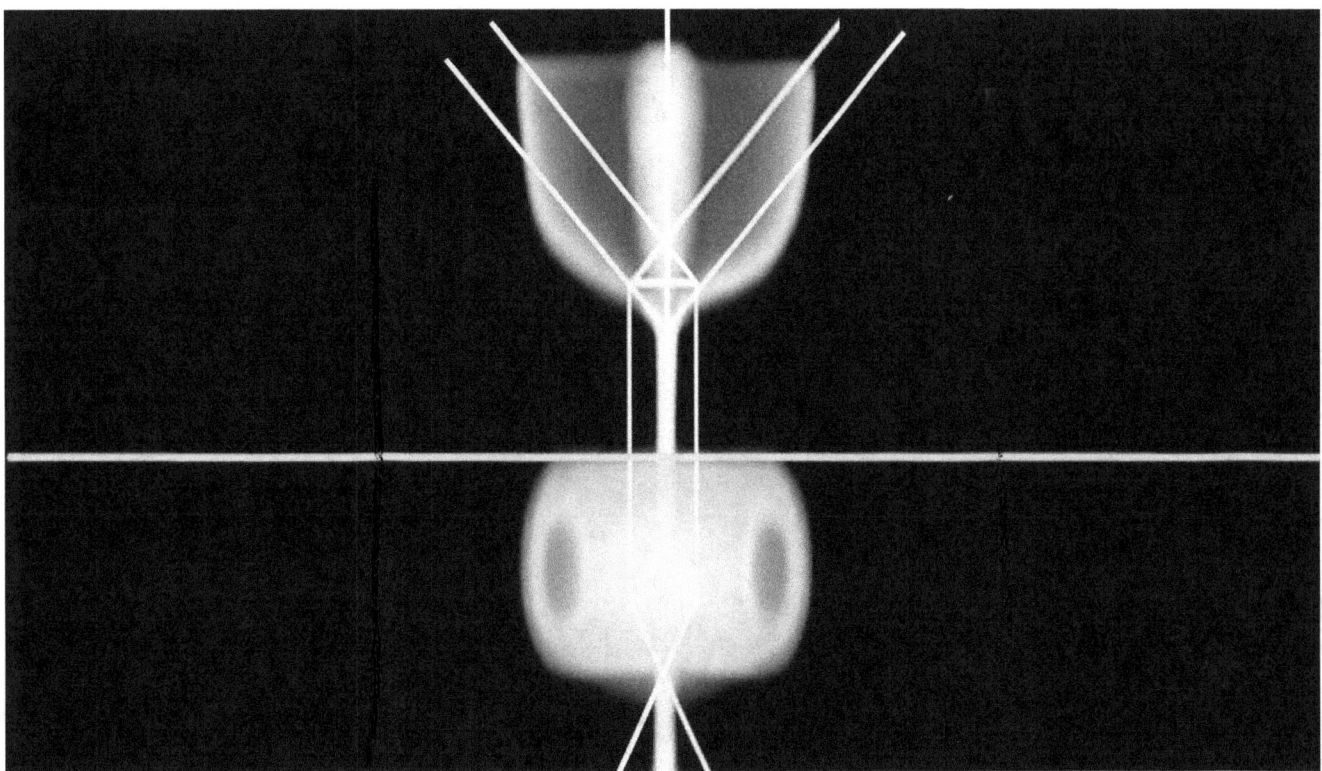

With the Sun positioned just below the horizon, the visible plasma geometry in the night sky, at the time when the geometry is perfectly vertical, would mark the direction towards the geographic pole of our planet, and with it the direction of the meridian, against which the pyramids have been aligned with this marvellous extreme accuracy that was been achieved.

Experiments have indicated that the plasma structure in the sky would have been perfectly perpendicular to the ecliptic plain, when it had been visible in times of strong plasma currents, thereby providing a celestial reference line. Of course, this reference line can no longer be seen today, in today's weak electric environment.

The idea for the building of the pyramids

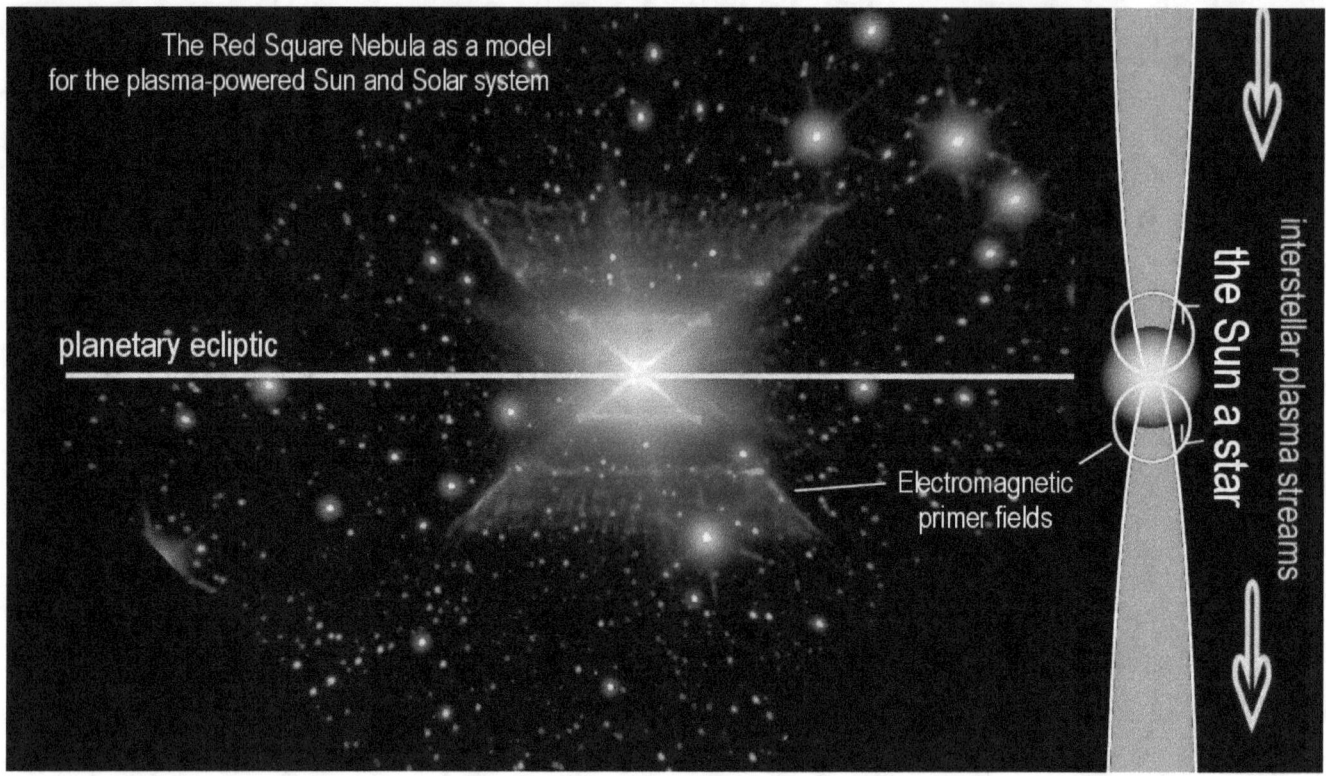

The idea, itself, for the building of the pyramids, might have been derived from features of the plasma structure that might have been incorporated into the geometry of the plasma-flow structure in some fashion.

Stone Henge

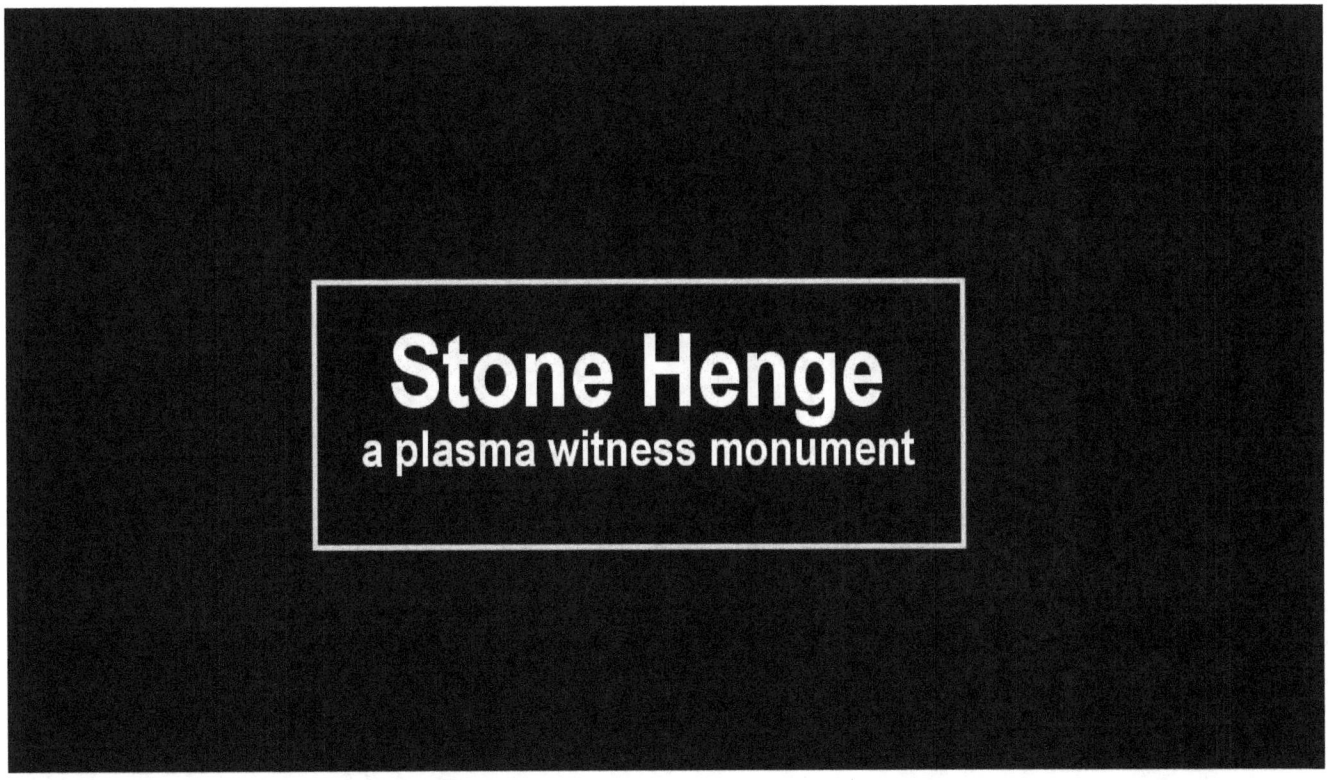

Stone Henge: a plasma witness monument

Replicated in the layout of the Stone Henge

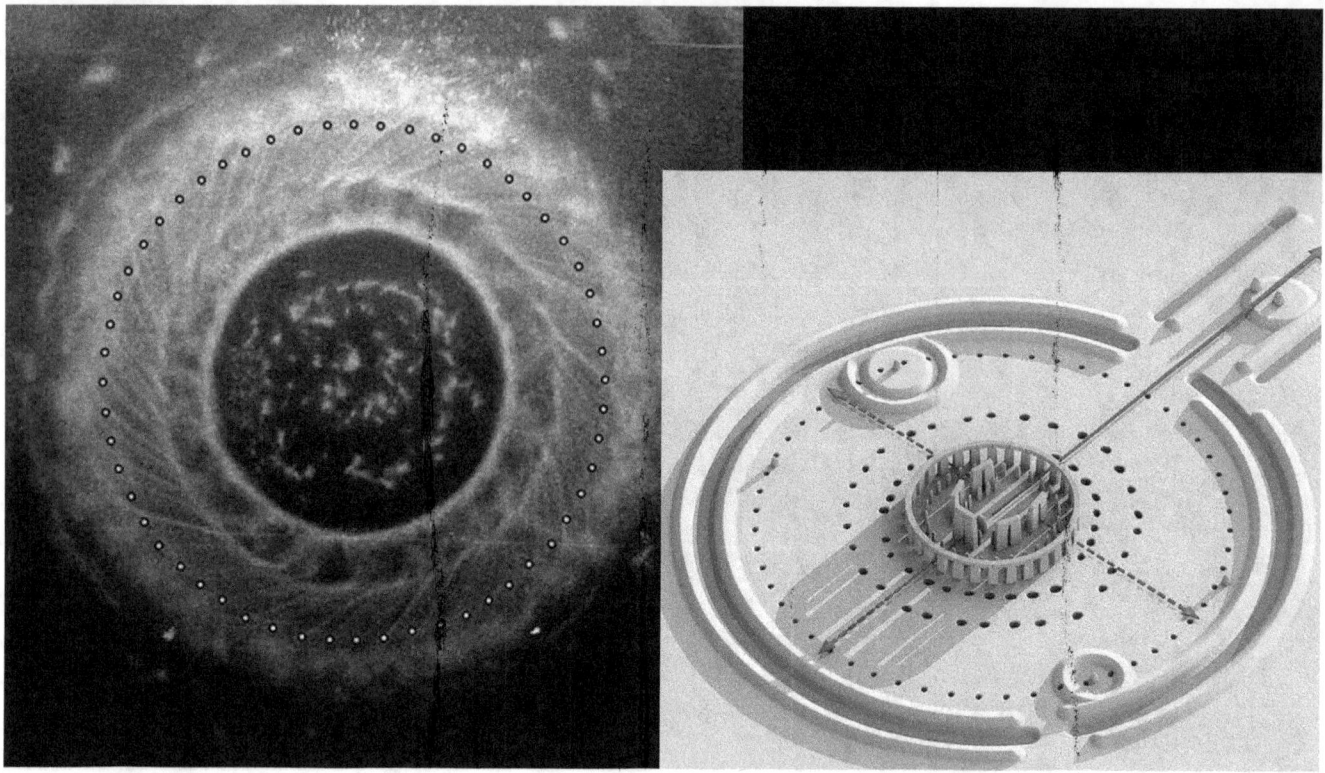

A further example of once visible plasma features is found in the more-recent historic monument, termed, Stone Henge.

The geometry that we see here matches the experimentally discovered geometry in plasma flows. We find this geometry amazingly accurately replicated in the layout of the Stone Henge monument.

The monument's features

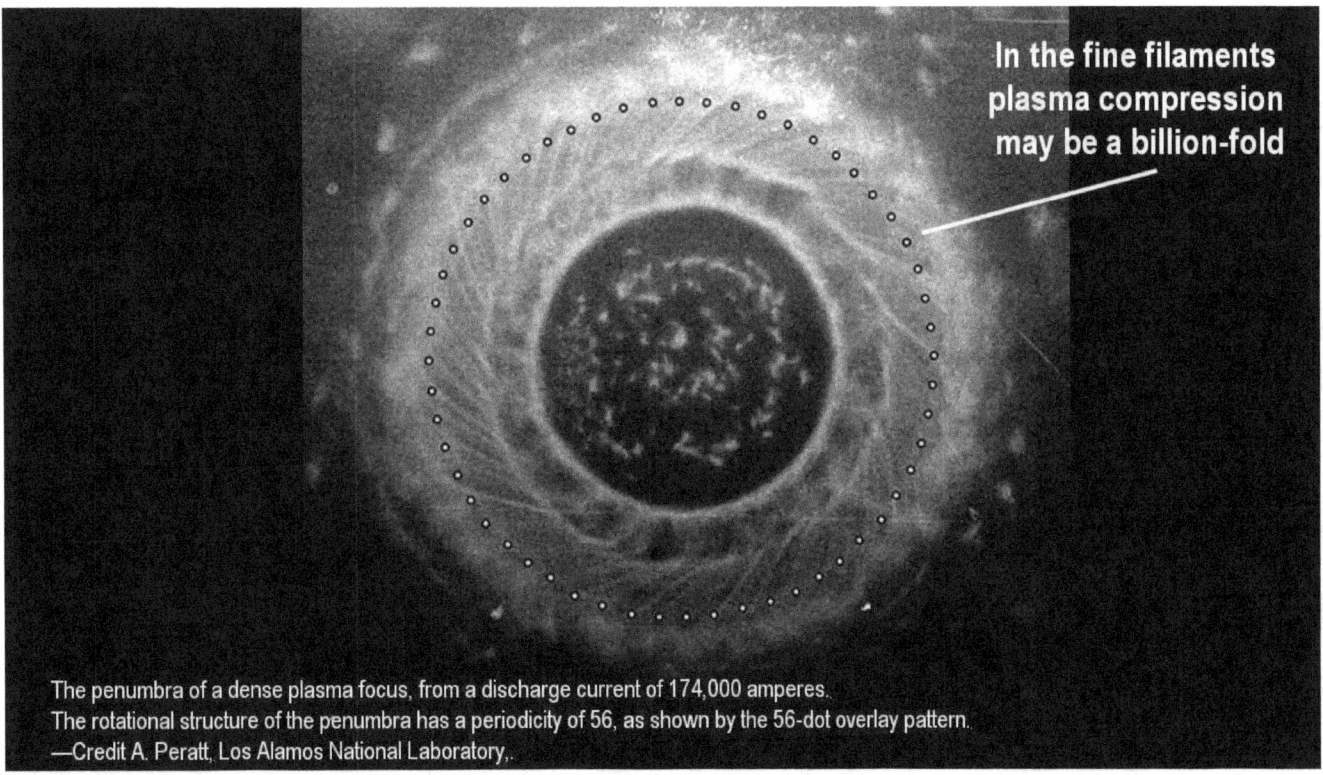

In the fine filaments plasma compression may be a billion-fold

The penumbra of a dense plasma focus, from a discharge current of 174,000 amperes.
The rotational structure of the penumbra has a periodicity of 56, as shown by the 56-dot overlay pattern.
—Credit A. Peratt, Los Alamos National Laboratory,.

The monument's features might have been visible during the interglacial optimum, from a high latitude position.

Replicated in the form of a large monument

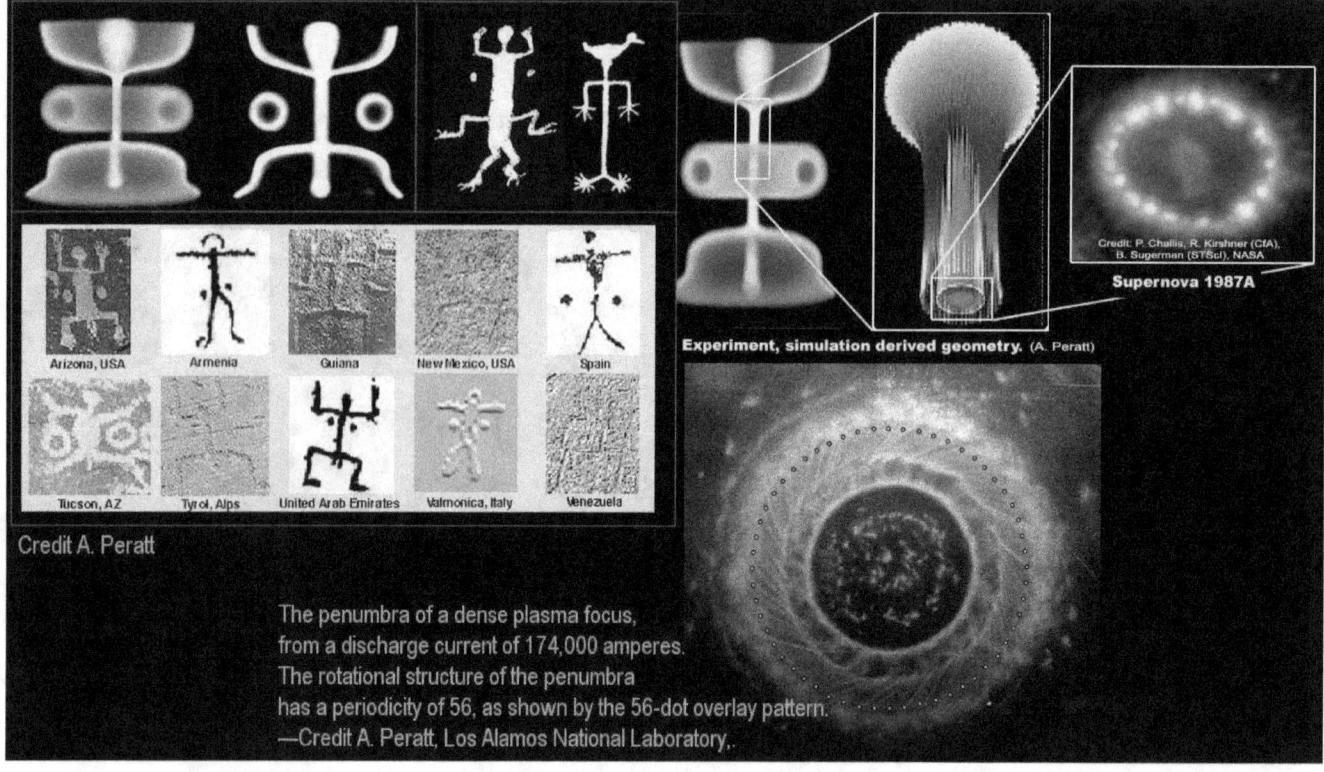

This means that the recently discovered features in high-energy plasma-flow experiments had been clearly visible in the sky, and have been replicated at the time in the form of a large monument for religious imperatives.

Aligned into 56 evenly-spaced filaments

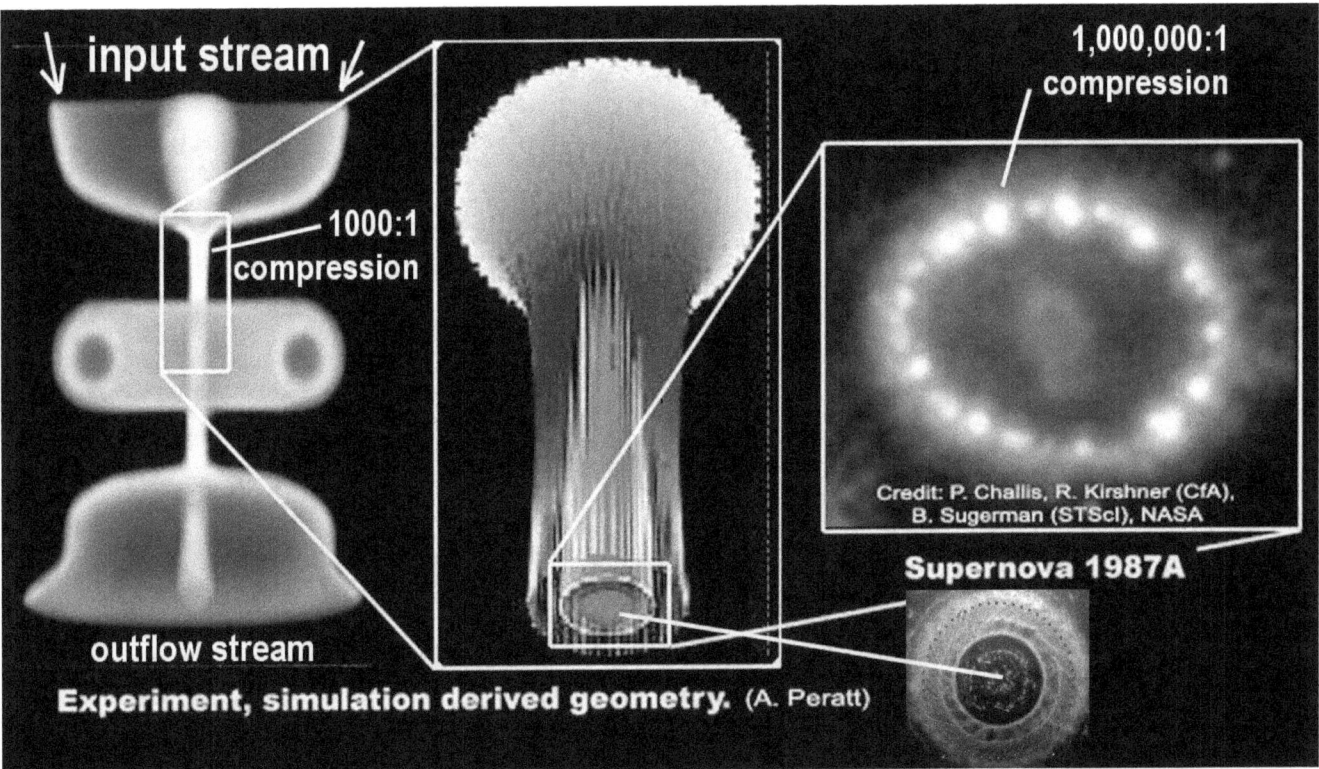

It has been discovered in high-power plasma-flow experiments that strongly moving plasma streams become magnetically aligned into 56 evenly-spaced filaments, or in lesser streams to fractions of this number, as it is seen in the case of the cross section view of a plasma stream that once caused what is erroneously termed Supernova 1987A.

The plasma-flow experiment

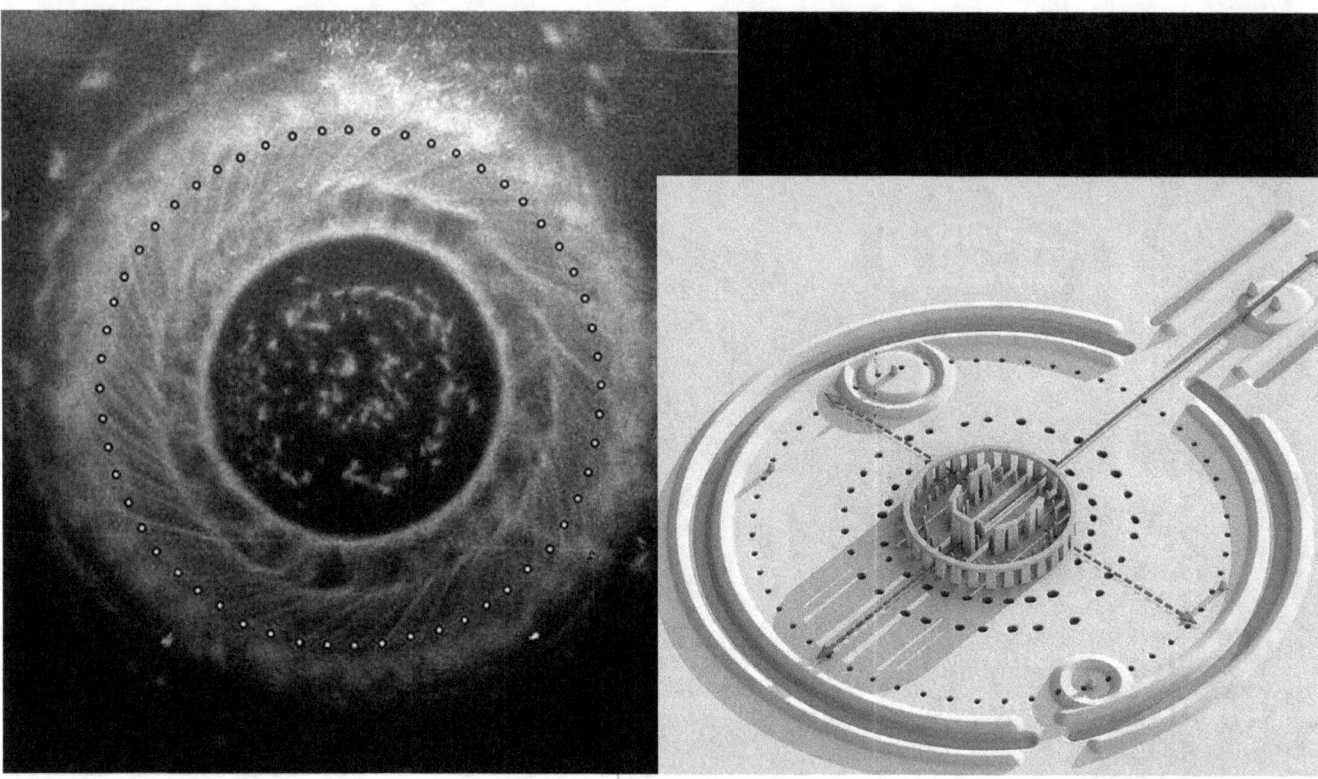

The plasma-flow experiment that had yielded the full complement of 56 filaments, matches amazingly closely the basic layout of the Stone Henge monument. The position of the 56 plasma filaments discovered in the experiments, matches the relative position of the monument's 56 Aubury holes amazingly accurately. The Aubury holes are pits that are believed to have served as sockets for a circle of large wooden poles.

These plasma features were once seen in the night sky

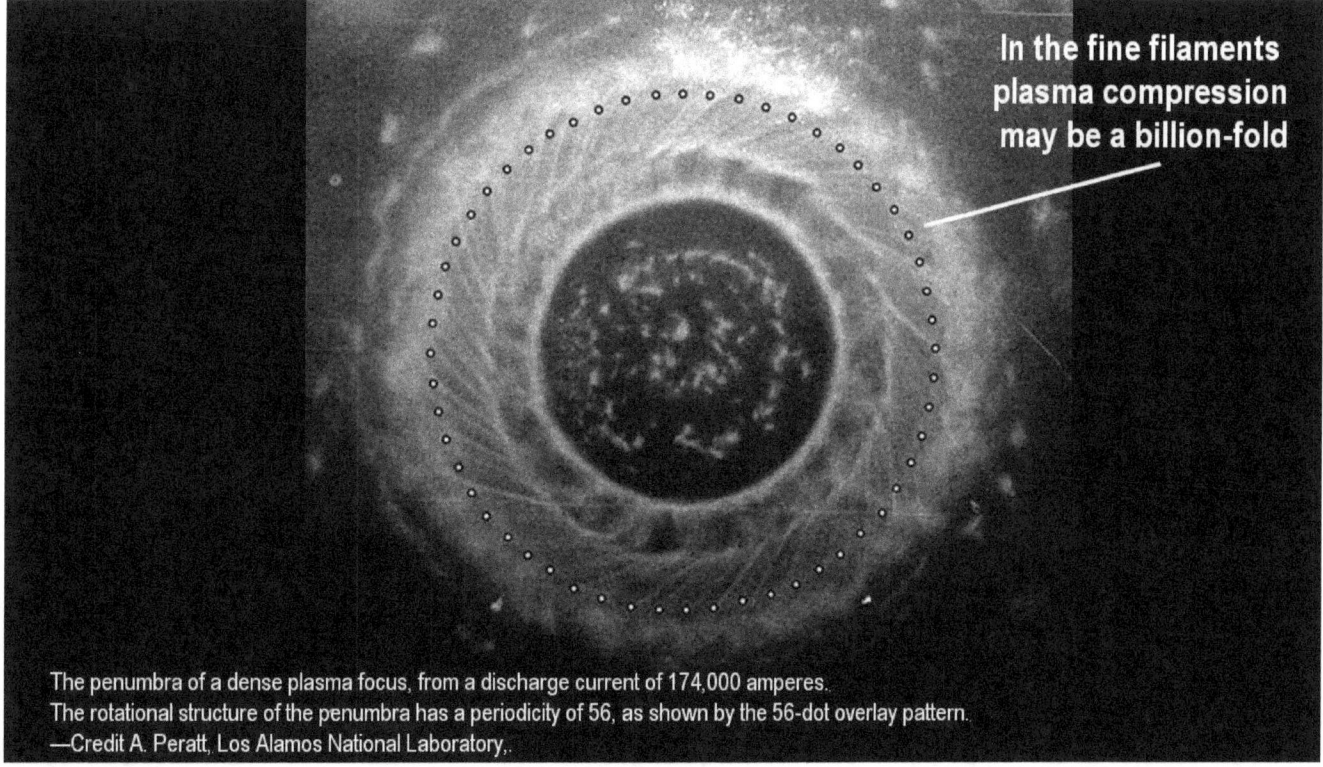

The penumbra of a dense plasma focus, from a discharge current of 174,000 amperes. The rotational structure of the penumbra has a periodicity of 56, as shown by the 56-dot overlay pattern.
—Credit A. Peratt, Los Alamos National Laboratory,.

In the fine filaments plasma compression may be a billion-fold

The close agreement of the construction of the monument with experiment-derived features in plasma-flow geometry, suggests that these plasma features were once seen in the night sky, possibly during the interglacial maximum, when the plasma-flow through the solar system would have been comparatively strong.

We are presently at a deep low point

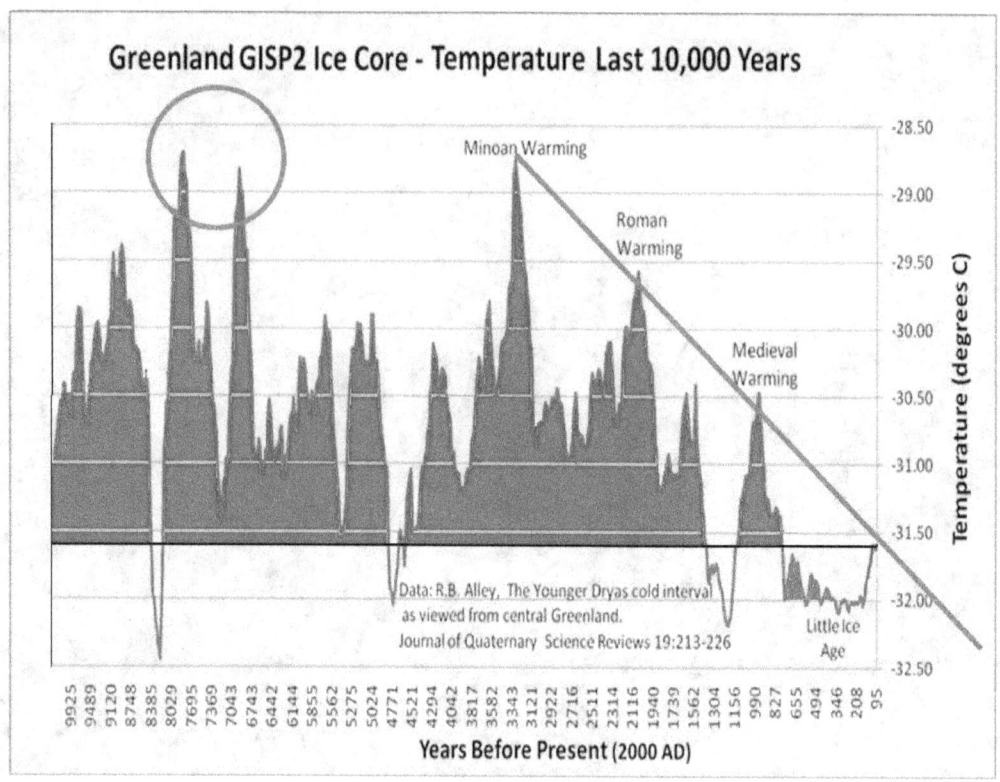

We are presently at a deep low point on the solar intensity scale, so that the plasma features that were evidently once plainly visible, are no longer visible.

Familiar features that are visible today

http://www.zam.fme.vutbr.cz/~druck/Eclipse/ - an example of the amazing solar eclipse photography of Milloslav Druckmueller

In fact, we are presently approaching a point in the ongoing electric weakening in the solar system, where familiar features that are visible today, such as the solar winds, may, in the near future, no longer be 'visible,' either.

Presently visible, large plasma structures

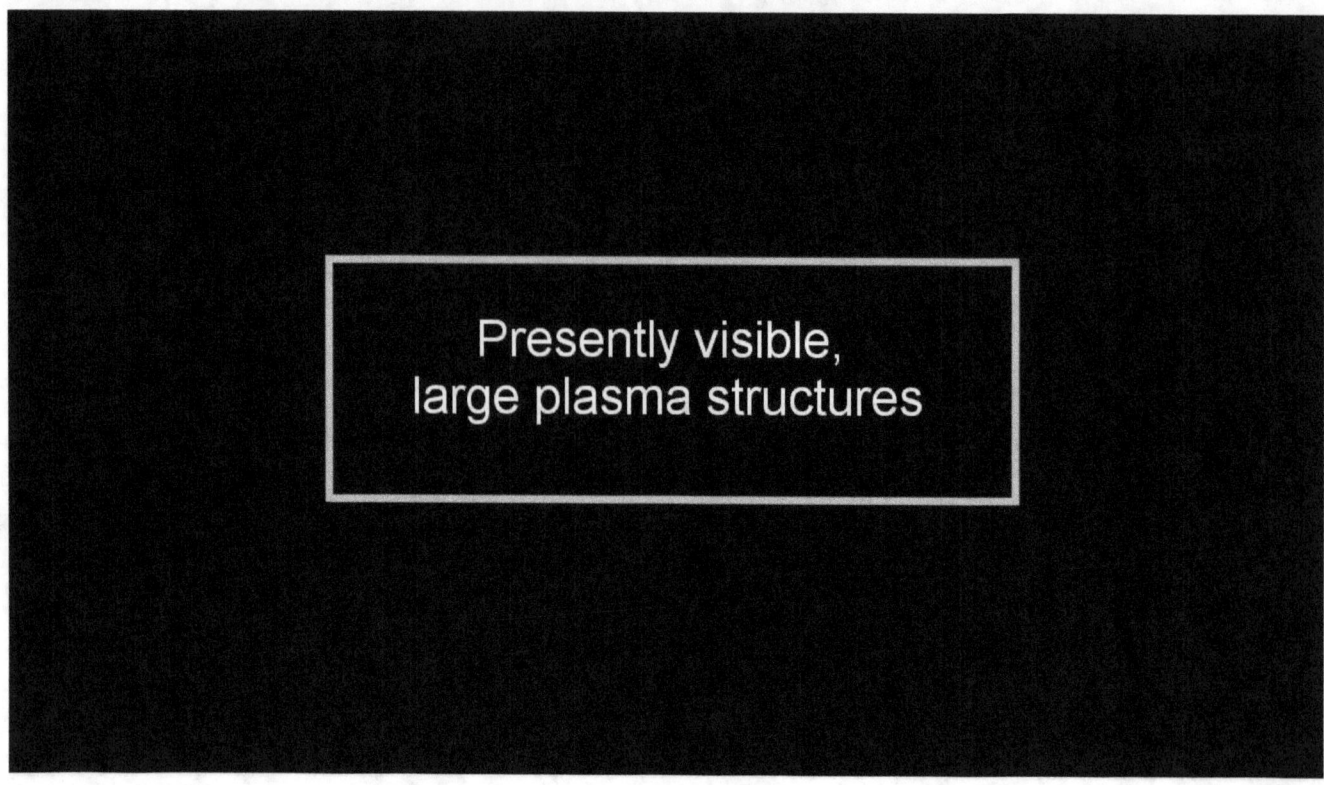

Presently visible, large plasma structures

Modern instrumentation has made it possible for gigantic plasma features to be seen that, in spite of their size, have remained invisible before.

Part 3: Historic evidence of large cosmic plasma structures

Two gigantic plasma structures from the heart of the Milky Way Galaxy

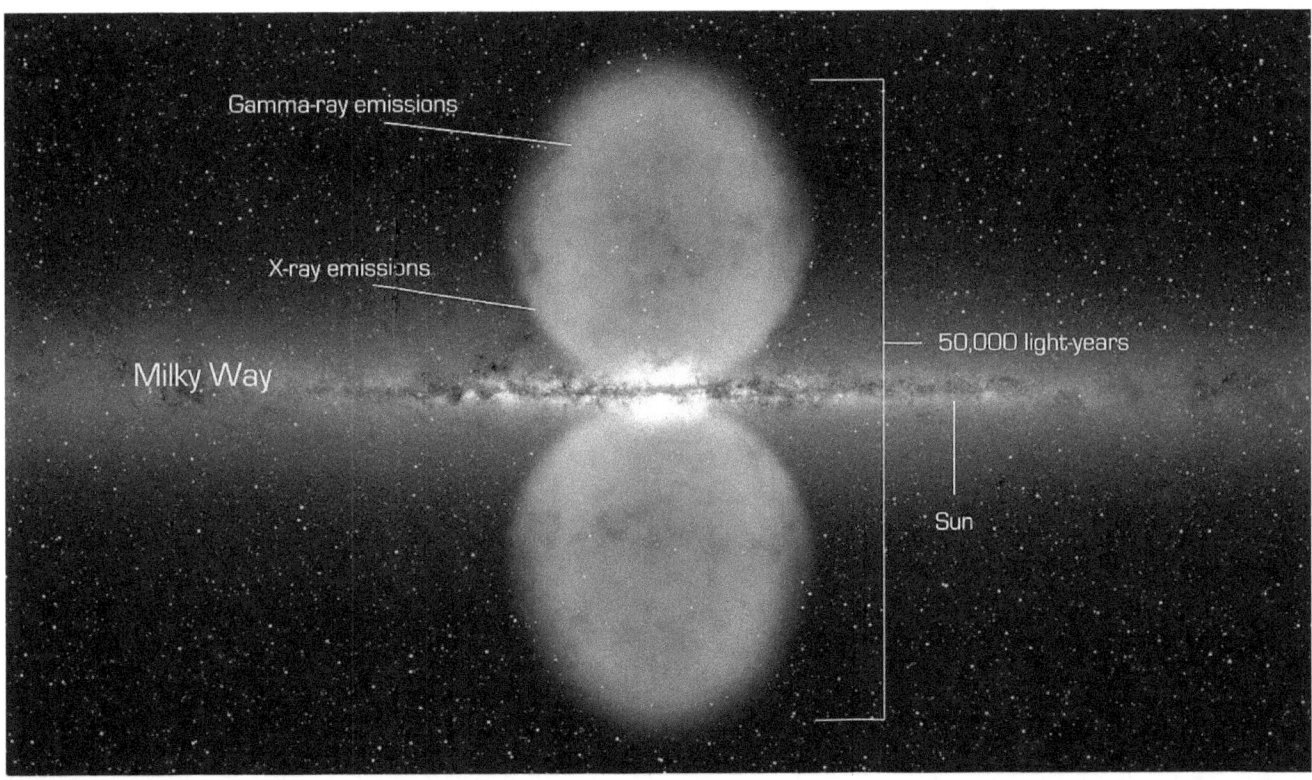

One of the newly detected features are two gigantic plasma structures that extend perpendicularly from the heart of the Milky Way Galaxy, across a distance of 25,000 light years above and below the galactic disk.

The plasma explorer Hannes Alfven had theorized

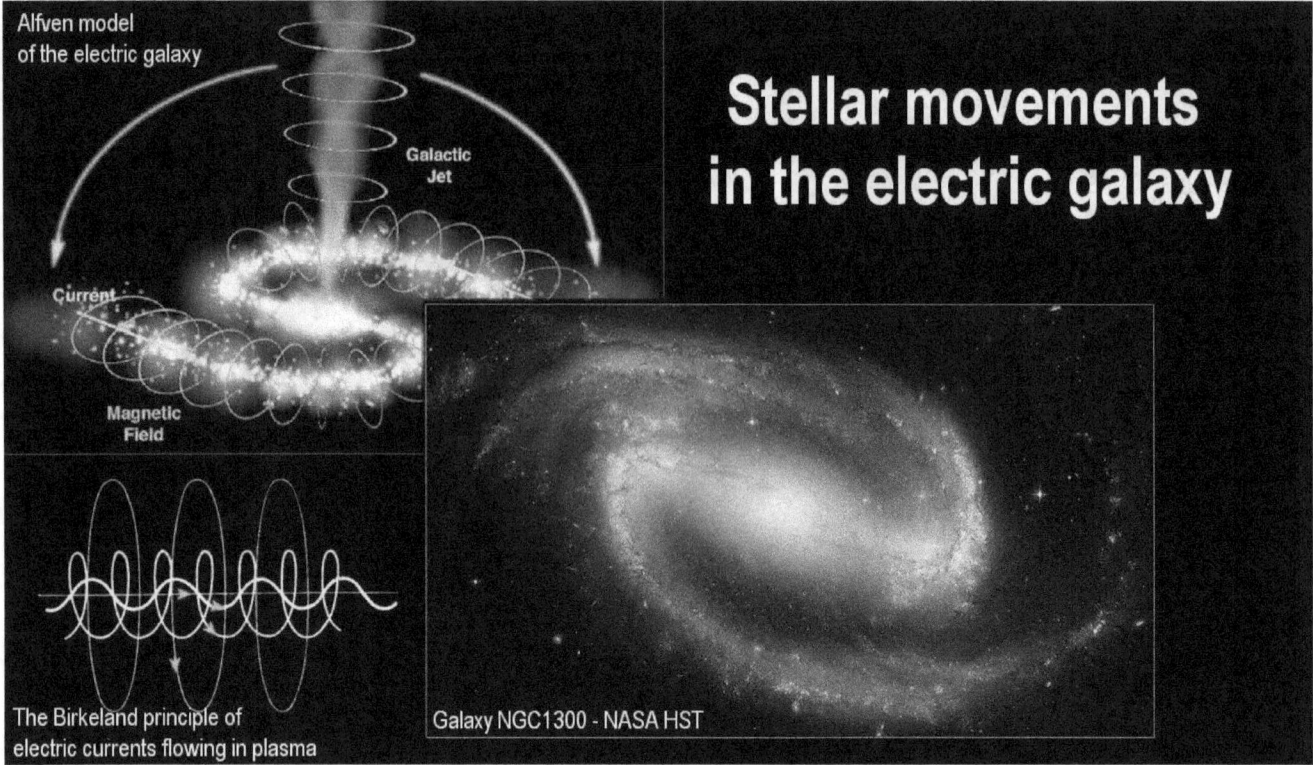

Long before this unique plasma structure became 'visible,' the plasma explorer Hannes Alfven had theorized that for the visible part of astrophysical dynamics, such as the galactic disk, must exist as a causative force that corresponds with the known geometric features in plasma physics, and that these must be expressed in the galaxies, including the invisible parts of the features.

The primer fields geometry derived from high-energy experiments

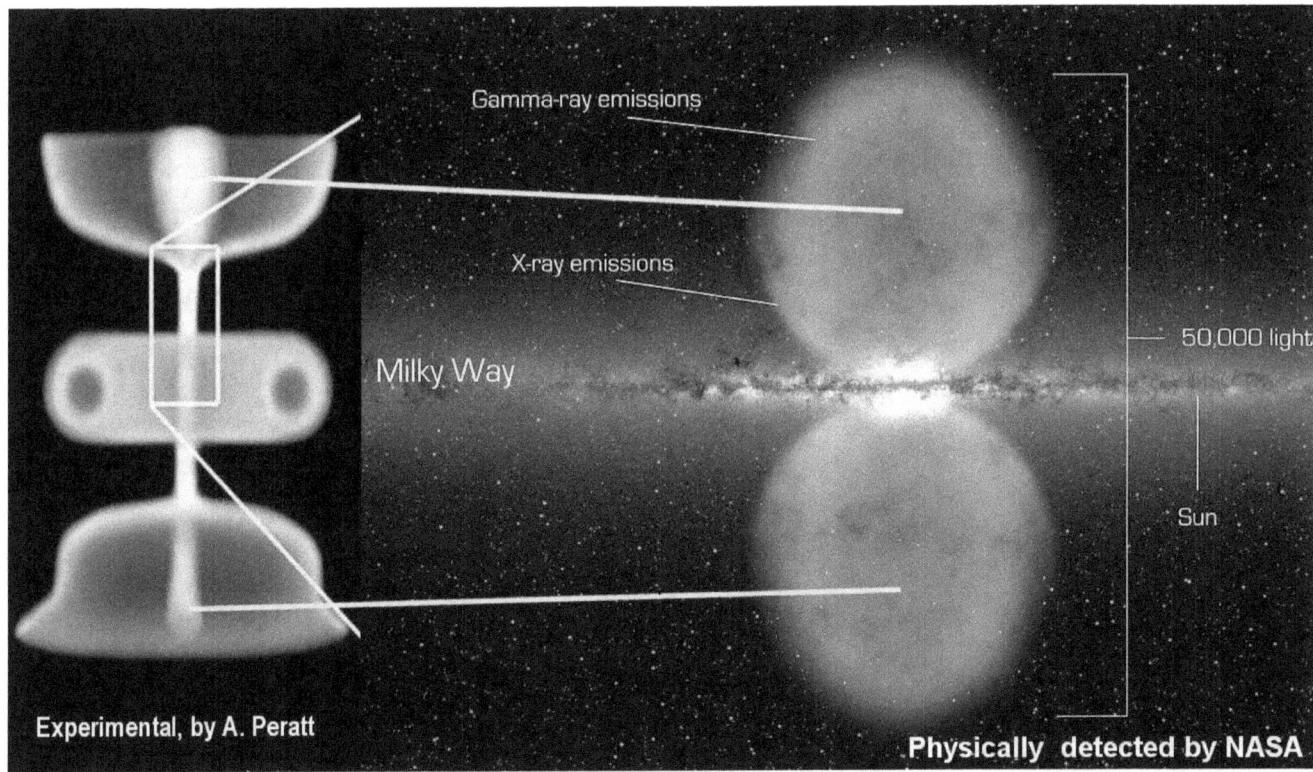

While many theories exist to interpret the phenomenon that NASA has discovered with advanced instrumentation, the only laboratory experiment that I know of that replicates in principle the observed galactic phenomenon in the small, is the primer fields geometry derived from high-energy experiments at the Los Alamos National Laboratory, conducted by Anthony Paratt.

The experiment-derived geometry illustrates

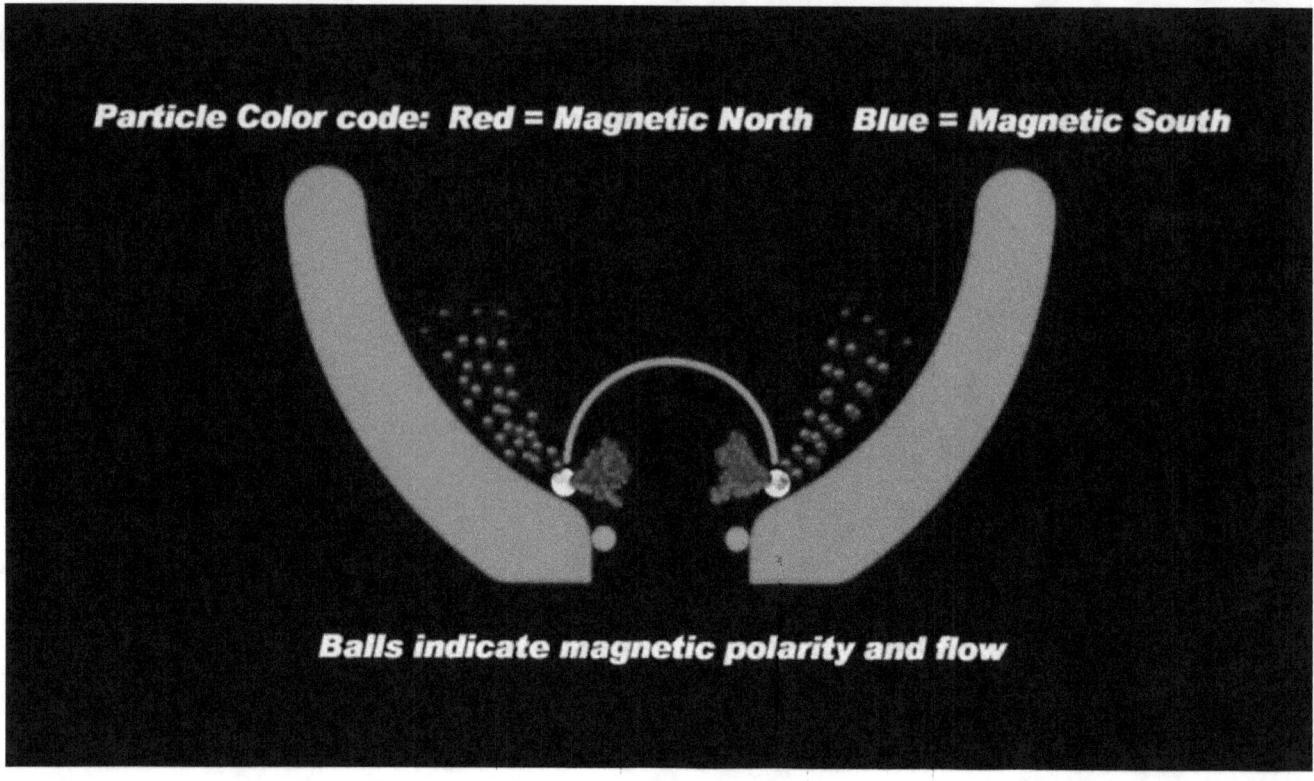

The experiment-derived geometry illustrates the galactic plasma structures to be magnetically confined concentrations of high-density plasma under, what David LaPoint refers to as, the magnetically created "confinement dome." He explored in details of it in 'static' laboratory experiments.

The magnetic bowl structure

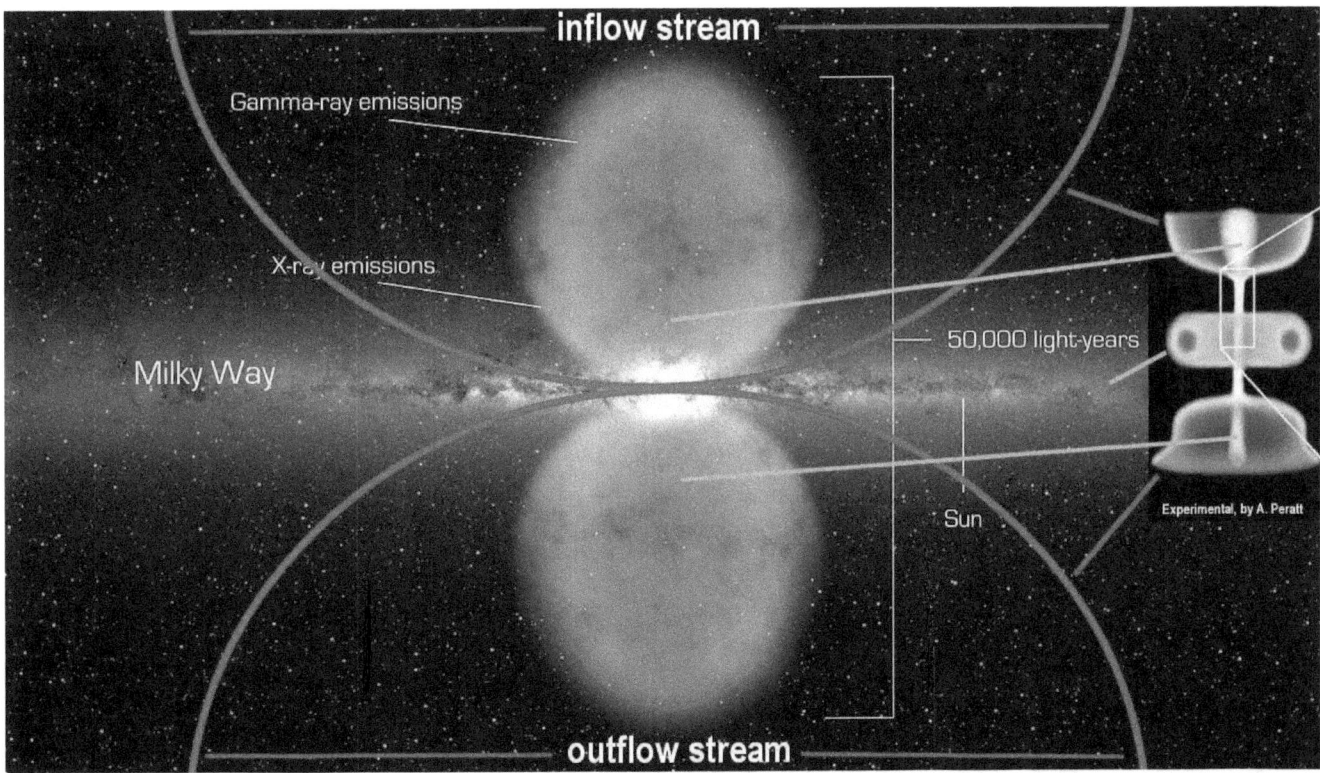

The magnetic bowl structure that channels intergalactic plasma streams under the confinement domes of the Milky Way Galaxy structure, is evidently too weak to be visible with the currently available instrumentation. However, the large magnetic field structure that creates the confinement domes, has become visible by its effect.

The huge scale of the plasma structure

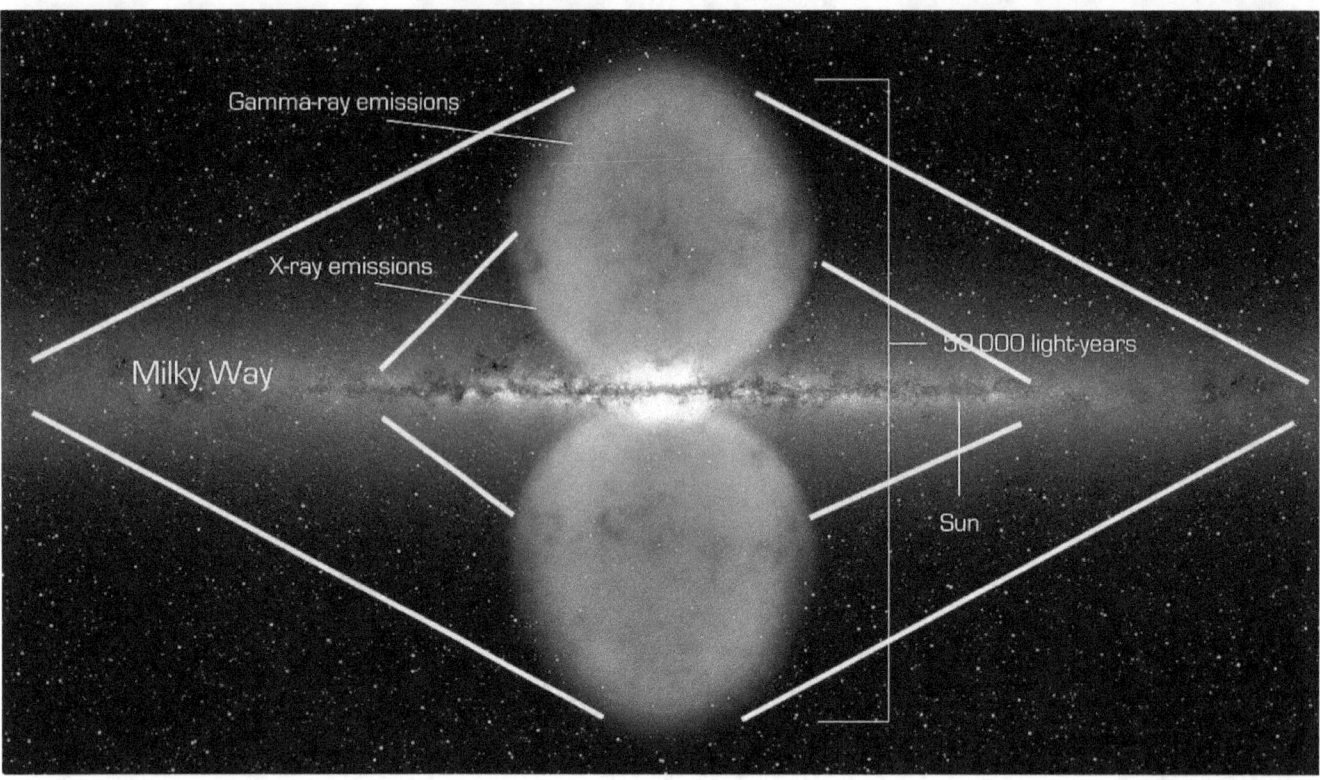

The huge scale of the plasma structure that has become visible, is evidently sufficiently large to have the effect that pinches the galaxies into an electrically aligned flat disk by electric repulsion from above and below, and by the magnetic fields that are associated with the gigantic plasma structures.

The bowl type magnetic field structure

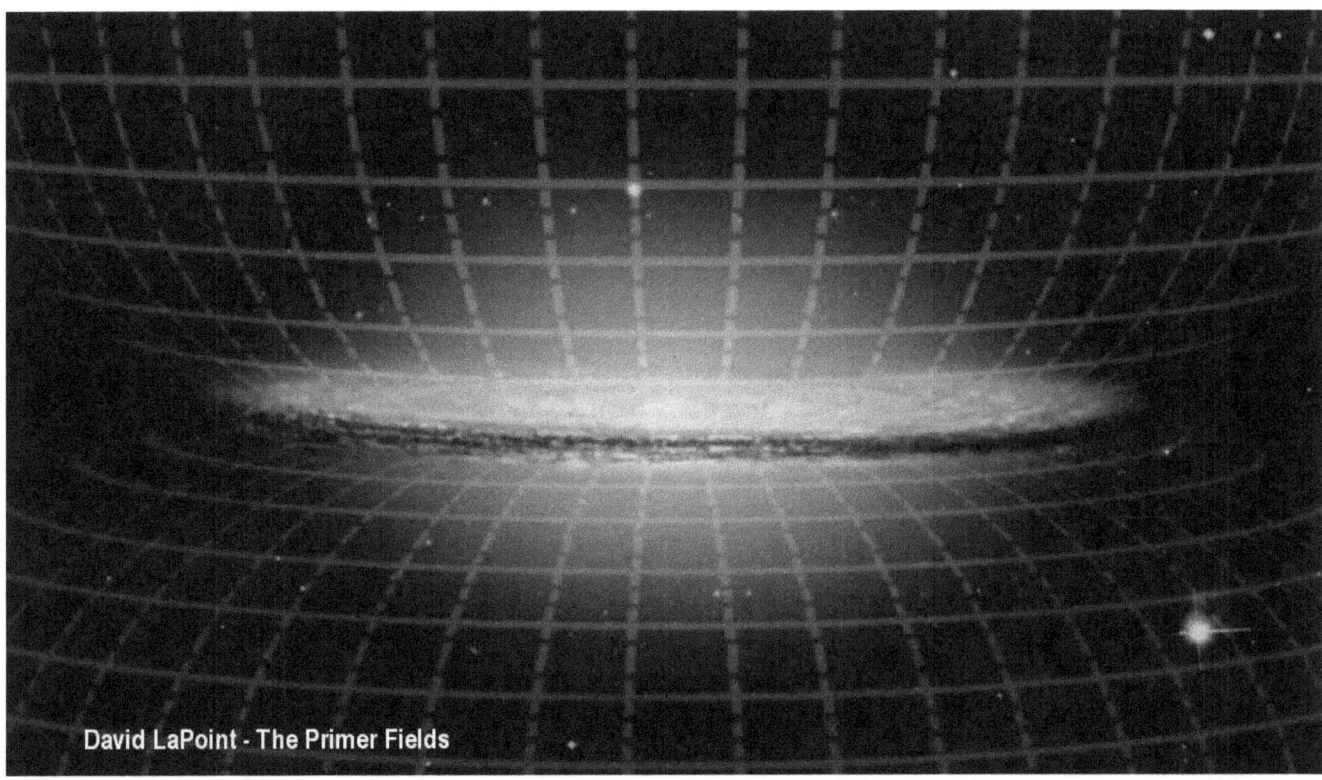

The bowl type magnetic field structure that the confinement domes are a part of, may be much larger than the experiment-derived geometry illustrates. The magnetic field may be as wide as the galaxy itself. The in-flowing plasma streams, of course, are presently invisible.

Intergalactic plasma streams may remain forever invisible

The intergalactic plasma streams may remain forever invisible. Plasma can only become visible in light when the plasma particles agitate atomic structures that it encounters. Plasma streams can also be recognized by their electromagnetic dynamic effect on the alignment of stars and galaxies with one-another. This alignment effect is discernable in the existence of groups of stars and galaxies that are frequently arranged into short or long strings, lined up like beads on a thread, often with equal spacing between them. These invisible, but discernable, interconnecting plasma streams, link small or large individual groups of galaxies into functionally connected larger dynamic phenomena, and so on.

String-bound groups, small and large

The string-bound groups, small and large, are visible 'everywhere,' as far as our telescopes can reach.

When dense plasma streams encounter atomic elements

ESO/VIMOS galaxy cluster ACO 3341

In some cases the interconnecting plasma streams are strong enough to be faintly visible as light as they encounter atomic material in their path.

When dense plasma streams encounter atomic elements, the electric interaction strongly affects the 'dance' of the atoms' electron swarms.

The interaction energizes the dance

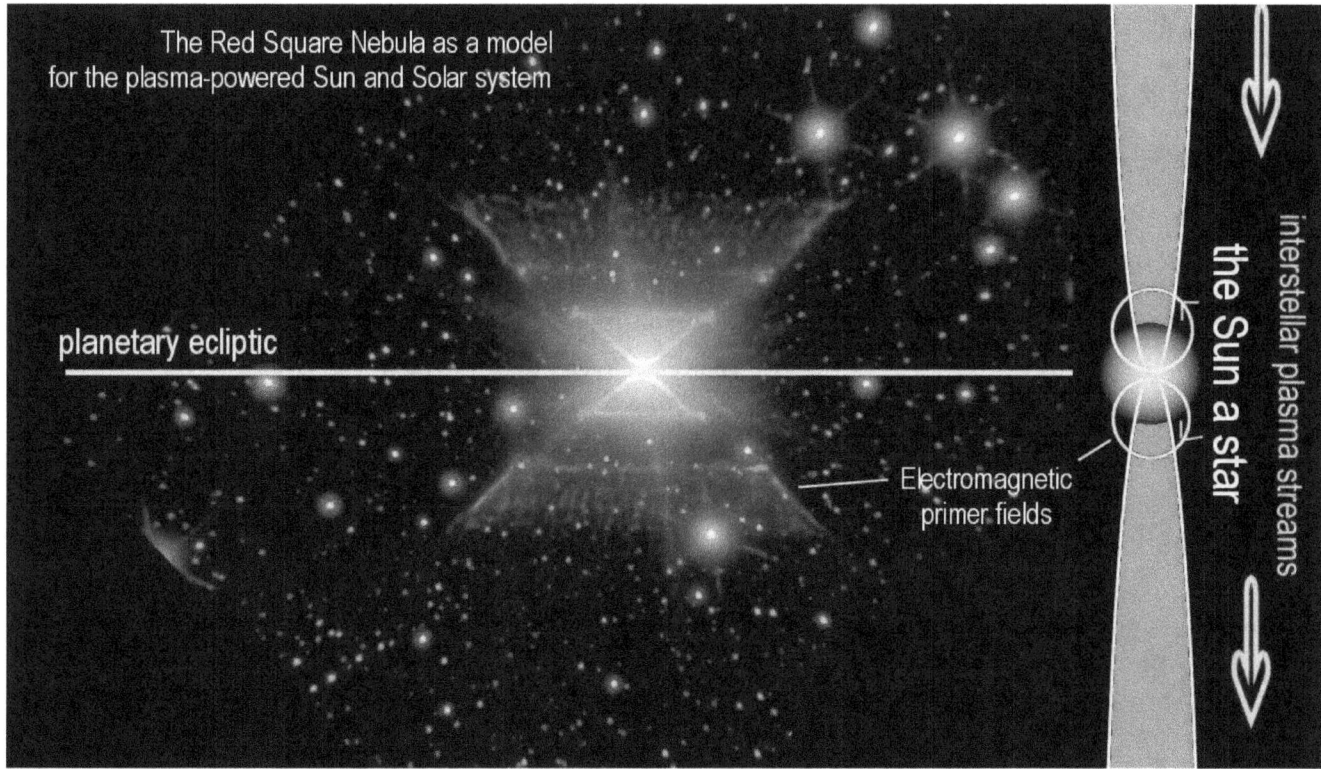

The interaction energizes the dance, whereby the atoms emit light, heat, and other forms of electromagnetic energy radiation. The agitating effect is the same as if the atoms had been heated to high temperatures by other forms of energy input.

The same type of effect in the form of lightning

On Earth we see the same type of effect in the form of lightning. The effect is the same as if the air had been heated to tens of thousands of degrees in the lightning stream, which evidently is not the case.

Part 4: The solar wind speaking of the coming Ice Age

Our Electric Cold Fusion Sun (Part 4) Solar winds and Ice Age

Solar wind, Ice Ages, and our future

Solar wind, Ice Ages, and our future

In the local theater, the solar wind is one of the major yardsticks that we have available, with which to judge how well-'funded' the electric fusion process on our Sun operates at a given time.

The solar synthesizing fusion

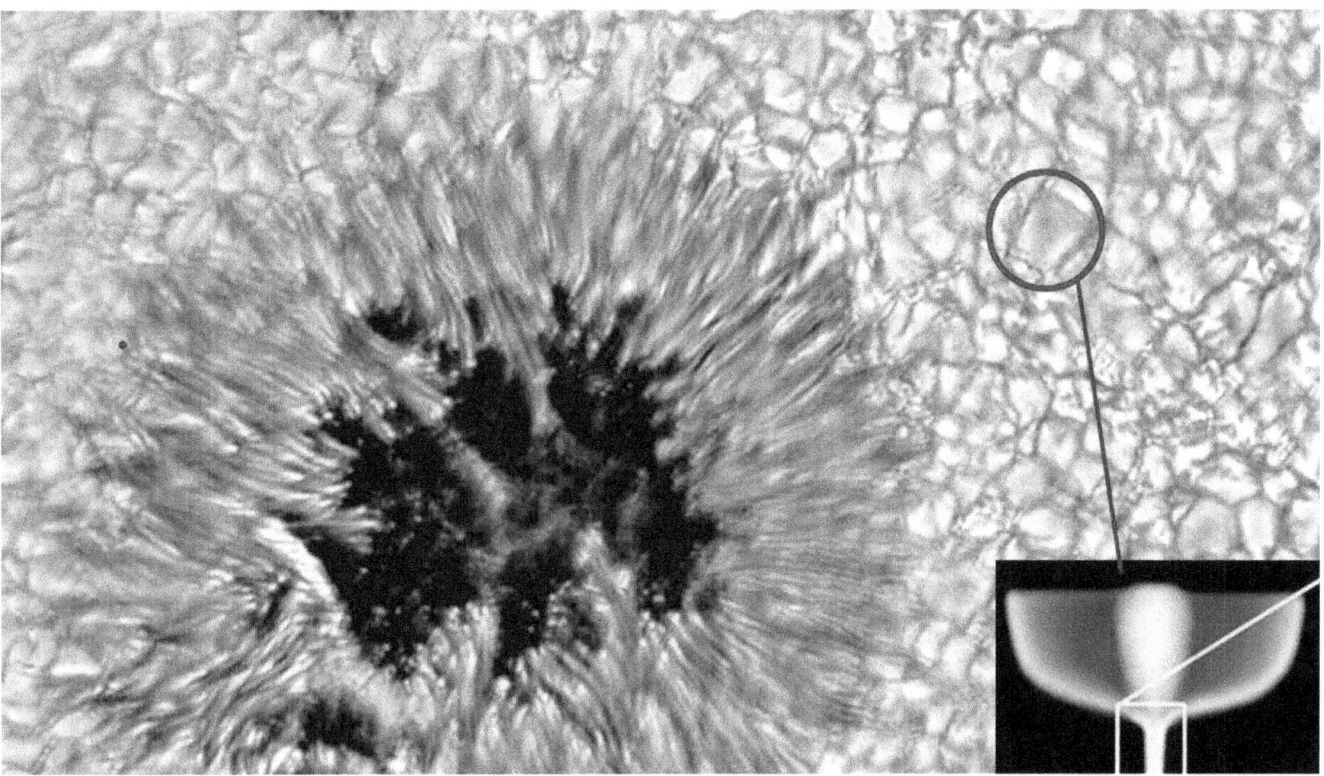

The process that synthesizes atoms from plasma, on the Sun, and provides the sink-effect for the plasma flow, also produces the solar wind. As I said before atoms are the only structures in the universe that are electrically neutral.

The solar synthesizing fusion, a form of cold fusion that binds plasma particles into electrically neutral structures, creates physical elements that no longer react as a part of the electric power system.

While they don't have a direct electric connection

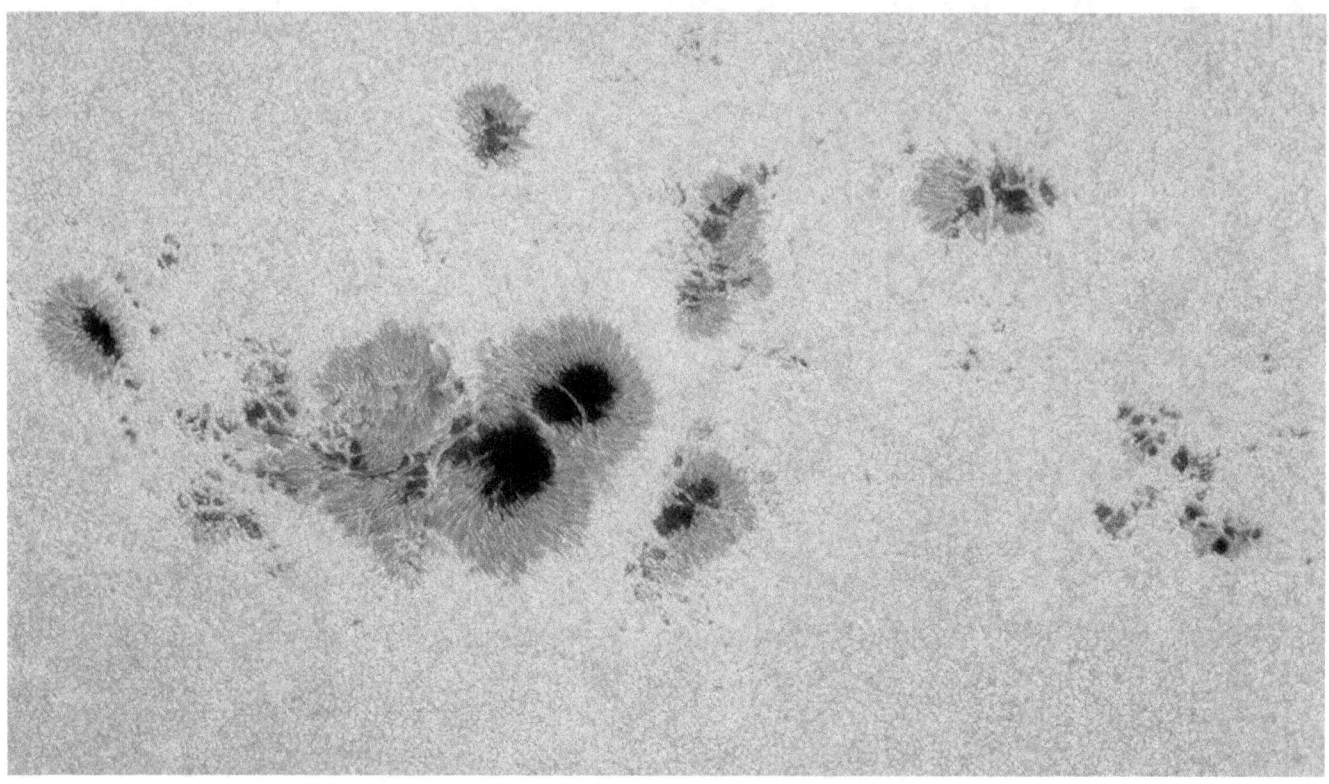

While they don't have a direct electric connection as with the solar wind, they do become caught up in its dynamic flow and move along with it.

The bound plasma particles that produce the atoms

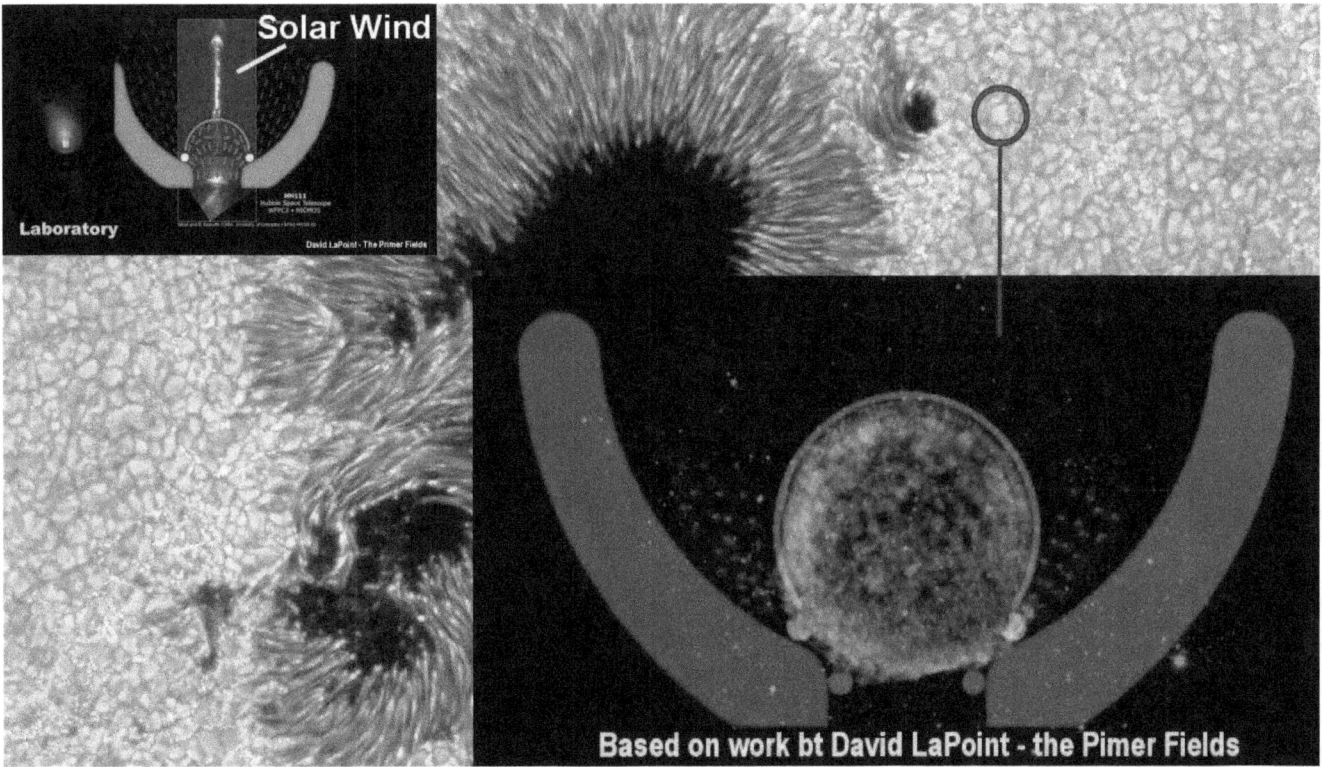

While the bound plasma particles that produce the atoms, that affectively disappear from the electric landscape as if they had vanished from the universe, which in a sense they have, physically remain in existence. So, what happens to these electrically-neutral fusion products, since not all of them flow away with the solar wind? Do they clog up the works? That's a question of principle.

Electrically neutral fusion products get blown along

- an example of the amazing solar eclipse photography of Milloslav Druckmueller
http://www.zam.fme.vutbr.cz/~druck/Eclipse/

The electrically neutral fusion products that do get blown along with the solar winds, fall out from the winds, slowed by gravity. In earlier times they formed the planets in a process of accretion. This process still happens, though to a lesser degree. To the present day, the solar wind provides the distribution service that makes all of this happen.

But what happens when the solar winds no longer blow? What happens when they no longer help to purge the fusion-reaction cells on the Sun, of the fusion product, and purge the corona of the synthesized atoms? Will this clog up the works?

In addition, the solar wind also fulfills a highly critical function as a part of an efficient plasma-pressure regulating system that keeps the Sun operating at a steady state.

The regulating feature

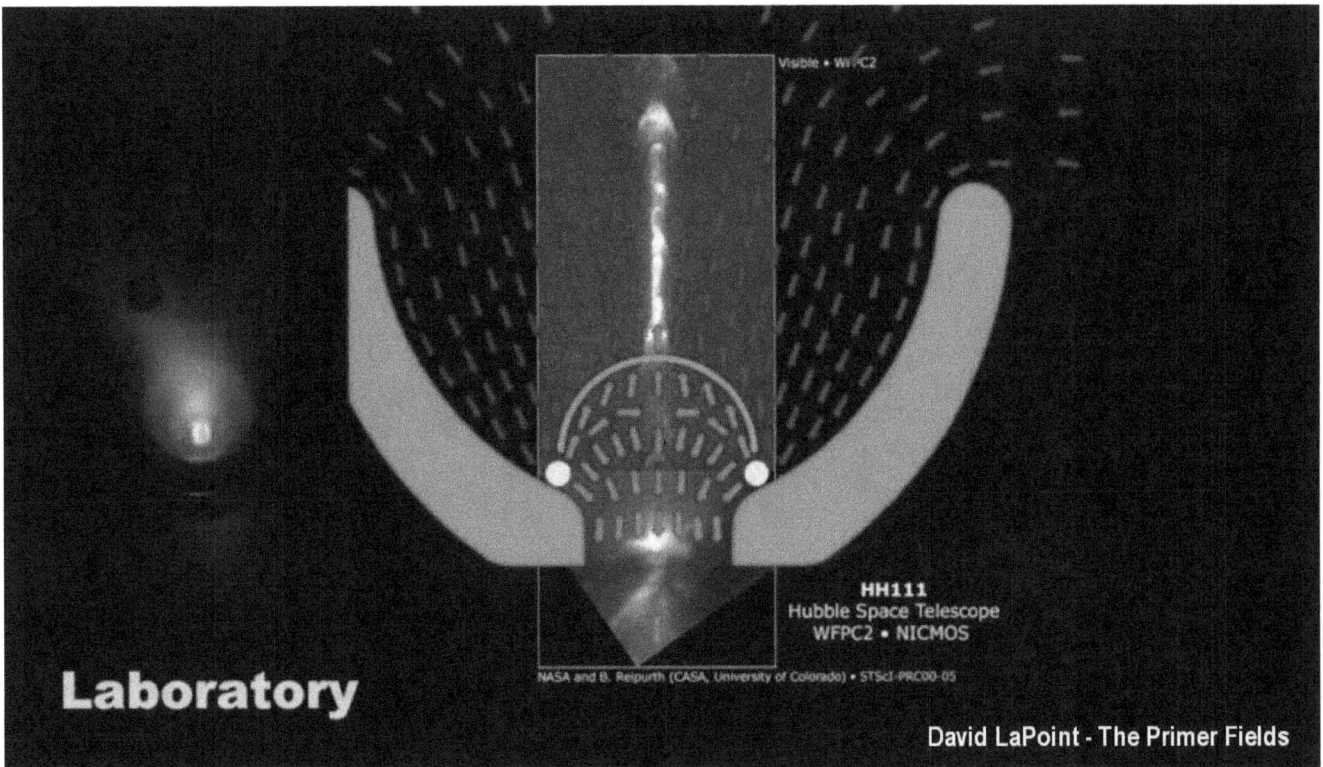

The regulating feature is the cause that produces the solar wind. When the plasma pressure under the confinement domes on the Sun's surface, exceeds the pressure that the magnetic dome can contain, the excess plasma escapes through the confinement dome in a fine jet, or numerous jets. The jets merging together become the solar wind. This means that for as long as the solar winds blow, there is enough plasma pressure under the confinement dome to keep the fusion process going. But when the winds diminish, what happens to the purging of the fusion products? These are all critical question for which no simple answers are presently available, though they all affect the efficiency of the fusion process that drives everything.

When the input streams diminish

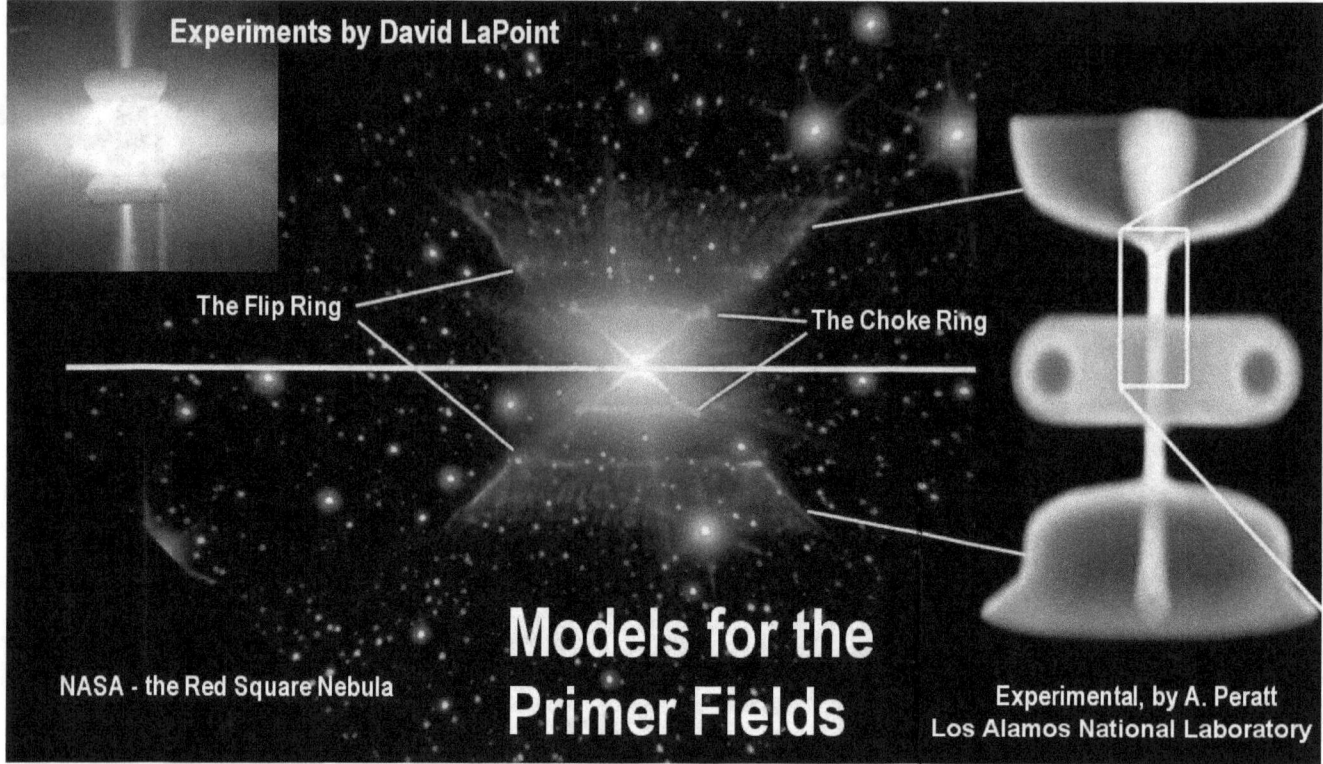

When the input streams diminish to below the minimal plasma pressure for the solar winds to form, then a lot of collapse-effects, evidently, begin to happen. The rate of the fusion reaction may diminish sharply. With it, the 'sink' effect becomes diminished. This reduces the plasma rate of flow, which in turn reduces the fusion reactions, which in turn reduces the plasma rate of flow, further. When the plasma flow diminishes, the primer fields diminish with it, which reduces the plasma flow still further. On this path the entire system can shut itself down quite rapidly, and almost without warning. How fast this may unfold is beyond our means to determine.

When the fusion products clog up the cells

The hardest of all, may be to determine the effect of the diminishing solar winds on the purging of the fusion products, of the synthesized atoms, from the fusion reaction cells. When the fusion products clog up the cells, they diminish the reaction process. It is unknown how critical this purging of the synthesized atoms from the reaction cells is. It may be more critical than we believe it to be, or hope it to be.

The longest duration of continuous fusion

One of the biggest problems that nuclear fusion-reactor experiments have encountered on Earth, is that their fusion product clog up the works as they dilute the fusion fuel that stops the fusion process. The longest duration of continuous fusion that has been achieved so far, is in the range of less than a second at full power, and 5 seconds at a third of the rated power output. That's presently the world record.

The giant ITER fusion reactor that is being built in Europe for a follow-on experiment, as an international project, is expected to a achieve a whopping 1000 second fusion burn by continuously purging the fusion product from the reaction chamber, which may not be possible.

The solar wind appears to fulfill the purifying function

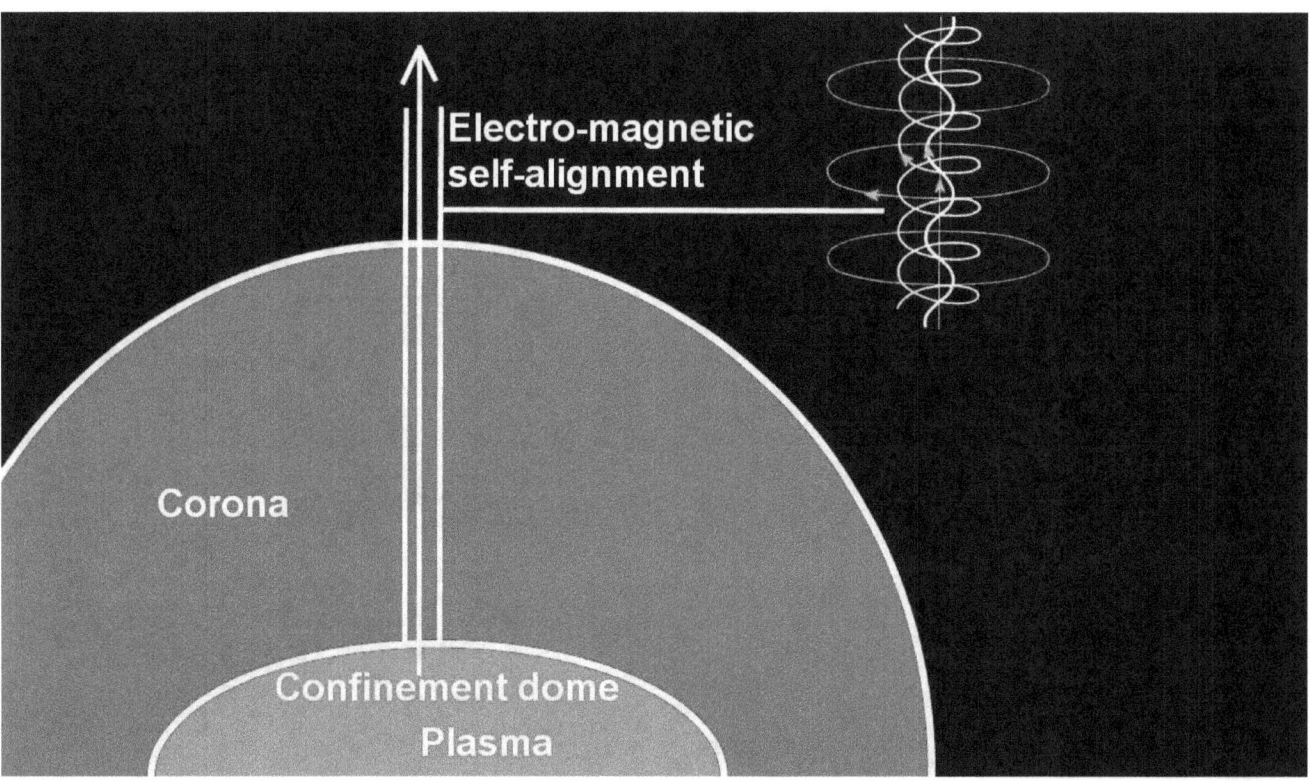

On the Sun, the solar wind appears to fulfill the purifying function. It is unclear to what degree the solar wind may diminish without it having a detrimental impact on the fusion reactions in the cells.

It may well be that the solar fusion system might collapse in a chain-reaction before the solar wind pressure diminishes fully to 0%. That's something to keep in mind.

In a chain-reaction negative feed-back loop

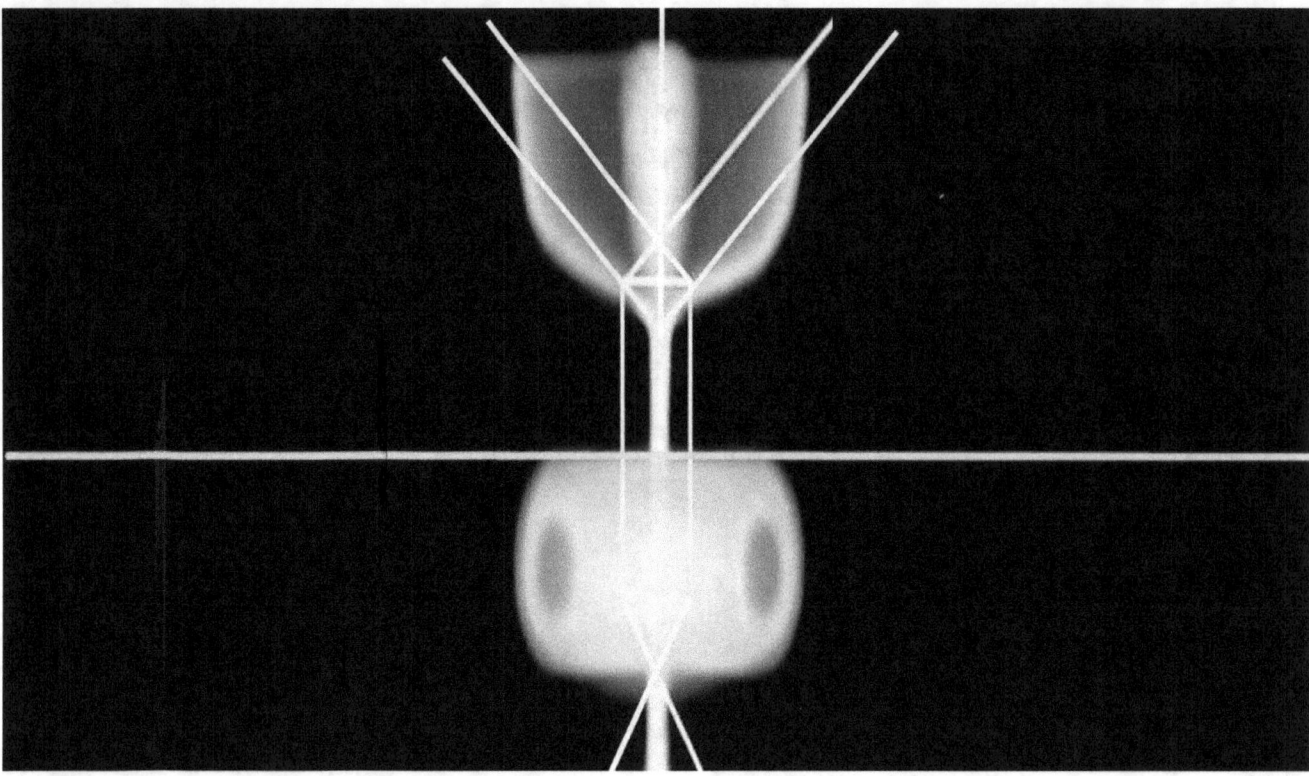

Whenever the atom-synthesizing process slows down, the sink-effect diminishes. In this case the reduced sink effect, reduces the plasma flow rate, which in turn reduces the magnetic fields. The entire system becomes affected thereby in a chain-reaction negative feed-back loop, whereby the entire, deeply interlocked electromagnetic system, may suddenly vanish as if it had never existed.

When the Primer Fields collapse

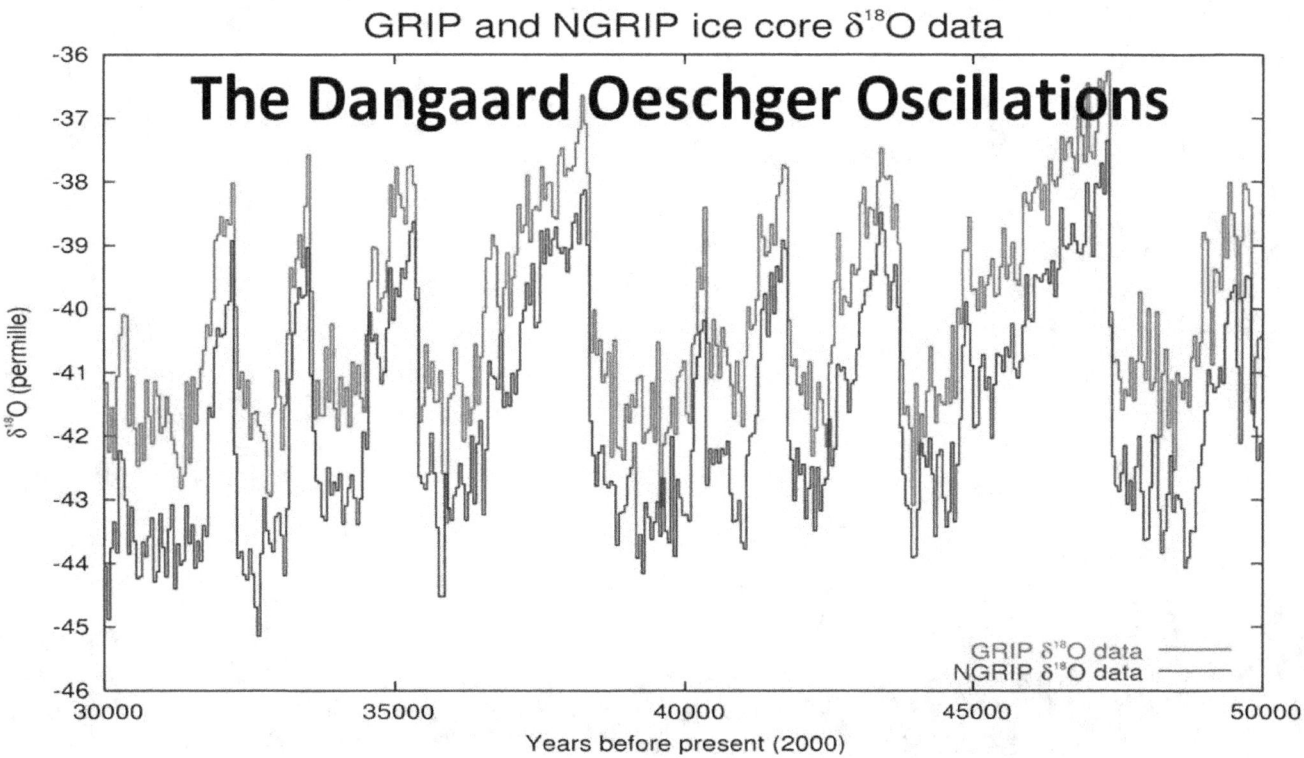

The Sun becomes inactive then, when the Primer Fields collapse. It looses 70% of its energy. The Sun goes dim. We have plenty of evidence of such radical events occurring in the past, preserved in the ice of Greenland from the last Ice Age.

The solar wind also tells us something else

http://www.zam.fme.vutbr.cz/~druck/Eclipse/ - an example of the amazing solar eclipse photography of Milloslav Druckmueller

However, the solar wind also tells us something else. As the carrier of the atoms synthesized on the Sun, the solar wind also tells us, that if we should ever care to utilize the solar-fusion sink process for energy production on Earth, we may be able to tune the fusion-part of the process to synthesize any atomic element we care to create.

The loss of Ulysses

The loss of Ulysses

NASA's Ulysses satellite gave us a big boost in understanding our universe. It gave us more than just a way of looking at the Sun from all latitudes.

Ulysses gave us the most 'pristine' view

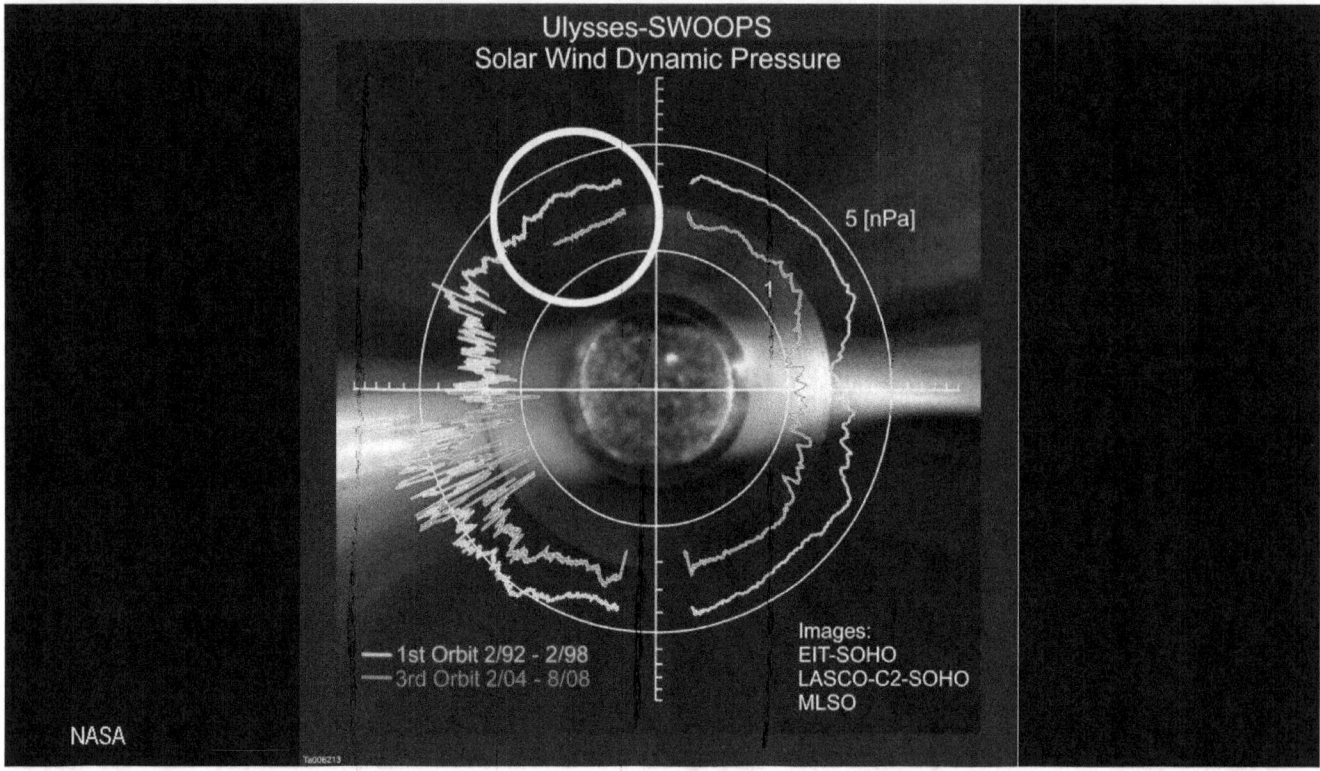

Ulysses gave us the most 'pristine' view of the solar-wind patterns that we can get. It took measurements in areas far away from the ecliptic plain where the heliospheric current sheet throttles the solar wind. It had measured the solar-wind speed outside the ecliptic at just below 800 kilometers per second, which we don't see in near-Earth space, or only rarely so.

This suggests that it would be wonderful if the Ulysses eye around the Sun had not been scrapped in 2009. Nevertheless, the measurements that it has provided, gave us enough valuable benefits for evaluating the solar wind measurements that we now get from satellites orbiting in near-Earth space, in ecliptic space, where the measurements are affected by the heliospheric current sheet.

Ulysses saw the solar wind pressure diminished

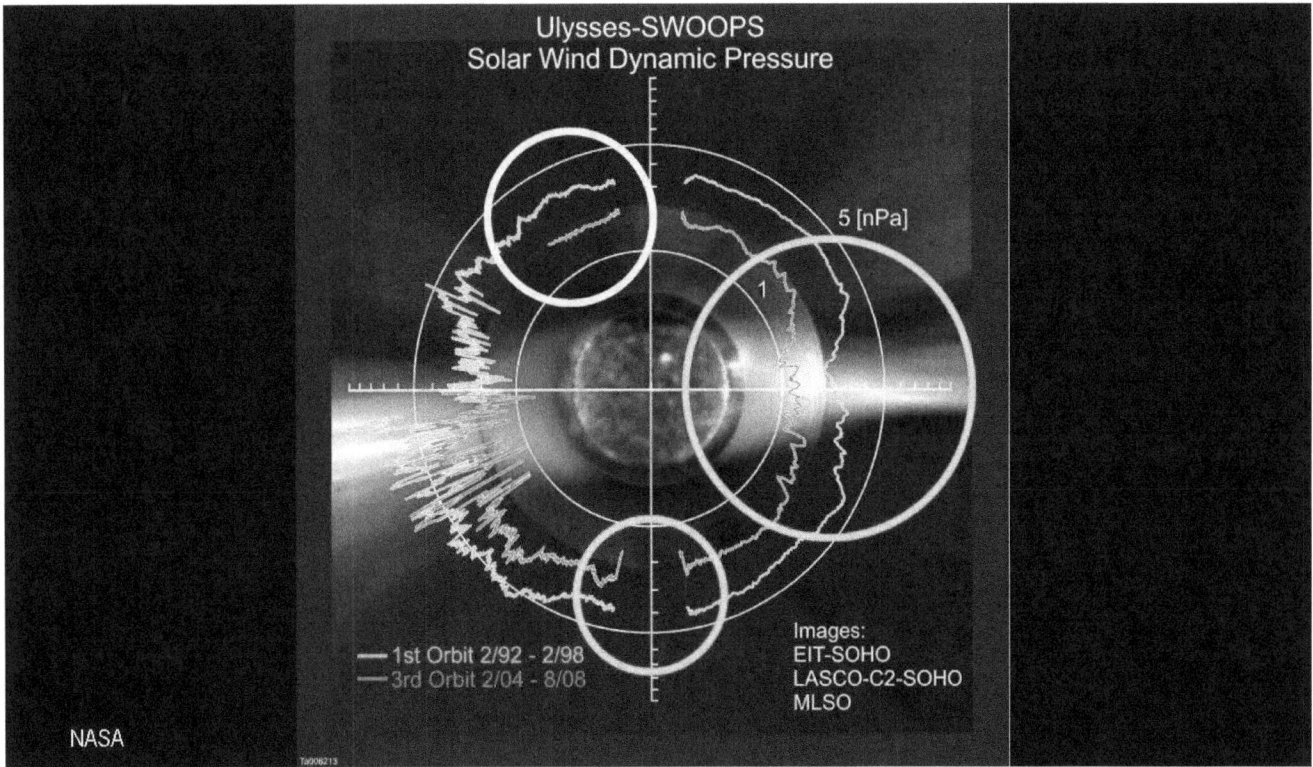

Ulysses saw the solar wind pressure diminished at the ecliptic, from four nano-pascals in the white circle, to two-and-a-half nano-pascals, in the green circle, for the 1st orbit, and from two-and-a-half nano-pascals to one, for the third orbit.

While the numbers are lower near the ecliptic in the green circle, the ratio between them remains essentially the same. This means that the solar-wind measurements that we are able to make in near-Earth space provide us with enough data, to render the solar wind data a useful thermometer for judging the health of the solar system. This means that the critical judgments that we must make in our time, to determine the start of the next Ice Age, are not severely impacted by the termination of the Ulysses mission.

For example, if the mission would continue, we would not see any changes in the polar region where the Sun has its connection with the plasma streams from the Primer Fields, which the solar wind is too puny to penetrate.

The diminishing trend towards the solar cut-off

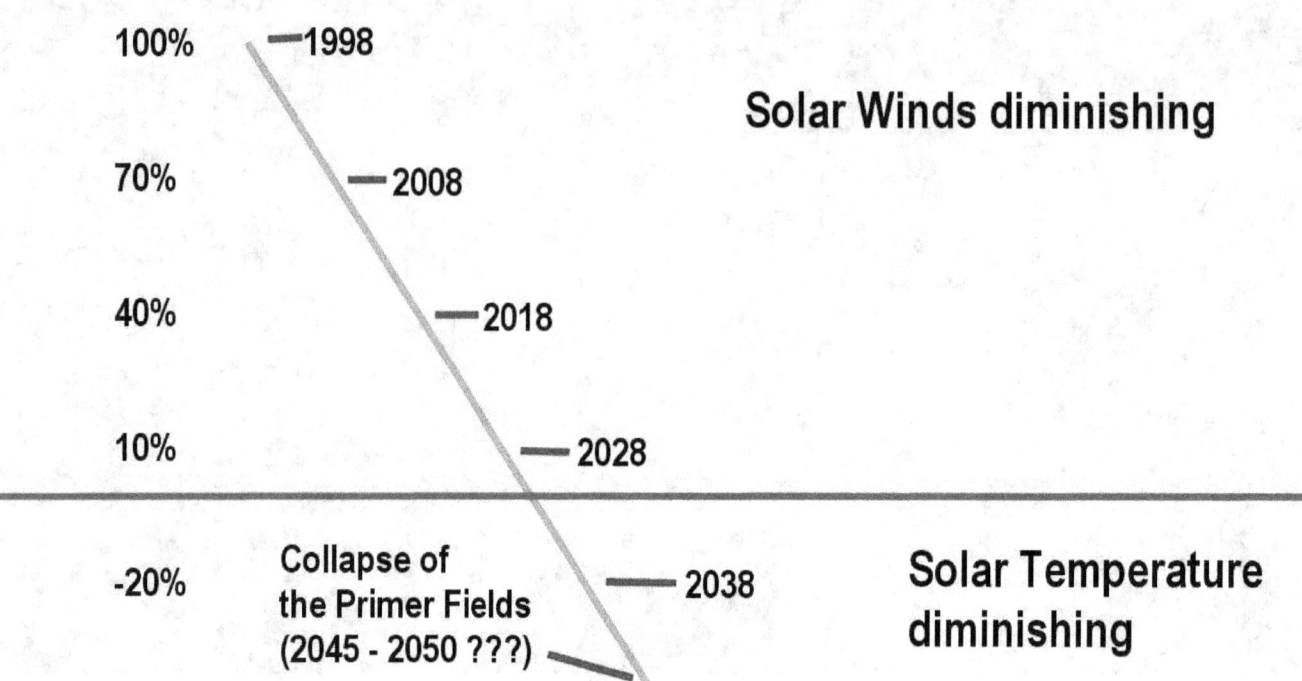

This also means that the diminishing trend of the solar winds towards the solar cut-off, that Ulysses saw the beginning of, can still be detected with contemporary instrumentation operating inside the ecliptic plane. The current observations have not significantly altered the potential that the solar cut-off might happen in the 2050 timeframe, and that the solar winds might cease before this time.

Ironically, the potential event itself, is actually not the most critical factor for our consideration. The principles are important. The principles make the potential events significant, because the recognition of the principles inspire the imperatives for action. The discovery of principles is rooted in the higher nature of man that is able to guide us today in uplifting our world in preparation for the needs of the future regardless of the unpredictable timing of events.

There are many principles in operation right now that are largely ignored. They are seen as insignificant events, but when they are seen in the context of the larger package, they illustrate the dynamics that precipitate events.

The dynamics of the sunspots can be seen in this manner, as aspects of sensitive electrodynamic phenomena.

When the sunspots speak to us

When the sunspots speak to us

We can observe the sunspot-principle in action

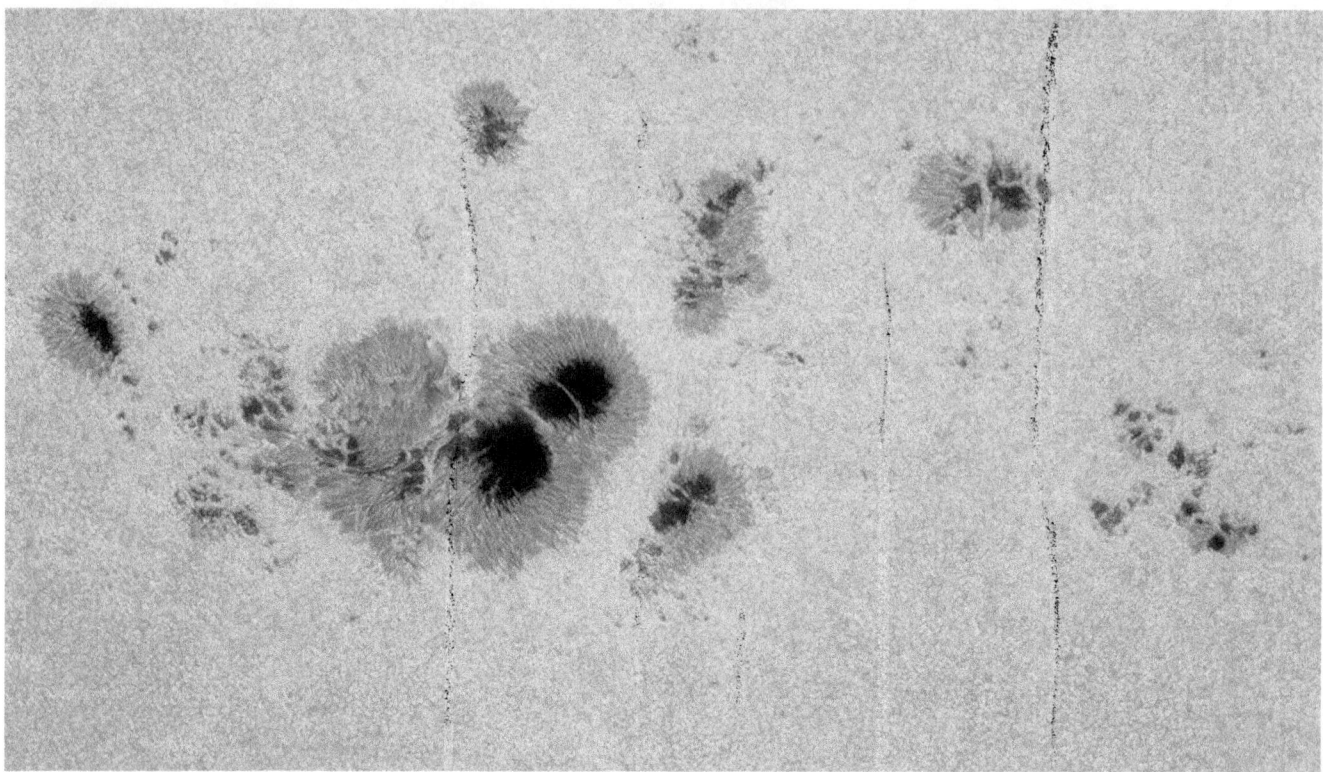

We can observe the sunspot-principle in action every day, the principle that produces sunspots.

Sunspots are voids on the solar surface. They result from localized disruptions of the dynamics of the Primer Fields that enable the reactive processes on the solar surface. The small cellular field structures that cover the entire surface of the Sun, sometimes breaks down, occasionally one cell at a time, but usually in groups, which leaves behind dark holes on the surface. The resulting dark areas are termed, sunspots. The sunspots frequently occur bunched together into groups across a region.

The reason for this localized collapse of reaction cells across a region, appears to be the building-up of 'backpressure' in that region below the cells. The backpressure may be caused by a regional high concentrations of synthesized atoms that tend to insolate the electric field connection below the reaction cells.

Thus, ironically, the occurrence of sunspots is an indication of high levels of fusion activity going on, which in the extreme is a danger in the process to itself.

Backpressure limits the plasma-current flow rate

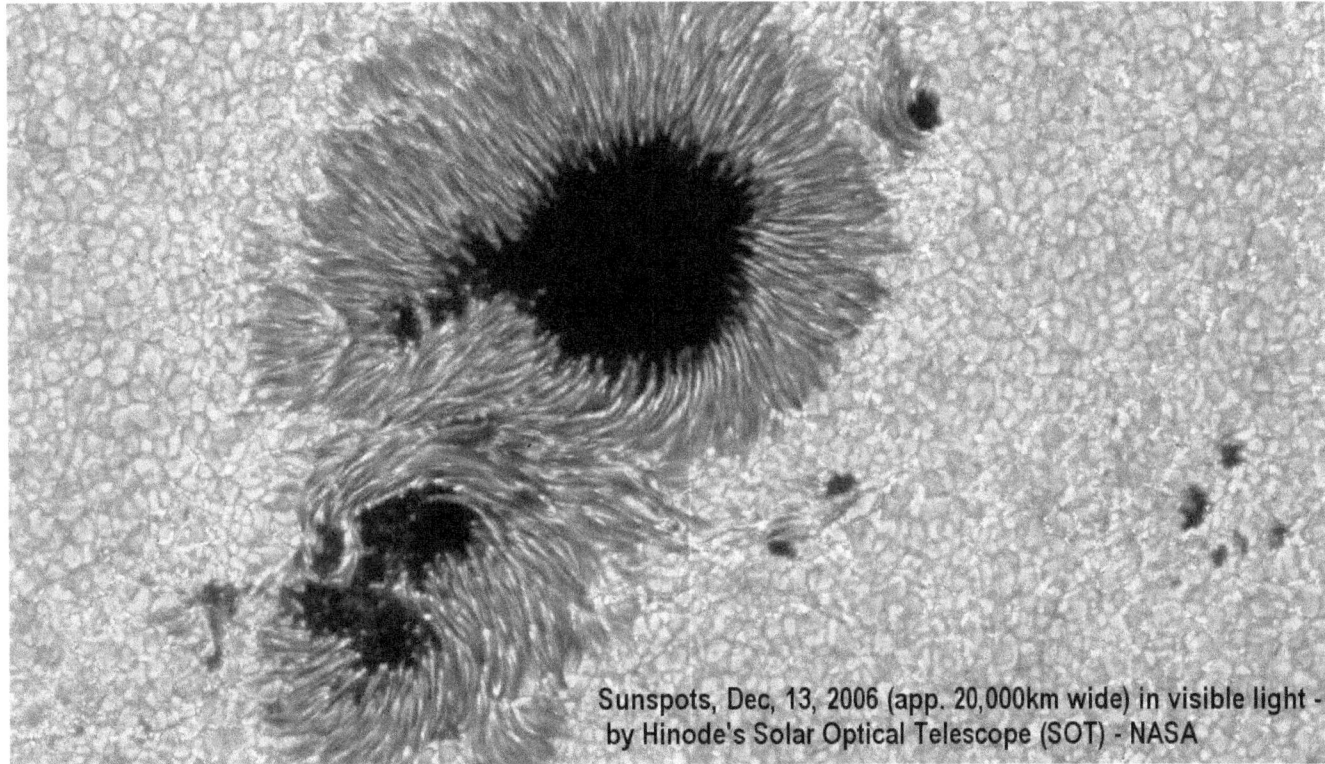

Sunspots, Dec, 13, 2006 (app. 20,000km wide) in visible light - by Hinode's Solar Optical Telescope (SOT) - NASA

That the backpressure limits the plasma-current flow rate, and thereby reduces the strength of the magnetic fields that concentrate the plasma and cause the fusion-reaction cells to function, is evident in the occurrence of regional sunspot groups, and the occurrence of single-cell failures in these regions or near large sunspots.

When active magnetic primer fields for the fusion cells diminish

Normally, when active magnetic primer fields for the fusion cells diminish, the plasma backpressure blows out around them, or it escapes in mass through the resulting hole in the magnetic 'carpet' after active cells have collapsed.

Escaping plasma streams are seen as plasma loops

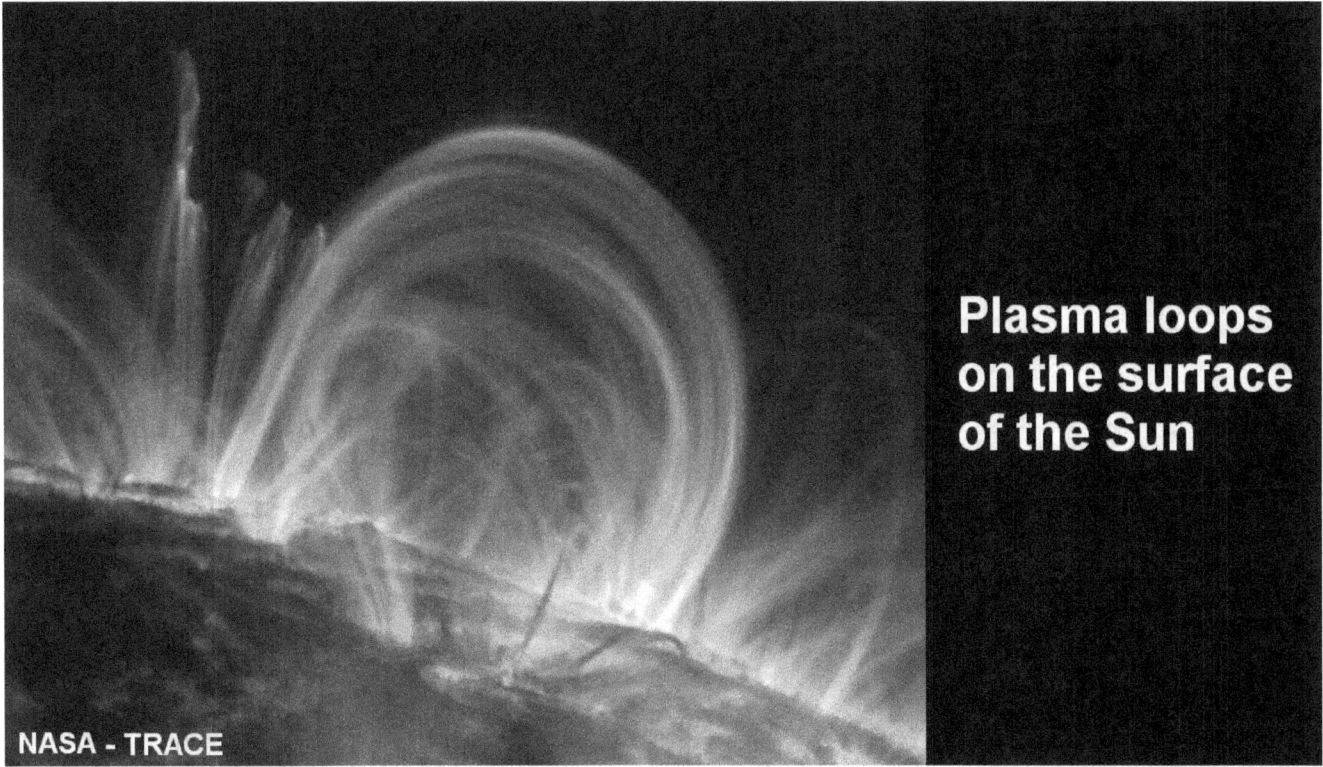

The escaping plasma streams are seen as plasma loops, and in larger cases, as giant prominences. They become 'visible' by the presence of atoms in their stream, that have caused the backpressure in the first place. Pure plasma is invisible. Plasma becomes visible as light, only by its effects on atomic elements. The atoms are 100,000 times larger than the plasma particles. They pervade the atmosphere and are moved along with the plasma streams, and become agitated by them.

When the escaping backpressure does not collapse the cells

In cases when the escaping backpressure does not collapse the cells completely, but escapes around them, the escaping plasma forms thin streams that loop back onto the Sun, guided by the magnetic fields that their flow is generating, which are then interacting with the much weaker background magnetic fields in a particular region of the Sun's surface. Active cells collapse rarely. Sunspots result only under extreme conditions.

Giant eruptions are extremely rare

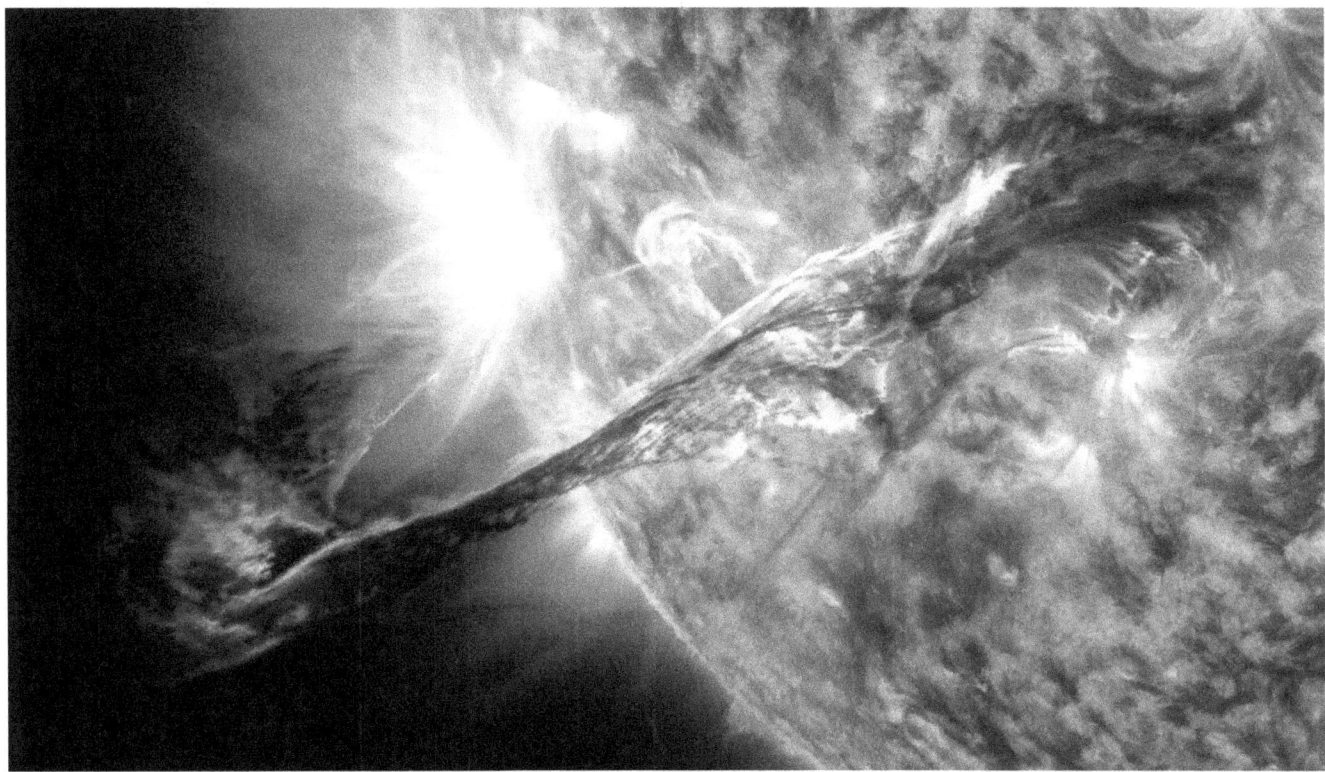

Giant eruptions, such as the prominence shown here, are extremely rare. They typically occur when the plasma pressure inside the Sun, as plasma is being pumped into it by the primer fields, exceeds the external plasma pressure produced by the process. When the build-up internal pressure is not vented continuously, or fast enough, but accumulated, the huge solar mass-ejection events tend to happen.

The occurrences of prominences, or solar flairs

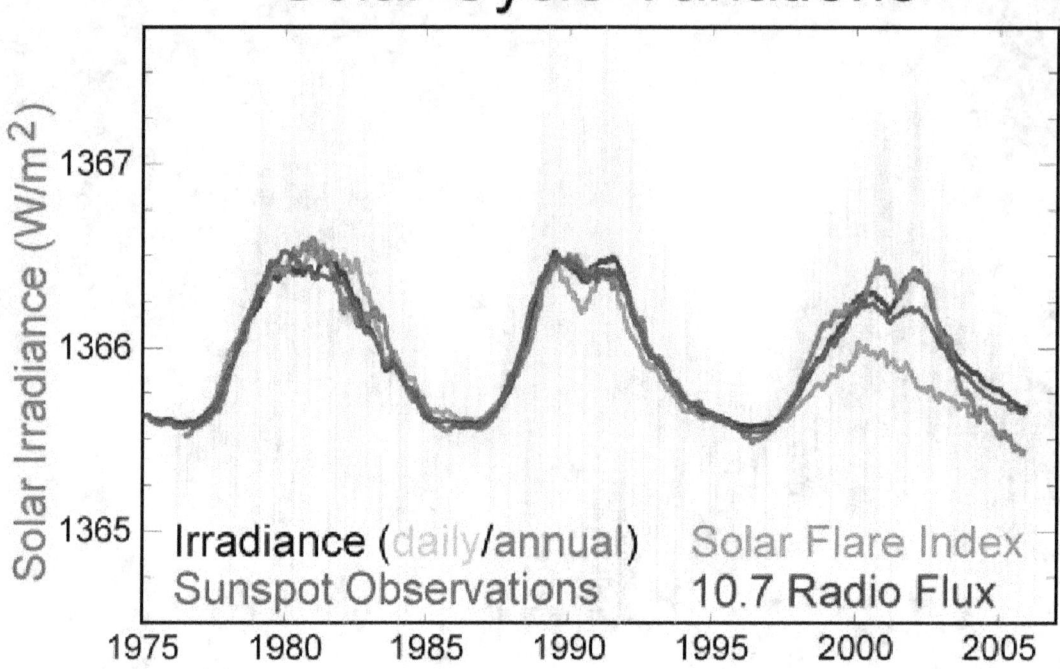

The occurrences of prominences, or solar flairs, typically follow the trend of the solar cycles. The reduced solar flair index, shown in green here, which was the most dramatically effected of all the indexes when the electric environment began the weaken, may well be the best 'thermometer' we have for measuring the electric health of the solar system, together with the measurements of the solar-wind pressure.

Plasma is a near-perfect electric conductor

Plasma is a near-perfect electric conductor. As a plasma sphere, the Sun soaks up plasma somewhat like a sponge that stores up whatever plasma streams are pumped into it, through the reaction cells, past the nuclear-fusion processes. The injection of plasma is a dynamic and widely distributed process. However, this injection of plasma increases the plasma pressure inside the Sun. The pressure, needs to be vented via small or large, solar mass-ejection events, or solar flairs as they are also called.

The details of the dynamics are still poorly understood, even though these ejection events are often feared on Earth as they disrupt radio communications, electricity transmission systems, and cause earthquakes.

When the solar flairs dramatically diminish

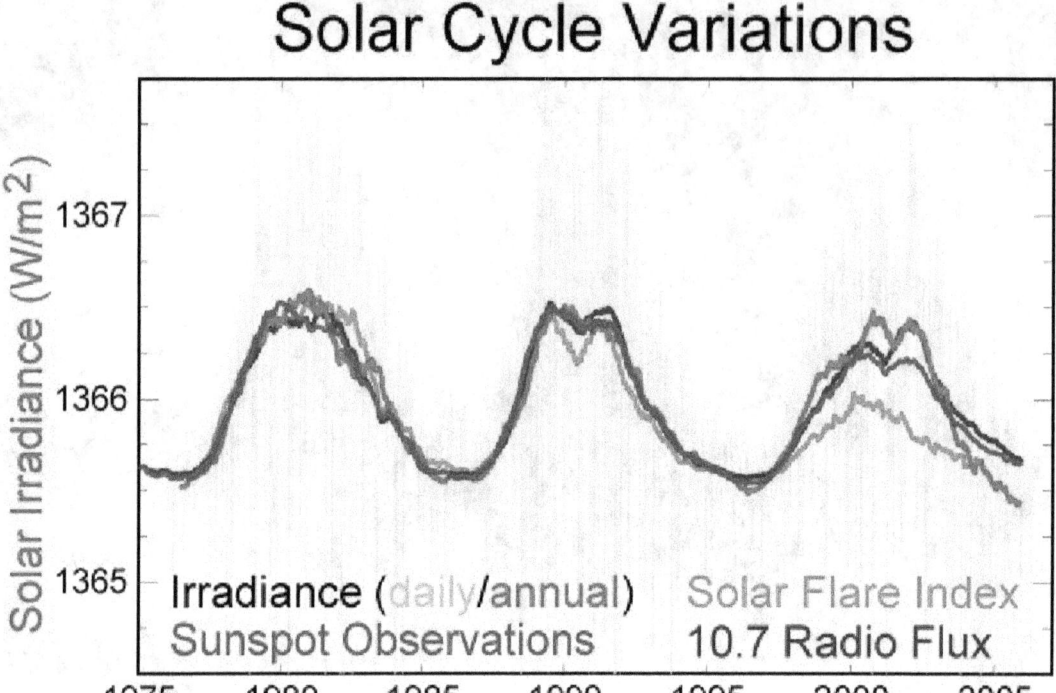

In real terms, however, solar-flairs, like sunspots, only more so, are a sign of good health for the Sun and for its operating dynamics. When the solar flairs dramatically diminish or no longer occur, then we have cause to be concerned. And they are diminishing as we see it in solar cycle 23, in the flair index, for the first time since the index was developed.

Part 4: The solar wind speaking of the coming Ice Age

The sunspots numbers are less dramatically affected

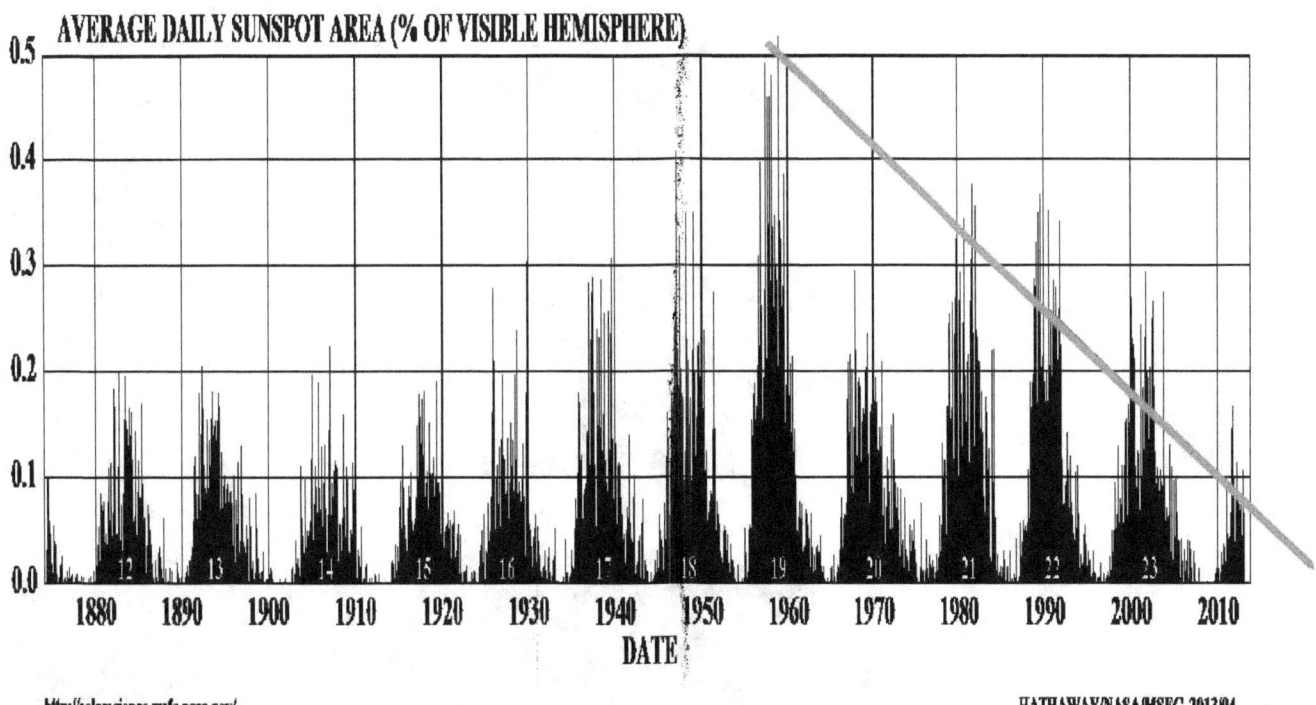

The sunspots numbers are somewhat less dramatically affected by the diminishing energy environment in the solar system. Still they are useful as a strong indicator of the unfolding trends.

From a science standpoint, the sunspots are valuable

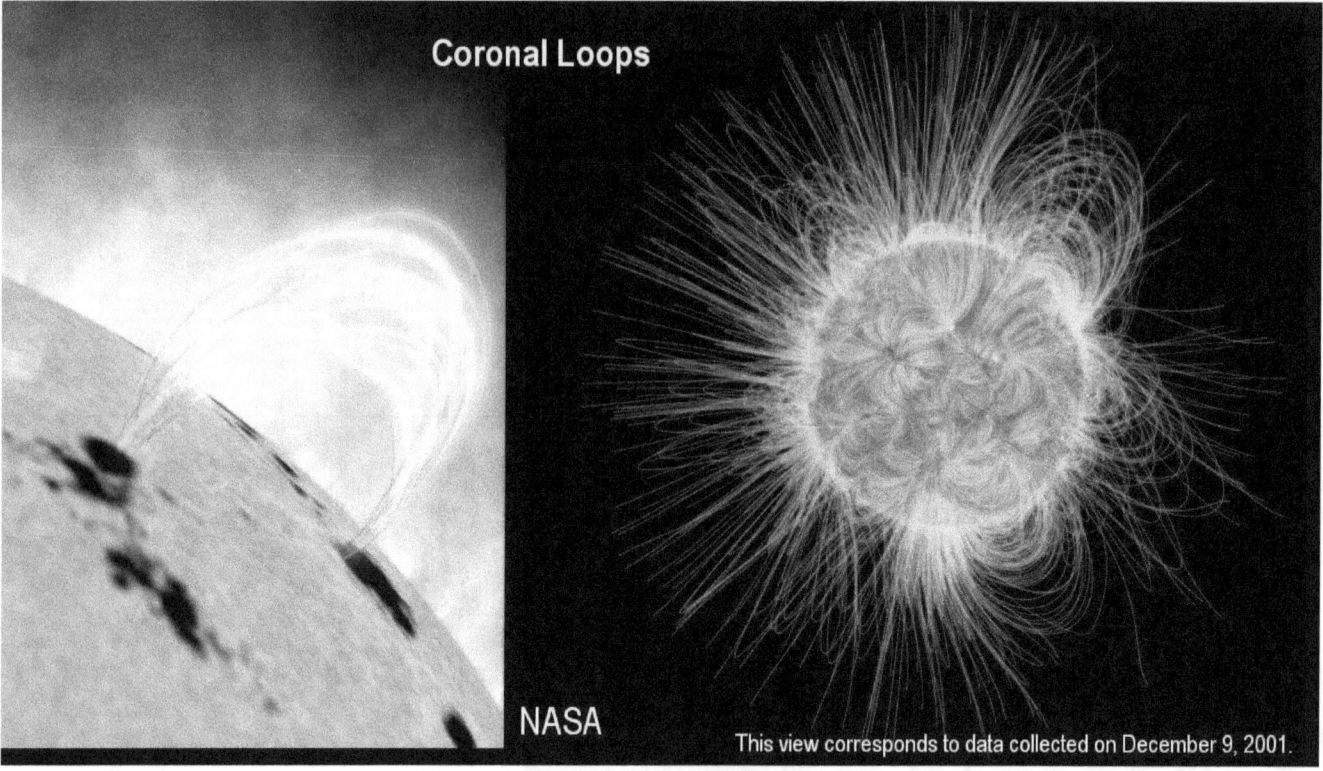

Nevertheless, from a science standpoint, the sunspots are valuable. They provide a portal through which we can see deep into the interior of the Sun, which, as I said in the beginning, is substantially darker below the active surface layer.

Sunspots also provide us a portal

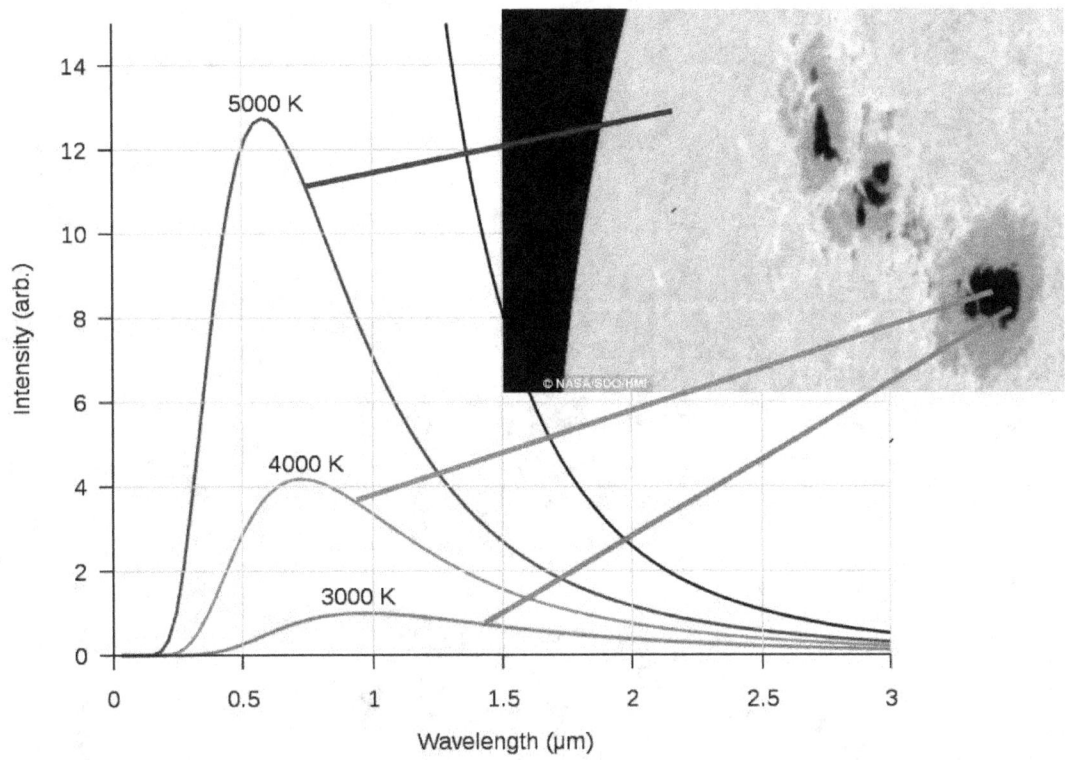

This means that the reaction process on the surface of the Sun is only skin deep. The dark interior reveals itself as essentially a sphere of plasma that furnishes merely the supportive electric environment for the surface reactions to happen.

The sunspots also provide us a portal to what the intensity of the inactive Sun may be when the reaction cells collapse, as the external primer fields collapse that focus plasma onto the Sun.

The recognition of the Sun as a dynamic plasma sphere, is critical for our determining its radiated energy in its inactive state, which we have to prepare the world for.

The time has come to get real

The time has come to get real

Science has become a trap

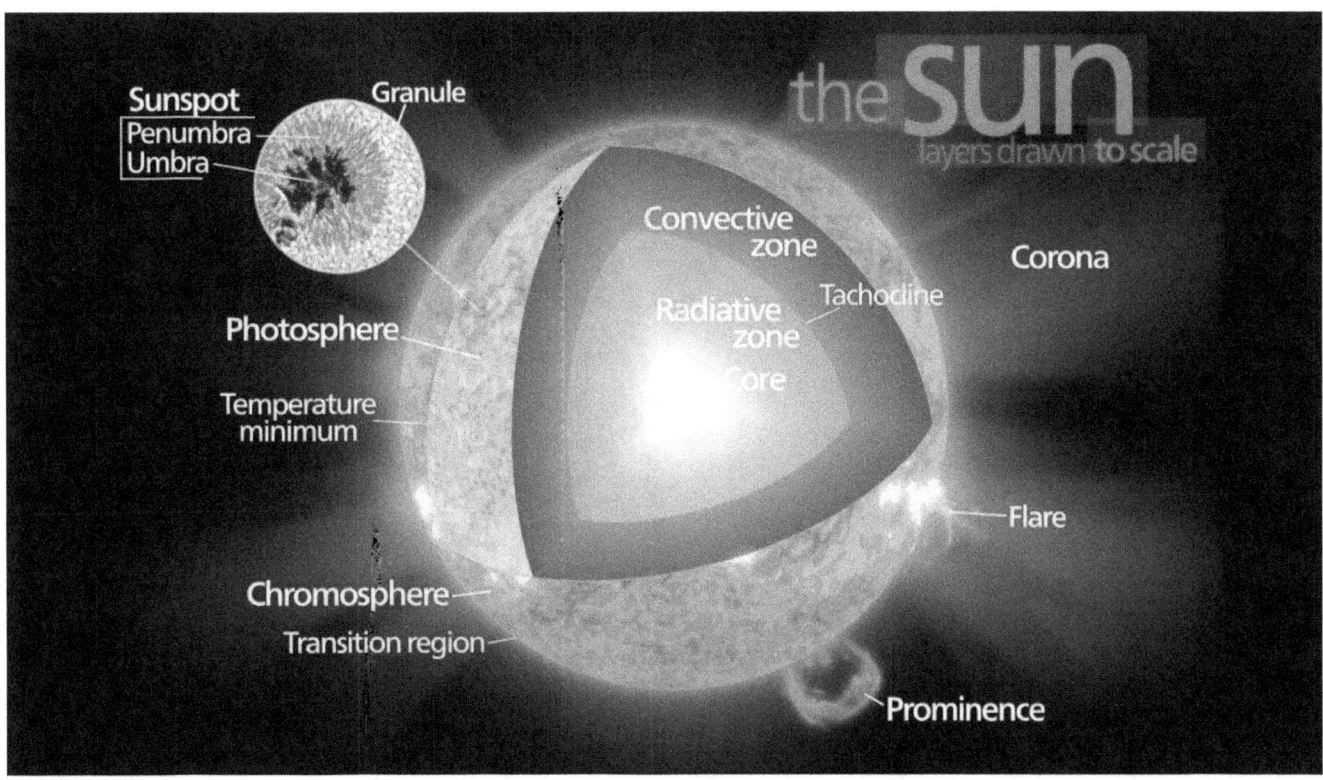

Science has become a trap. It resists reality in the name of doctrines.

When doctrines require that one set reality aside

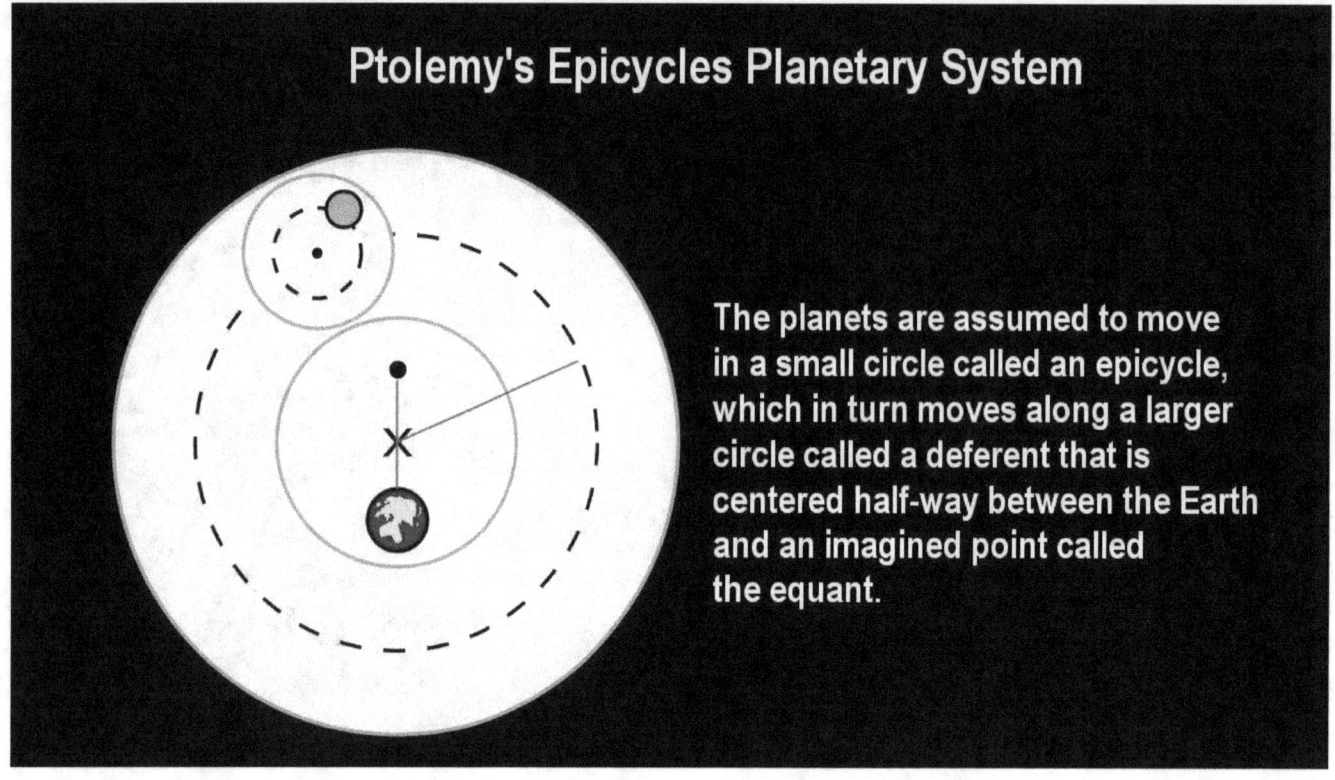

The moment when doctrines require that one set reality aside, exotic 'epicycles' become necessary to rescue the doctrines, like Ptolemy had done when he invented epicycles, with which he proved scientifically what doesn't actually exist.

The Big Bang theory was developed as a counter-theory

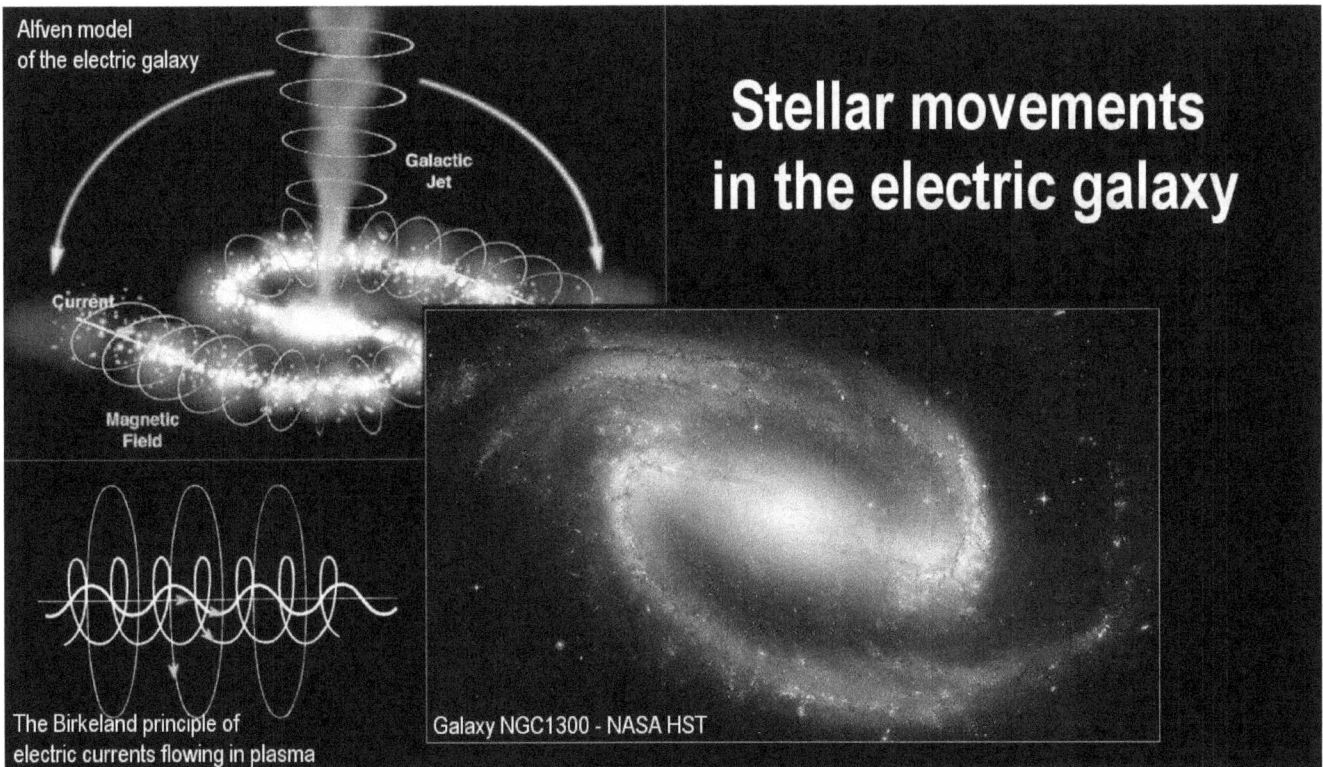

It may have been for this reason, that as soon as the electric universe theory was put onto the table of humanity, that the Big Bang theory was developed as a counter-theory. The Swedish electrical engineer, plasma physicist, and winner of the 1970 Nobel Prize in Physics, Hannes Alfven, had landed a bombshell that didn't fit the political doctrine.

Similar to more modern Global Warming theory

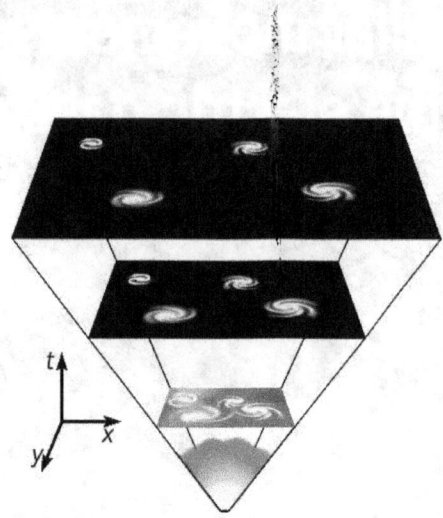

The Big Bang creation myth refuted by the electric solar fusion model

The Big Bang theory was set up as a cover-up doctrine, similar to more modern Global Warming theory.

Under the weight of countervailing doctrines

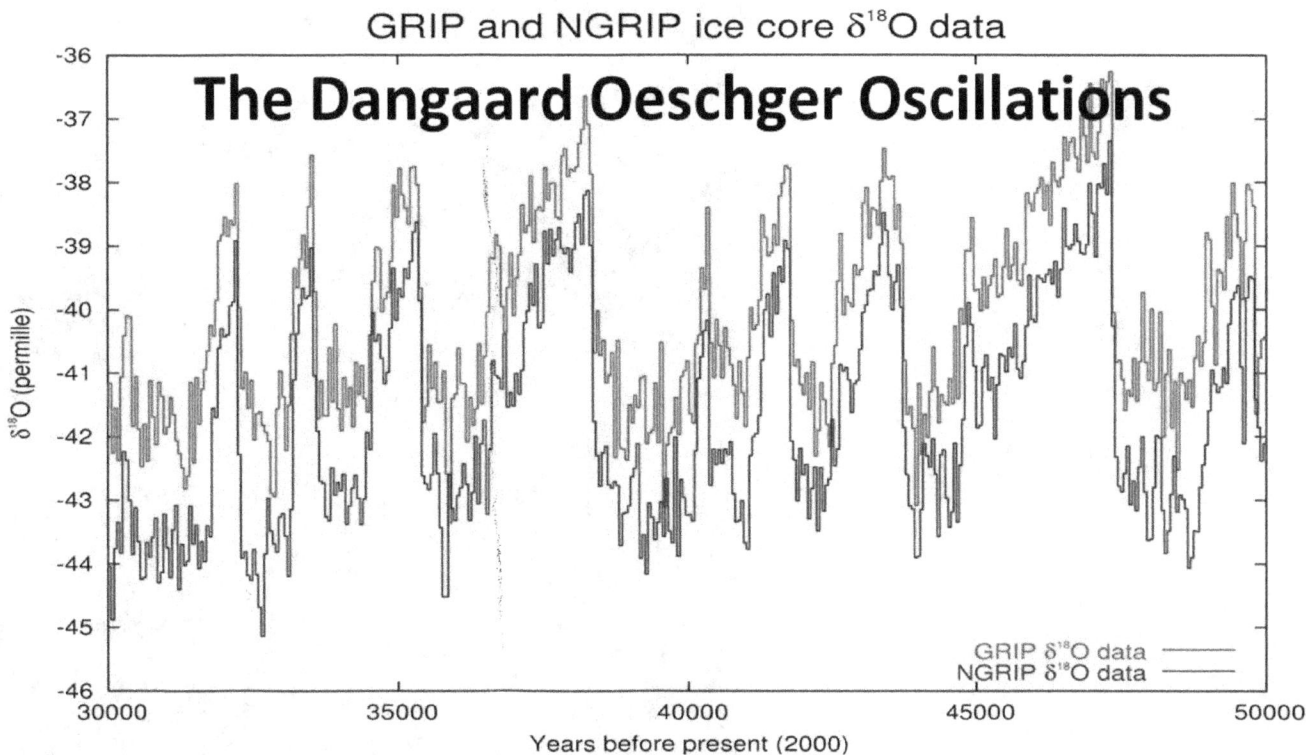

Under the weight of these countervailing doctrines, it became nearly impossible to formulate a rational Ice Age theory. The electro-astrophysical basis for it, became denied to exist so that the entire subject remains left hanging to the present day, as if it was irrelevant.

Science still plays with epicycles in solar physics

Margarita Philosophica by Gregor Reisch, published in 1508

Science still plays with epicycles in solar physics, to defend the doctrine that plasma in space does not exist; that plasma streams do not exist; that electric plasma interactions do not happen - asserting that an electrically powered sun is therefore not possible. Thus science is being put to the task in solar physics to relate all observed phenomena to a platform of the fantasies where reality is disallowed. That's the same type of prison that Ptolemy was stuck in where he was 'forced' to explain scientifically what does not exist, for which his exotic epicycle theories was invented to maintain the imprisonment.

How do we get out of the paradoxical trap?

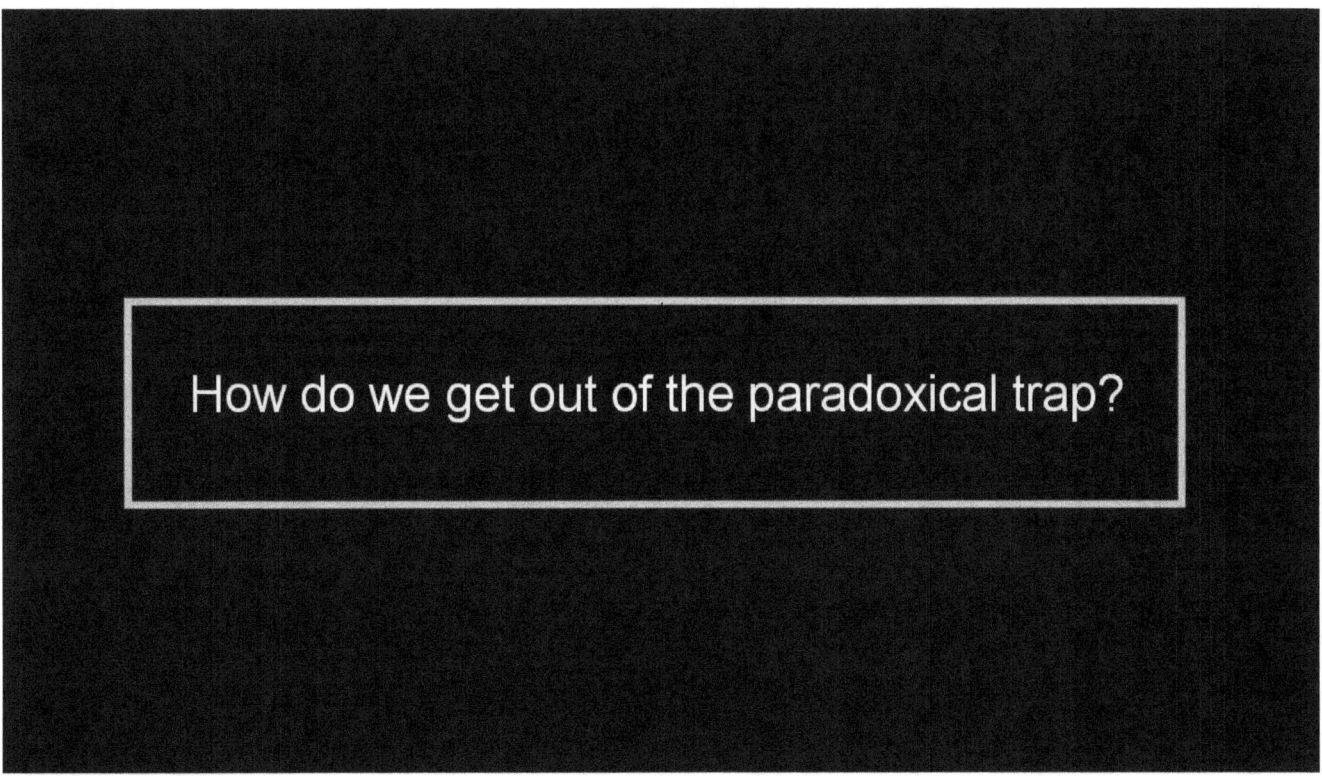

How do we get out of the paradoxical trap?

We get out of the trap like Johannes Kepler did

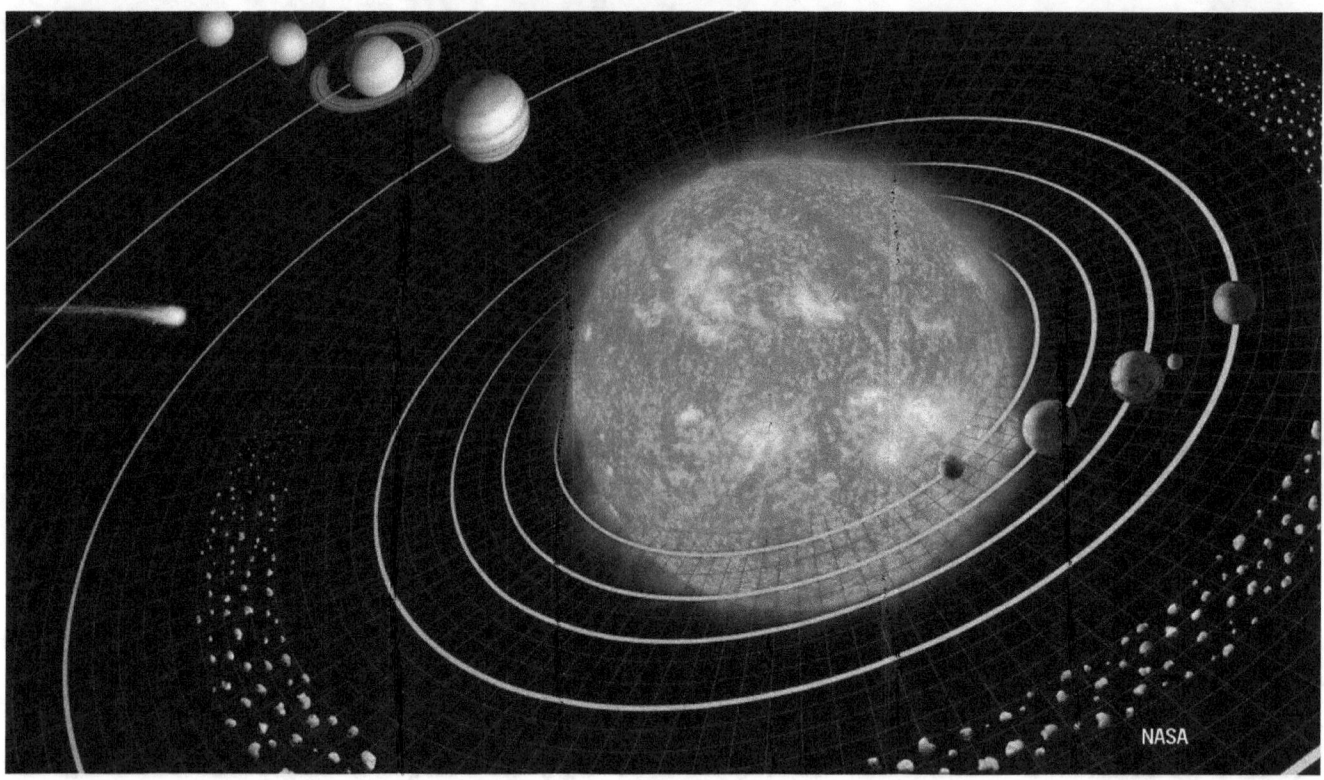

We get out of the trap like Johannes Kepler did in astronomy, by stepping away from the doctrines, to exploring what is actually real.

Some people did this in modern time half a century ago.

The mission named Ulysses

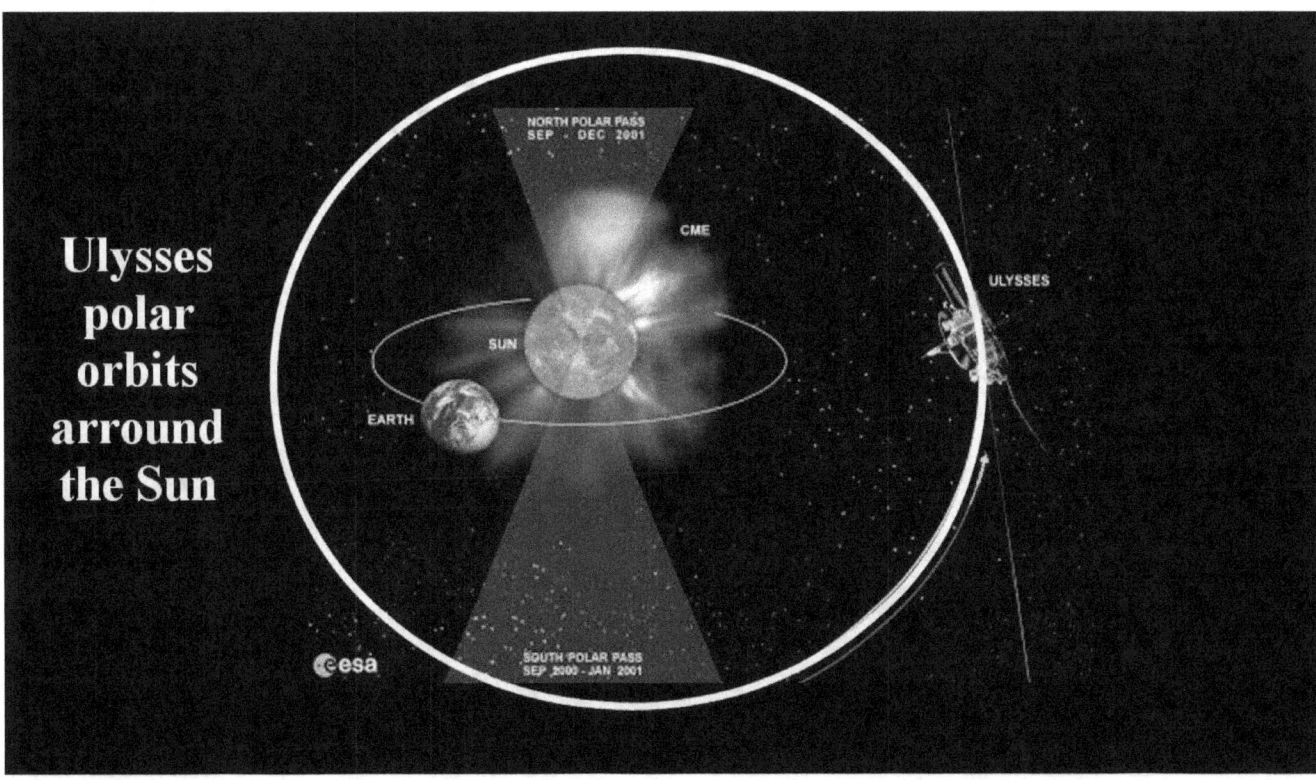

They designed a mission in the early days of the American space program, that would launch a spacecraft into a wide polar orbit around the Sun, to measure the Sun's electric characteristics, such as its solar wind speed, pressure, temperature, density, and so on, outside the ecliptic plane where electric measurements are distorted by the heliospheric current sheet that flows there. The mission was named Ulysses.

The Ulysses spacecraft was launched from the shuttle Discovery on October 6, 1990. In order to reach high solar latitudes, the spacecraft was aimed close to Jupiter so that Jupiter's large gravitational field would accelerate the spacecraft out of the ecliptic plane towards the high solar latitudes. Ulysses' encounter with Jupiter occurred 14 months later, on February 8, 1992.

The Ulysses satellite flew three orbits around the Sun

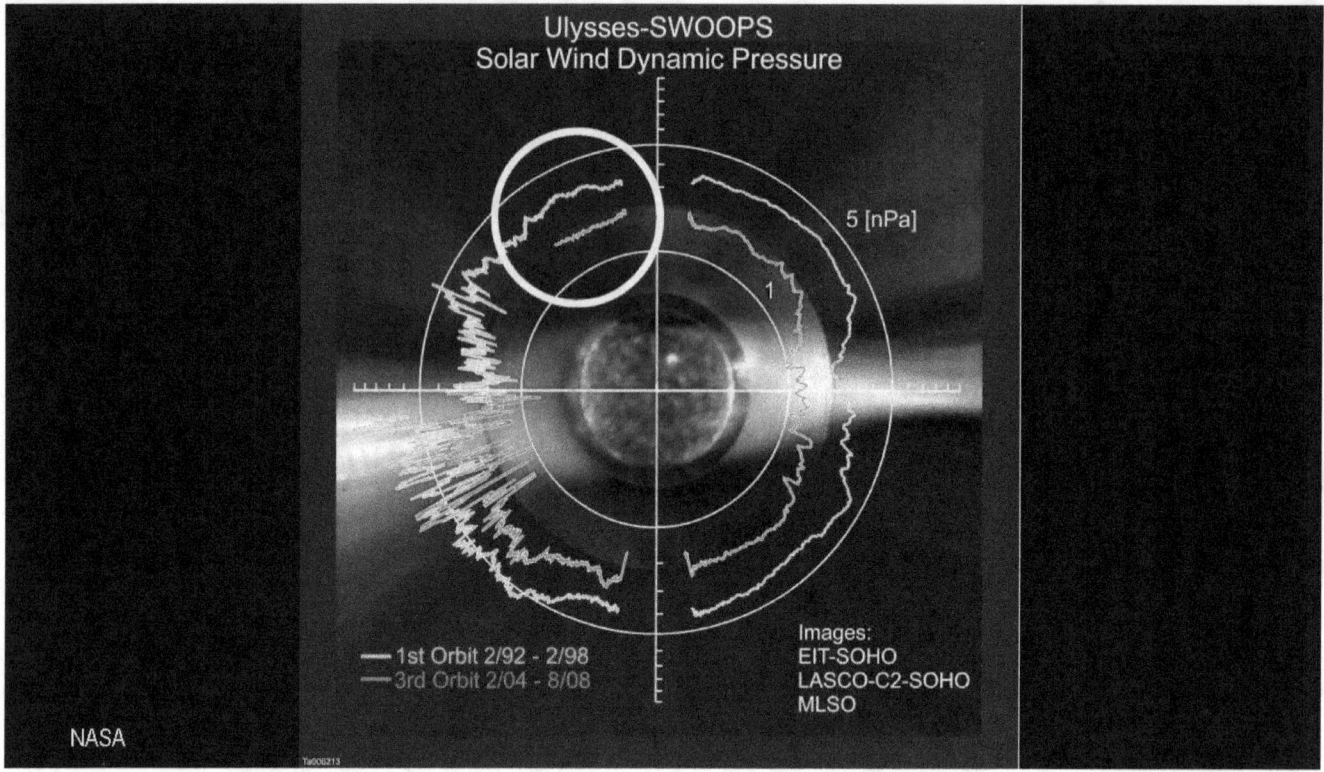

The Ulysses satellite flew three orbits around the Sun perpendicular to the ecliptic. The first orbit began when the spacecraft encountered Jupiter. The resulting first orbit around the Sun was completed 6 years later, in February 1998. The third orbit began 6 years after that in February 2004. The spacecraft was turned off in August 2008, four and a half years into the third orbit.

What we got back as data from the satellite gives us the first physically measured verification that our sun is NOT an internally heated fusion-powered star, but is an electrically powered star.

What the Ulysses spacecraft has measured

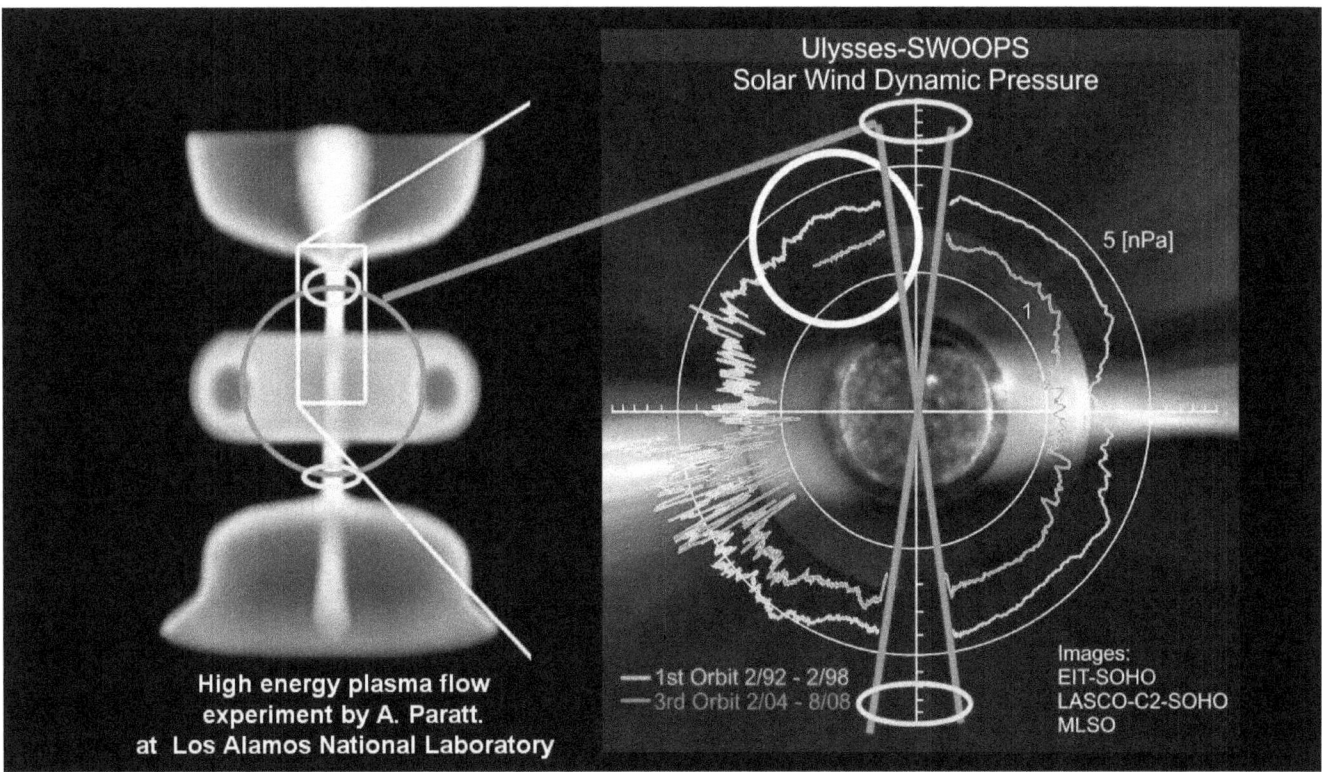

What the Ulysses spacecraft has measured, confirms physically in space the observed geometric characteristics of magnetically self-shaped plasma into primer fields, as observed in high-energy plasma-flow experiments.

If the Sun existed as an internally heated star, the abrupt loss of solar wind in the polar regions should not occur. But with the Sun being powered by focused plasma streams, with a geometry observed experimentally at the Los Alamos National Laboratory, one finds the lack of the solar wind in the Sun's polar region not surprising, but one finds it instead a confirmation of principles that have been experimentally discovered.

The resulting verification by the Ulysses spacecraft, of a fundamental principle in plasma physics, physically detected to be operating on the large scale in cosmic space, places the Sun into an entirely different category than that of an internally heated self-powered star. It places the Sun into the science category of high-energy plasma physics, as an electrically powered star.

When sunlight from our Sun is expanded

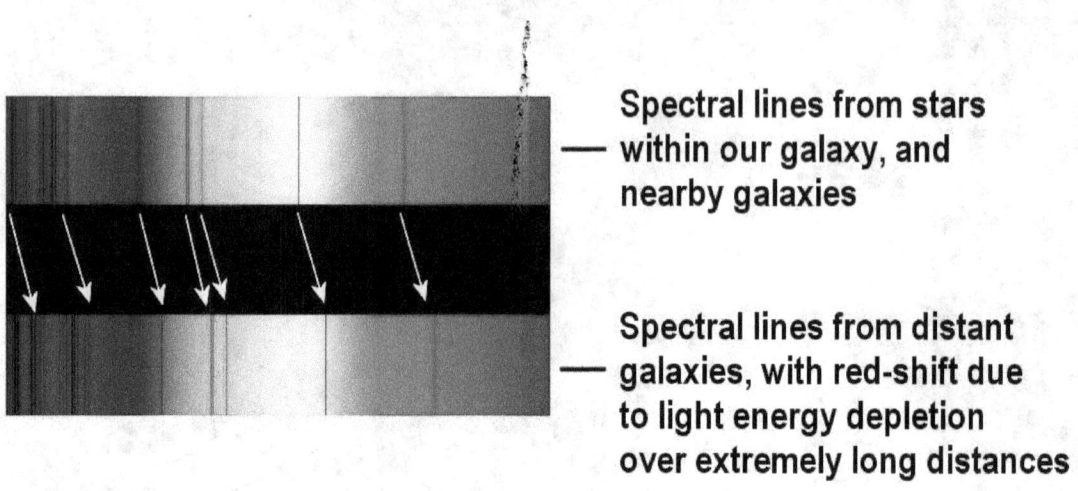

Spectral lines from stars within our galaxy, and nearby galaxies

Spectral lines from distant galaxies, with red-shift due to light energy depletion over extremely long distances

The electrically powered star that our Sun is, which has synthesized all the atomic elements of the planets in the solar system, also proves that every star in the universe is likewise so powered, and that all planets are similarly composed as our own.

When sunlight from our Sun is expanded with a prison into its various bands of color, we see a number of lines drawn of different intensity. The lines represent individual resonance characteristics of specific atomic elements in the solar corona that the light from a star has to pass through, which absorb light-energy at specific wavelengths. By recognizing the lines, and the width of the lines, it becomes possible to determine the atomic composition and temperature of the corona of each star. Researchers were surprised at first that the light from all the different stars in the galaxy, and even the light from different galaxies, includes the same spectral pattern. The evidence tells us that a single type of electric nuclear fusion occurs everywhere in the entire universe, with a similar cosmic abundance of elements being produced everywhere. The spectral lines vary only by the width of the lines according to the temperature of an individual stars' corona.

Part 5: The plasma solar system evident in CO2

Our Electric Cold Fusion Sun (Part 5) Evidence in CO2

Evidence that disputes the Big Bang theory

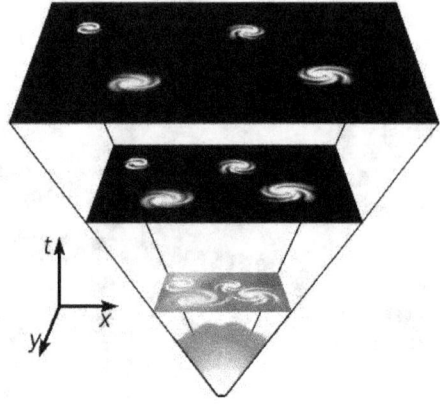

The Big Bang creation myth refuted by the electric solar fusion model

A part of the evidence that disputes the Big Bang theory, and proves the electric universe, is actually located in the Big Bang the theory itself.

The theory states that the entire universe exploded into being 13.8 billion years ago, and expanded from a single point as a cloud of dust that spread across cosmic space with ever-increasing speed.

The proof that the expansion is not happening

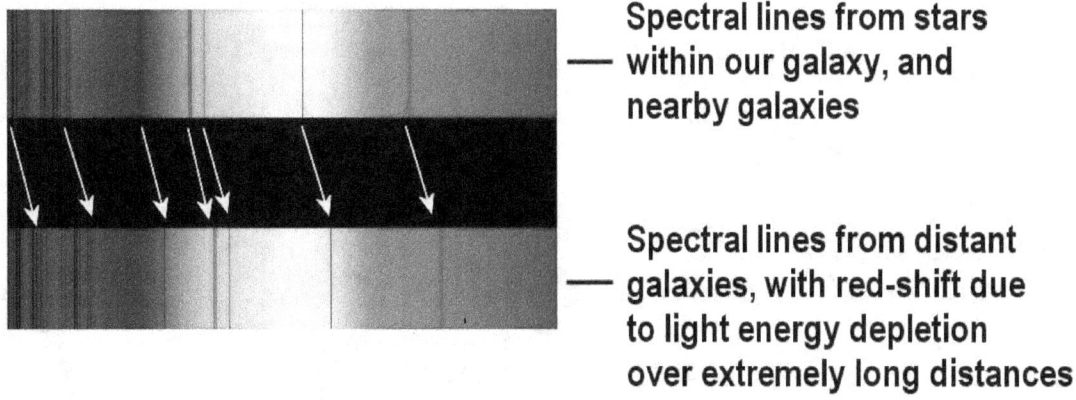

Spectral lines from stars within our galaxy, and nearby galaxies

Spectral lines from distant galaxies, with red-shift due to light energy depletion over extremely long distances

The proof that the expansion is not happening becomes evident in the amazing uniformity of the spectral lines in light from close-by stars, with the light from the most distant galaxies.

When light passes through atmospheres of atomic gases, the various elements contained in them resonate at specific frequency, where they absorbed light energy. From the location of the abortion lines it becomes possible to determine the types of elements contained in the gases, including their ratio.

While the energy-depletion in the propagation of light from extremely distant galaxies, on the order of hundredth of million of light years, has caused a red-shift in the spectrum, resulting from the energy depletion of photons*, an amazing uniformity of elementary composition comes to light.

The spectral uniformity tells us

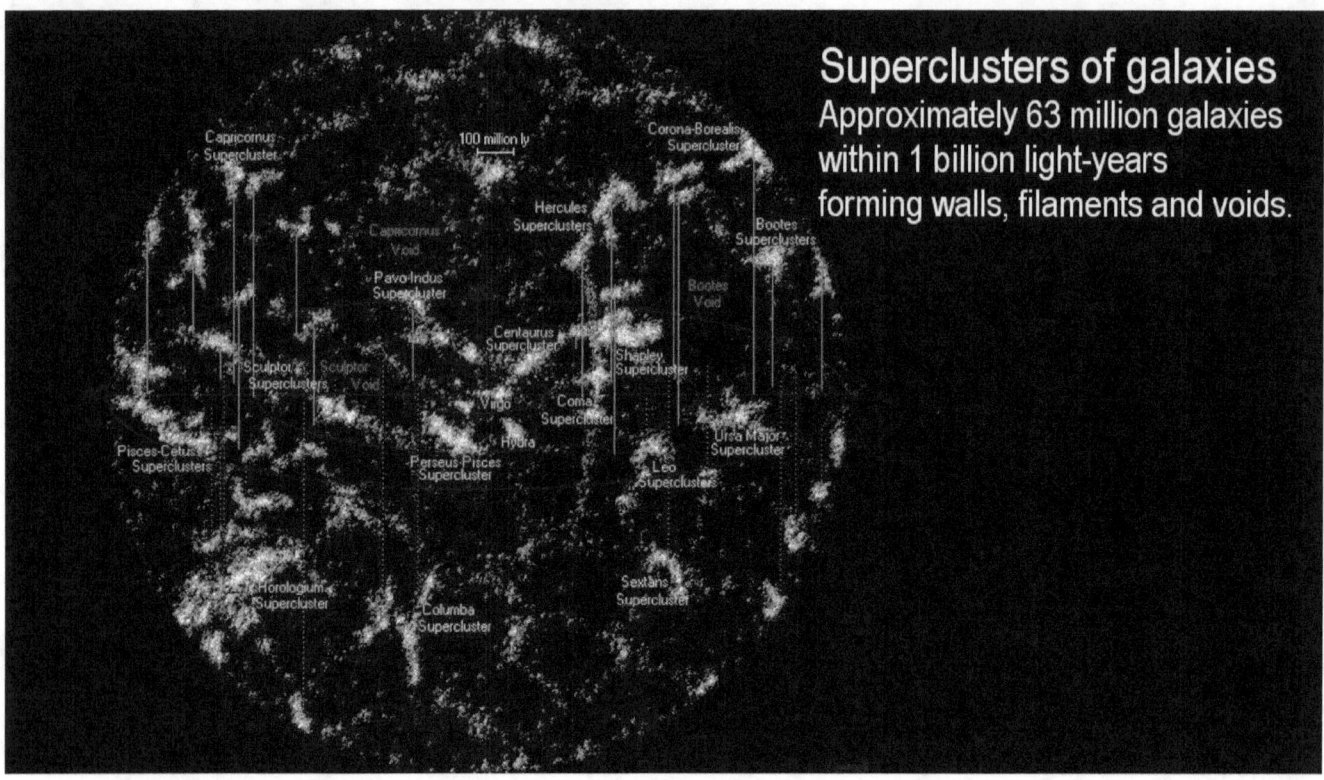

The spectral uniformity tells us that all the stars in the universe are enveloped by gases with the essentially same elementary composition; no matter how near; no matter how far. This uniformity is not possible on a Big Bang basis. * (see part 2)

When a gas cloud expands for billions of years

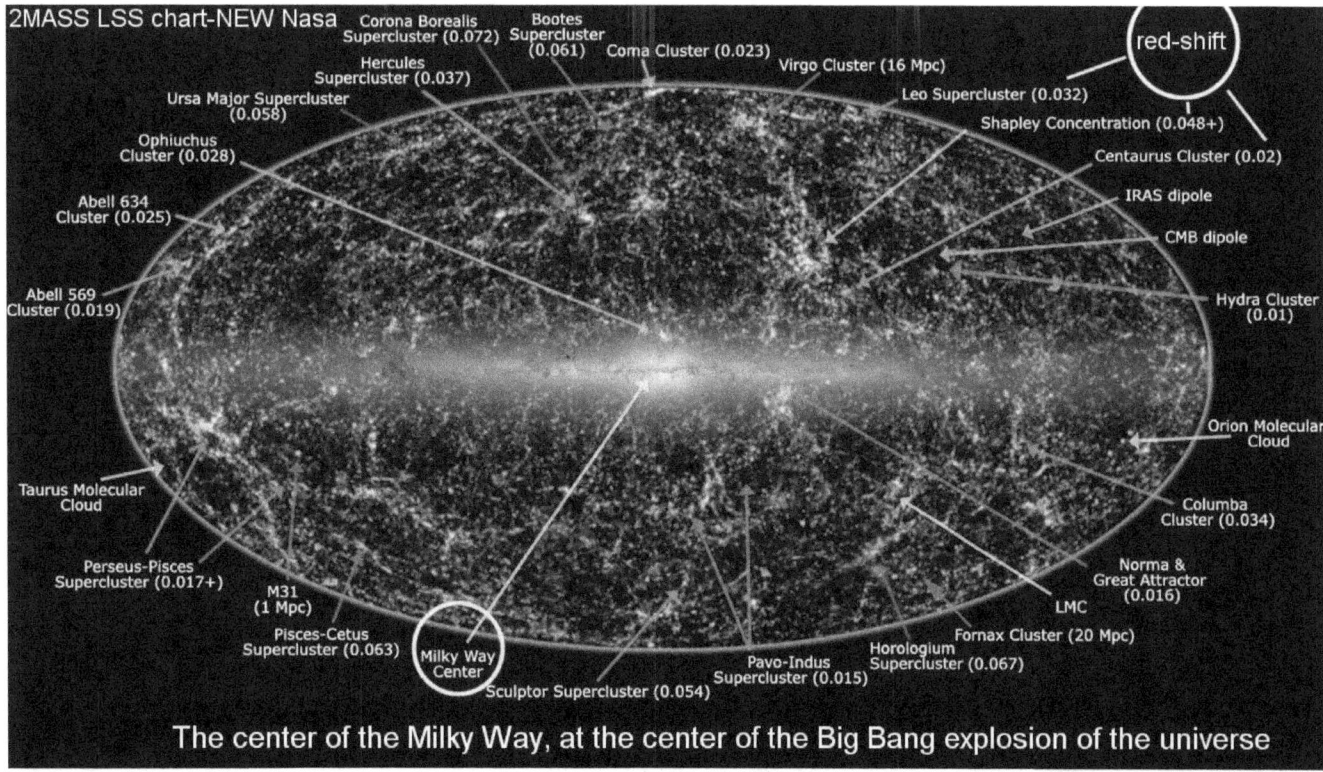

The center of the Milky Way, at the center of the Big Bang explosion of the universe

When a gas cloud expands for billions of years, it is not possible that its elementary composition is not radically altered as it expands, considering the vast differences in gravitational attraction, up to 240 to 1, which exists among the various atomic elements. The composition would have been radically altered during the 13 billion years of the supposed expansion of the dust and gases.

The atomic elements that cause the similarity

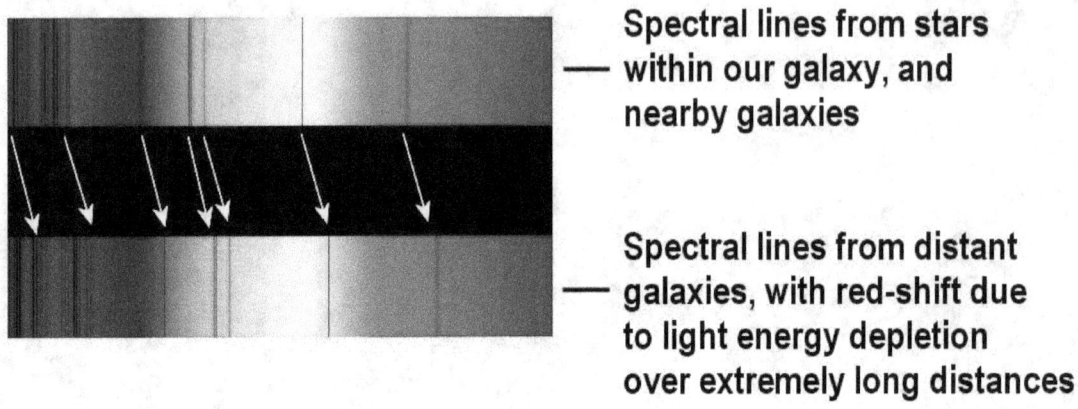

Spectral lines from stars within our galaxy, and nearby galaxies

Spectral lines from distant galaxies, with red-shift due to light energy depletion over extremely long distances

The similarity of the elementary composition in distant galaxies, with the composition surrounding the stars nearby, proves that the atomic elements that cause the similarity were evidently created everywhere locally, by each star itself, resulting from the same synthesizing electric fusion, reflecting the same principle, and being expressed in the same process, no matter where they the stars would be located in the universe, near or far.

This means that plasma alone is the singularity

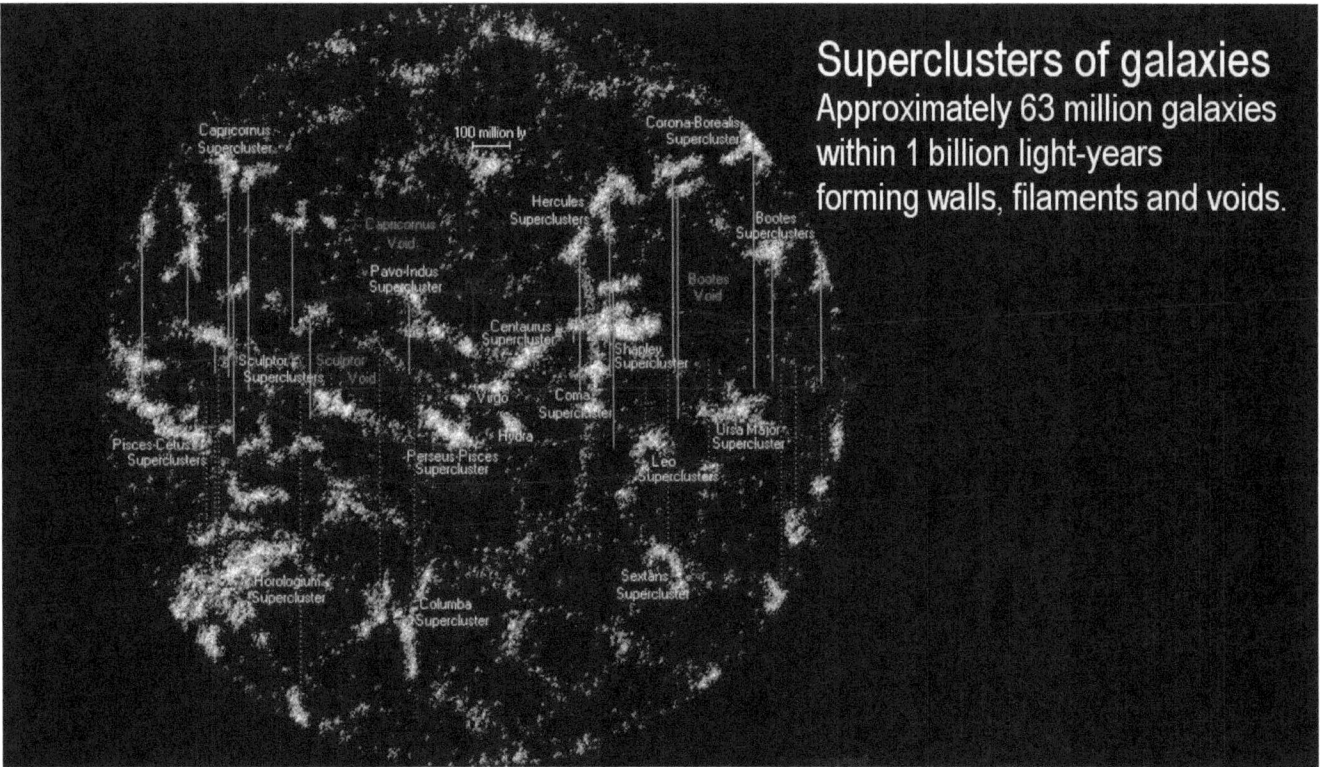

This means that plasma alone is the singularity in the equation that we see the result of, which of course is universally abundant as its is an element of space itself, according America's most advanced theoretical physicist, David Bohm.

David Bohm named Einstein's successor

- David Bohm named Einstein's successor by Einstein himself. For reference see part 2 of the series.

The similarity of the spectral lines

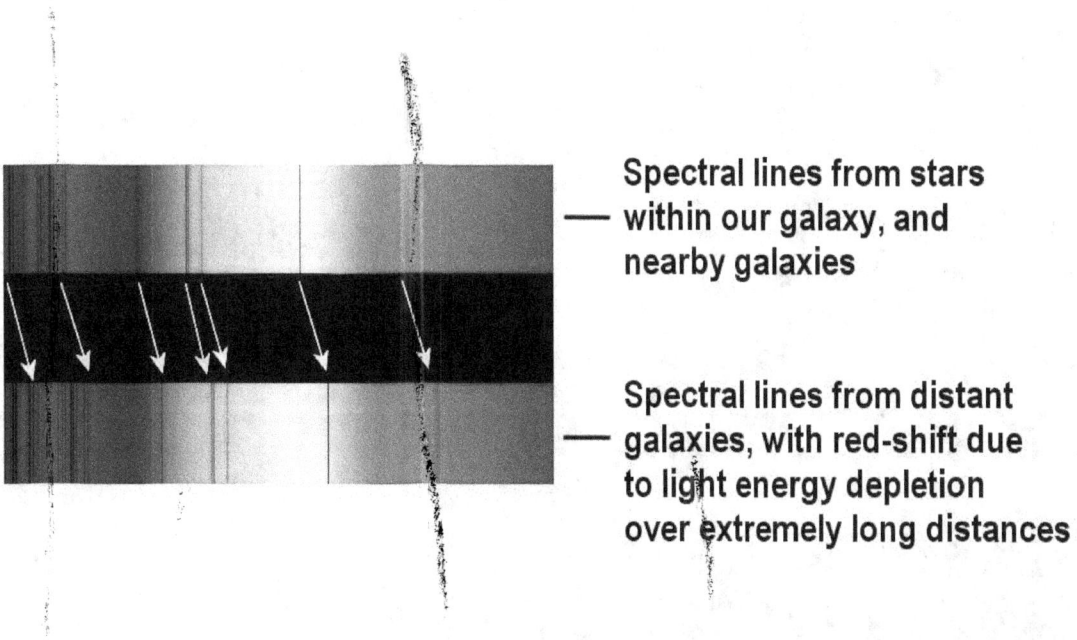

The similarity of the spectral lines, also solves a long-standing paradox regarding the CO2 concentration on Earth.

Cold Fusion Evidence in CO2

Cold Fusion Evidence in CO2

If the Big Bang composition is universally similar

If the Big Bang composition of created elements is universally similar, it shouldn't be possible that the Earth has extremely small amounts of CO_2 in its atmosphere, while the atmosphere of its neighbour planets is made up almost entirely of CO_2. This shouldn't be possible. It is a paradox.

The paradox becomes resolved when we recognize that all the carbon and the oxygen for the CO_2 in the solar system where synthesized on the Sun from plasma by electric fusion. With this being recognized, the paradox becomes resolved as but a local distribution characteristic.

The heavy CO2 molecules

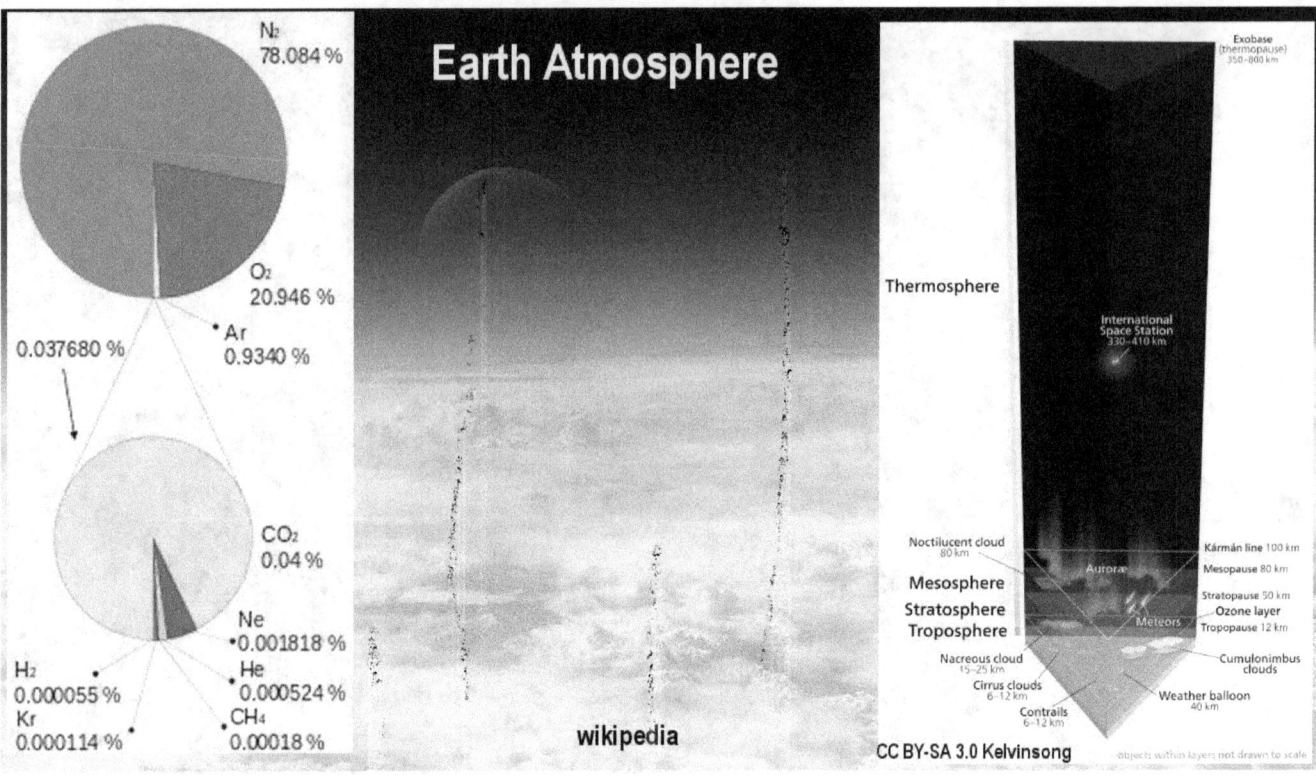

In the same manner as the water vapor in the high stratosphere, for the noctilucent clouds, was deposited into the outer atmosphere of the Earth by the solar winds, other atomic elements are being showered onto the Earth, from the Sun. The heavy CO2 molecules that we also find existing at high altitudes, were evidently likewise so deposited there.

High concentrations of CO2 have been measured in the stratosphere, in the 600 parts per million range, at an altitude of 30 kilometres above the surface. The large concentrations of this heavy gas, which are nearly twice as dense as on the ground, should not be found at such this extreme altitude, unless they were deposited there from the Sun, carried by the solar wind.

Evident by their distribution of the elements in space

That all the atomic elements of the solar system were synthesized on the surface of the Sun is evident by their distribution of the elements in space around the Sun. Since the heavy elements are not as easily carried along by the solar wind, they are first to fall out. This would result in the forming of heavy planets at close distances to the Sun, which we term the inner planets, ranging from Mercury to Mars.

The same can be said of the atmospheres of the planets. The closer the planets are, the heavier their atmosphere is. For this reason, the innermost planet with an atmosphere, which is Venus, has an atmosphere that is made up almost entirely of carbon dioxide, termed CO_2, which is a heavy gas with an atomic weight of 44. For this reason, the farthest planet is made up almost entirely of hydrogen gas with an atomic weight of 1.

The atmosphere of the Earth is lighter than that of Venus, for reasons of it being further away from the Sun. Our atmosphere is made up of gaseous oxygen with a weight of 16, and nitrogen with a weight of 14. Most of the oxygen near the Earth has been bound to hydrogen, to form water, that has a weight of 34. Only a tiny portion of the synthesized carbon atoms, when they became combined with oxygen into carbon dioxide, has made it as far as the Earth. The present CO_2 concentration on Earth is a mere 390 parts per million.

CO2 concentration on the Earth is presently at the lowest level

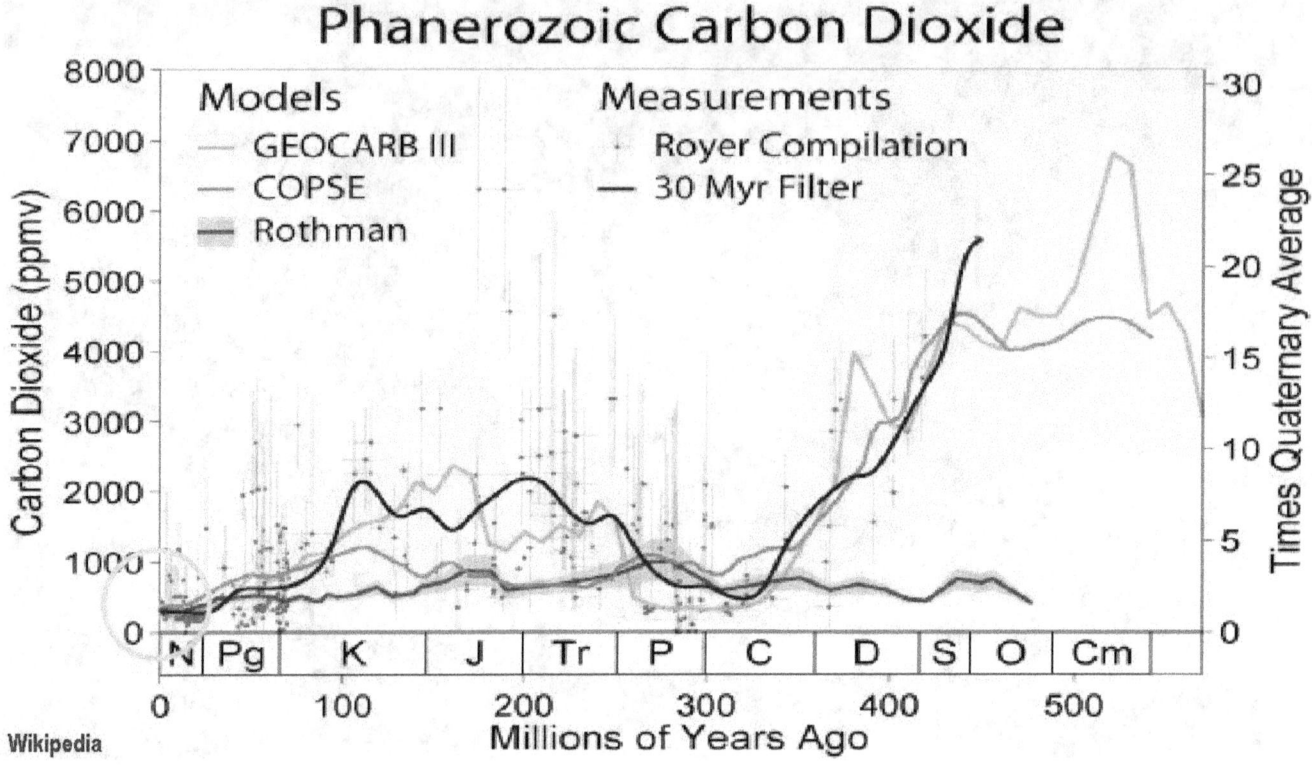

The averaged CO2 concentration on the Earth is presently at the lowest level it has been during the entire time-span of abundant life on our planet. It may have been more than 20 times as dense 500 million years ago, according to a number of major studies.

CO2 never stays in the atmosphere for long

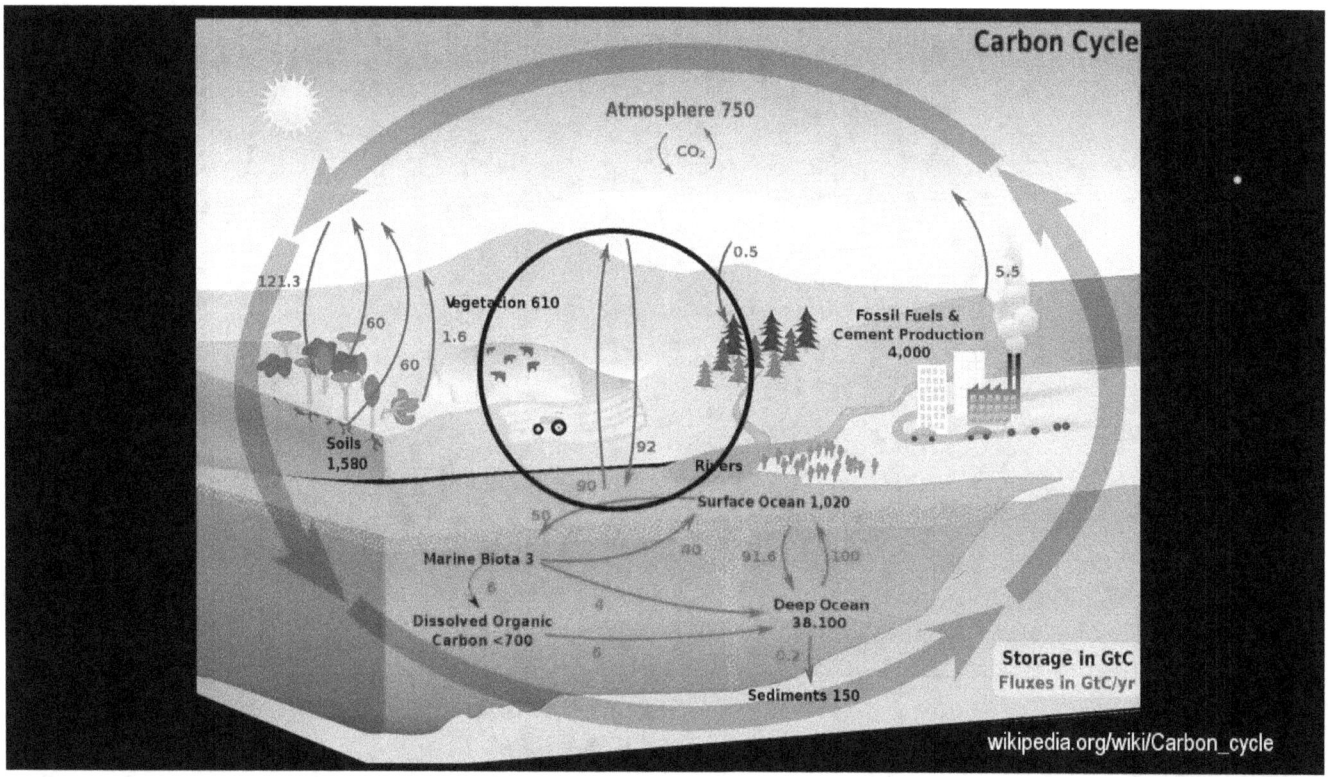

CO2 never stays in the atmosphere for long. Roughly 30% of it is recycled every year, through the oceans and plants. In the recycling process much of the CO2 is lost over time, as it remains dissolved in the oceans.

Because of the recycling loss

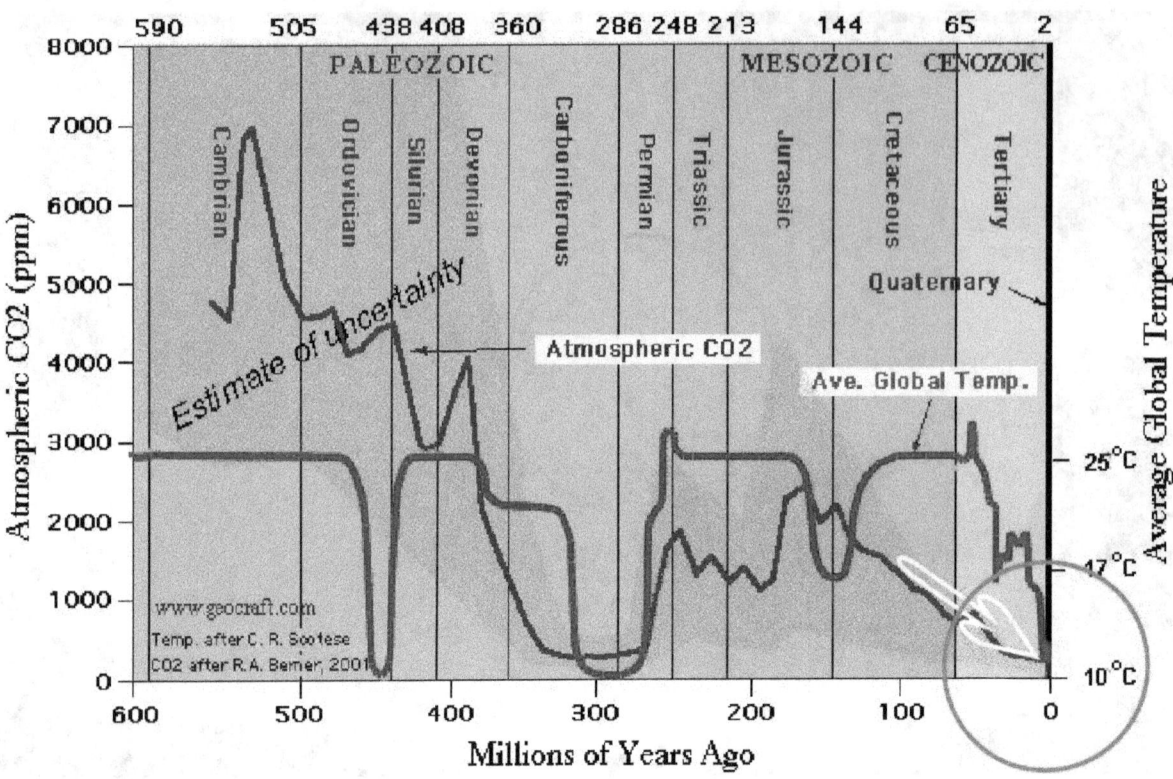

Because of the recycling loss, the atmospheric CO2 had reached extremely low levels around 300 million years ago, as we see it here in this simplified diagram. The evidence tells us that the solar fusion synthesis of oxygen and carbon, and their flowing in the solar wind, is typically insufficient to offset the recycling losses. This loss had made life on Earth rather poor during the late Carboniferous Period and the early Permian Period, until a major CO2 uplift had rescued the biosphere.

CO_2 is extremely critical for all life on our planet

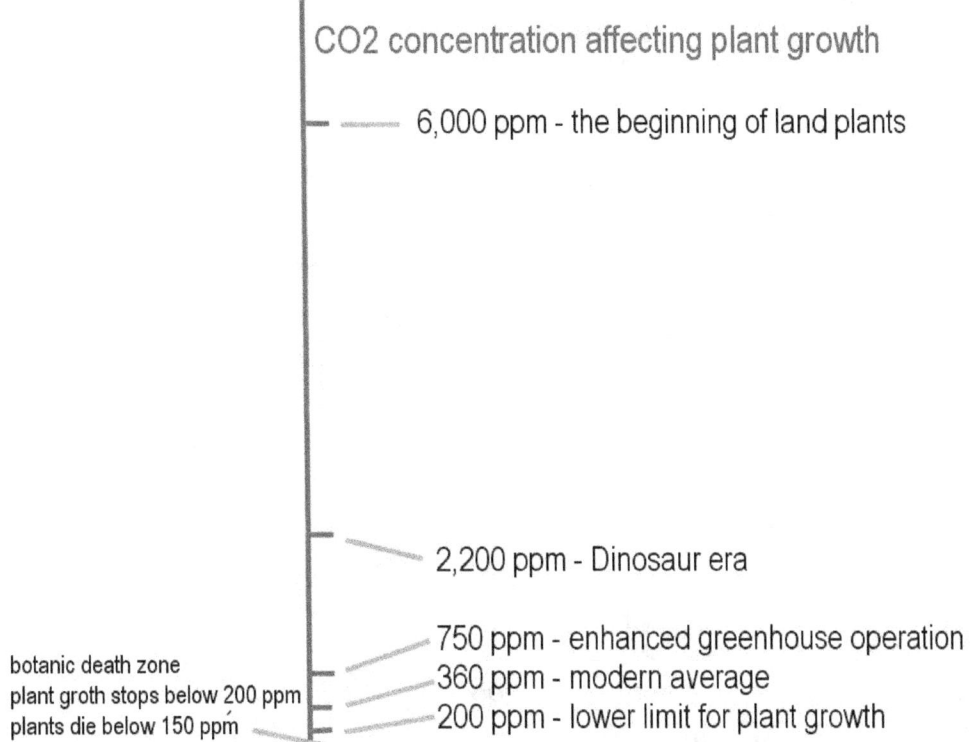

CO_2 is extremely critical for all life on our planet. Plants can't exist without it, and we ourselves, cannot exist without the plants that our entire food chain is build on. CO_2 is so critical, that if the present concentration in the air is cut in half, the plants die, and if it is doubled, a 50% increase in plant-growth results.

If the CO2 loss had continued past the Permian Period

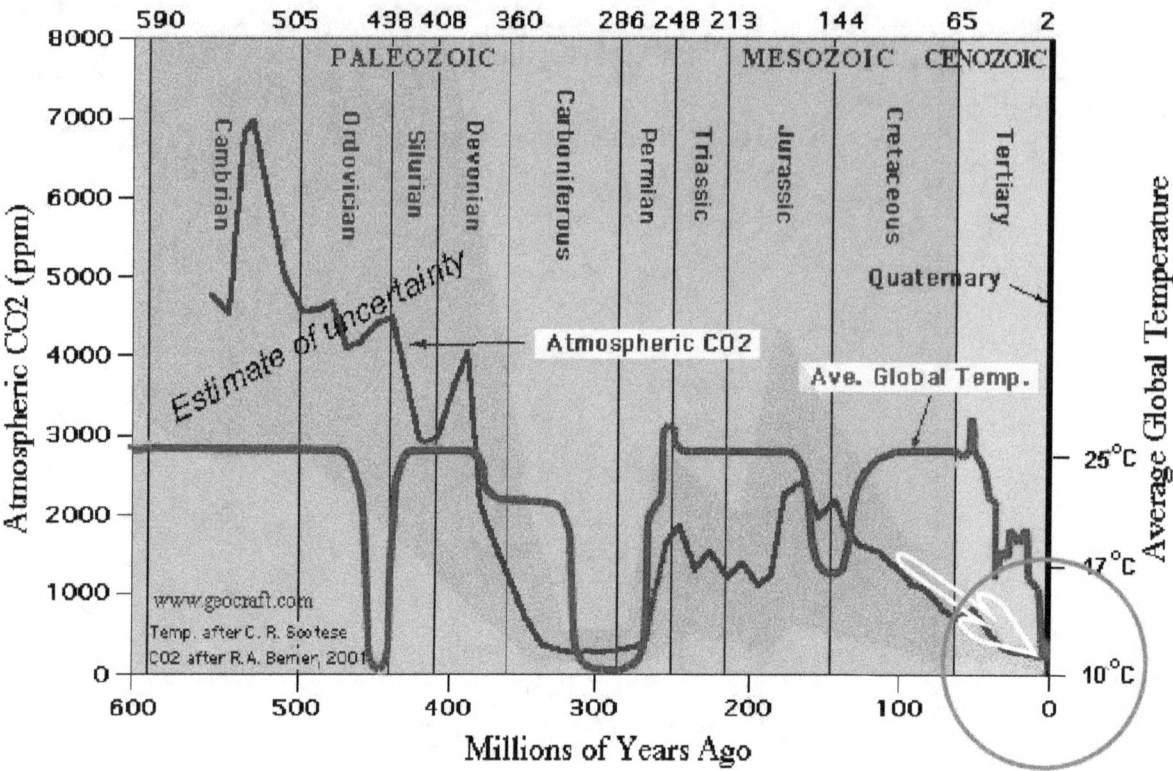

If the CO2 loss had continued past the Permian Period, life might have become extinct for the lack of CO2. This didn't happen. Life was rescued by a gigantic solar event that gave life a huge boost.

The solar event occurred during a peak period

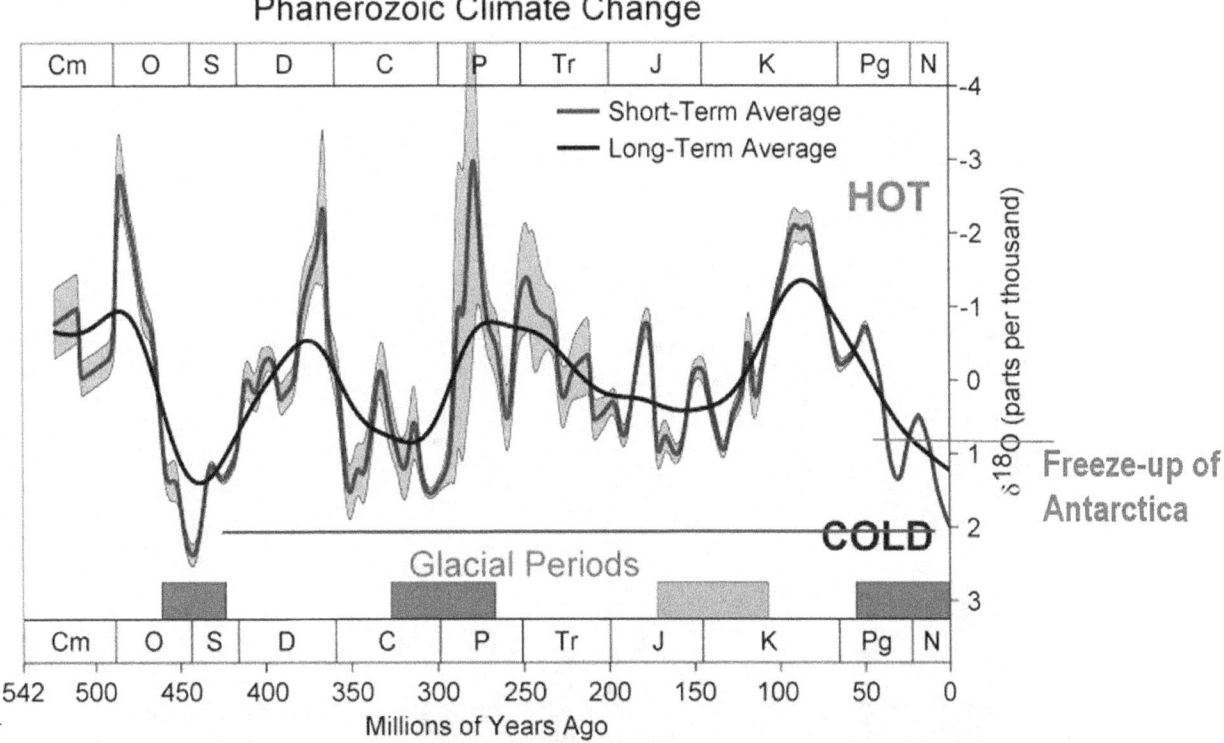

The solar event occurred during a peak period in solar activity. The plasma density was extremely high at the time, which was reflected as a much hotter Sun with denser and faster solar winds. What we see measured here reflects the surface temperature of the oceans measured in variations of a heavy isotope of oxygen in the waters of the oceans. The 'normal' abundance of the heavy oxygen, named O18, is 20 mill, or 20 parts per thousand, or 2 tenth of a percent. When the sunlight is hotter, the lighter water is more readily evaporated than the heavy water, which changes the abundance ratio at the surface. The ratio is preserved in the calcium shells of microorganisms that lived at the surface region of the sea. By examining the accumulations of their remains in sediments, the global climate can be reconstructed for the last half a billion years, as shown here.

A type of solar lightning

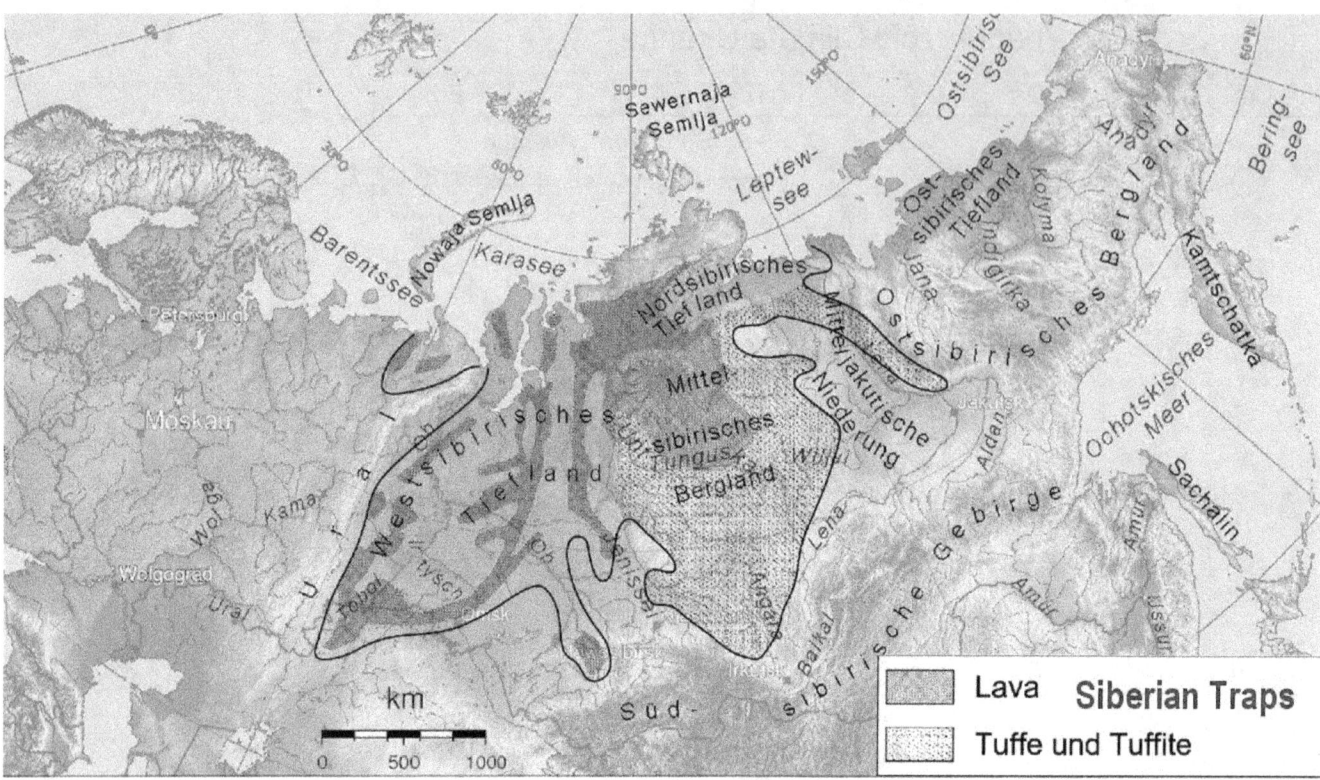

During the peak period of extreme solar activity in the Permian Period, an extremely large event occurred, which evidence suggests would have been an electric discharge event between the Sun and the Earth. The solar wind is basically an electric wind, made up primarily of plasma. Plasma is a near perfect conductor of electricity. In times of extreme solar-wind density, an efficient electric connection between the Earth and the Sun would likely have resulted, to cause a type of solar lightning. It may have been a gigantic discharge event of this type that with electric stress effects pulverized much of the mantle of the Earth across large areas of what is presently northern Siberia, enabling enormously large magma flows, which created the flood basalt province, termed the Siberian traps. It is estimated that the flood basalts cover up to seven million square kilometres of land and contain up to 4 million cubic kilometers of basalt.

The single event raised the global CO2 level 5-fold

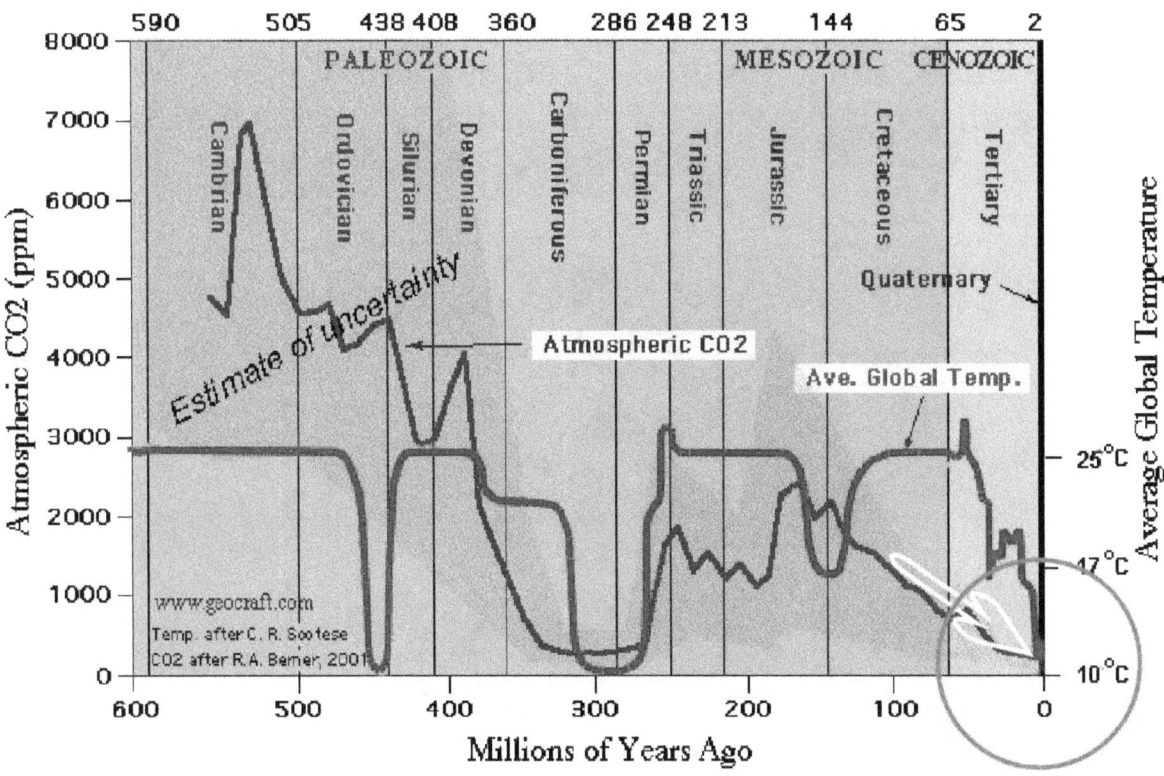

The single event raised the global CO2 level 5-fold at the time, as you can see here, near the Permian boundary, 250 million years ago. For this increase close to 5 trillion tonnes of CO2 would have been injected into the Earth's atmosphere. It may be argued that this increase resulted from the volcanism of the Siberian flood basalt event. It is far more likely, however, that this huge increase flowed directly from the Sun, which is surrounded by a heavy gaseous atmosphere, and would have been even more surrounded during the hot times. The solar lightning strike may have lasted less than half an hour.

A similar lightning strike occurred during the Jurassic Period

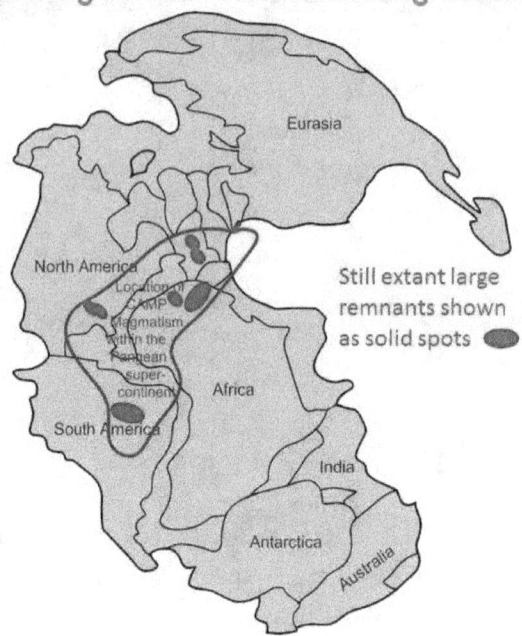

Evidence suggests that a similar lightning strike occurred during the Jurassic Period. In this case the gigantic Central Atlantic Magmatic Province was created, with lightning offshoots across the super-continent. The event injected another 5 trillion tonnes of CO2 into the atmosphere. The resulting electric stress induction, during the event at the time, may have started the breakup of the super-continent, termed Pangaea.

Massive Australian Precambrian/Cambrian Impact Structure

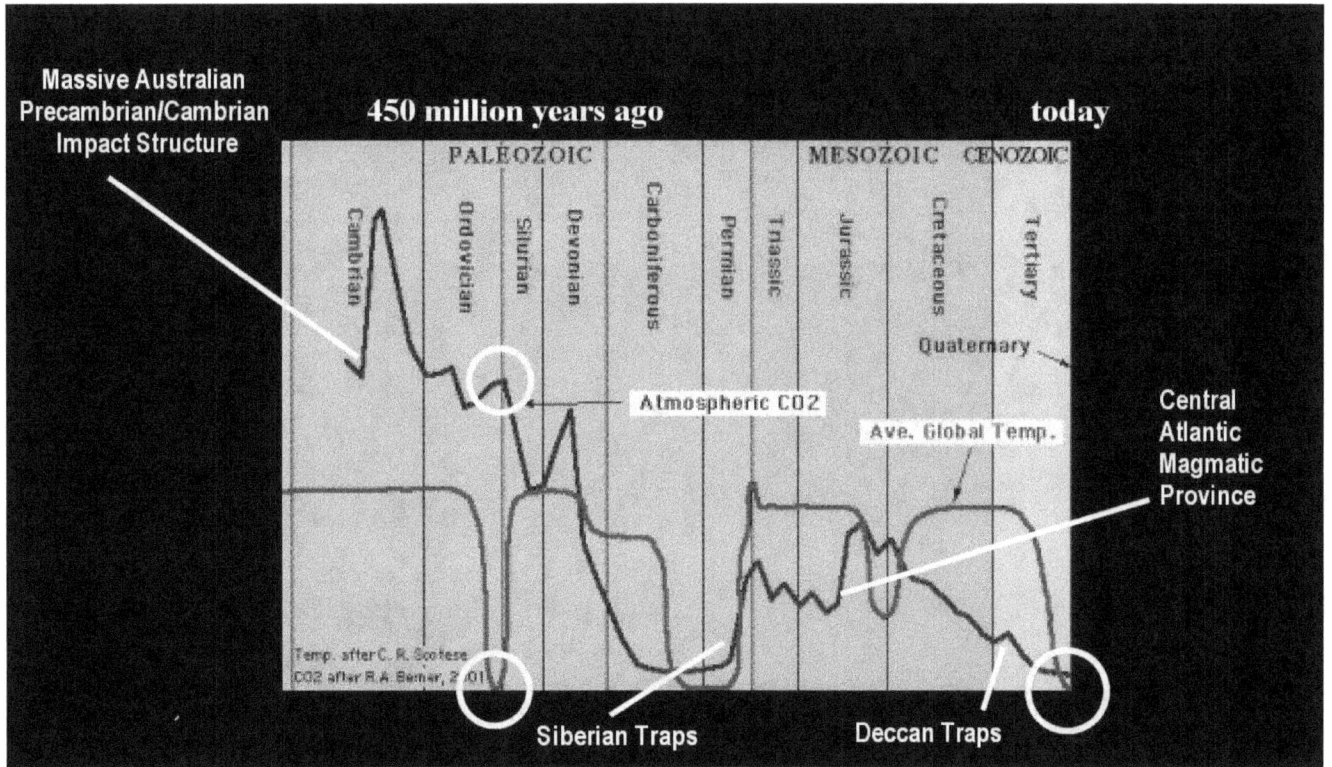

A still larger solar discharge event may have produced the enormous CO2 uplift, which evidence suggests has occurred in early Cambrian time, more than 500 million years ago. The existence of an enormous lightning discharge crater has been discovered, located in north-central Australia. it is termed the "Massive Australian Precambrian/Cambrian Impact Structure."

Lightning impact crater was carved 2000 kilometers wide

Lightning over Las Cruces, New Mexico - photo by Edward Aspera Jr. - United States Air Force, VIRIN 040304-F-0000S-002

The huge lightning impact crater was carved 2000 kilometers wide. Can you imagine the immense type of lightning that creates an impact crater spanning 2000 kilometers? Actually, several of these exist. A large one of this type exists on the Moon. Another one exists on Mars.

The big discharge-impact crater on Mars

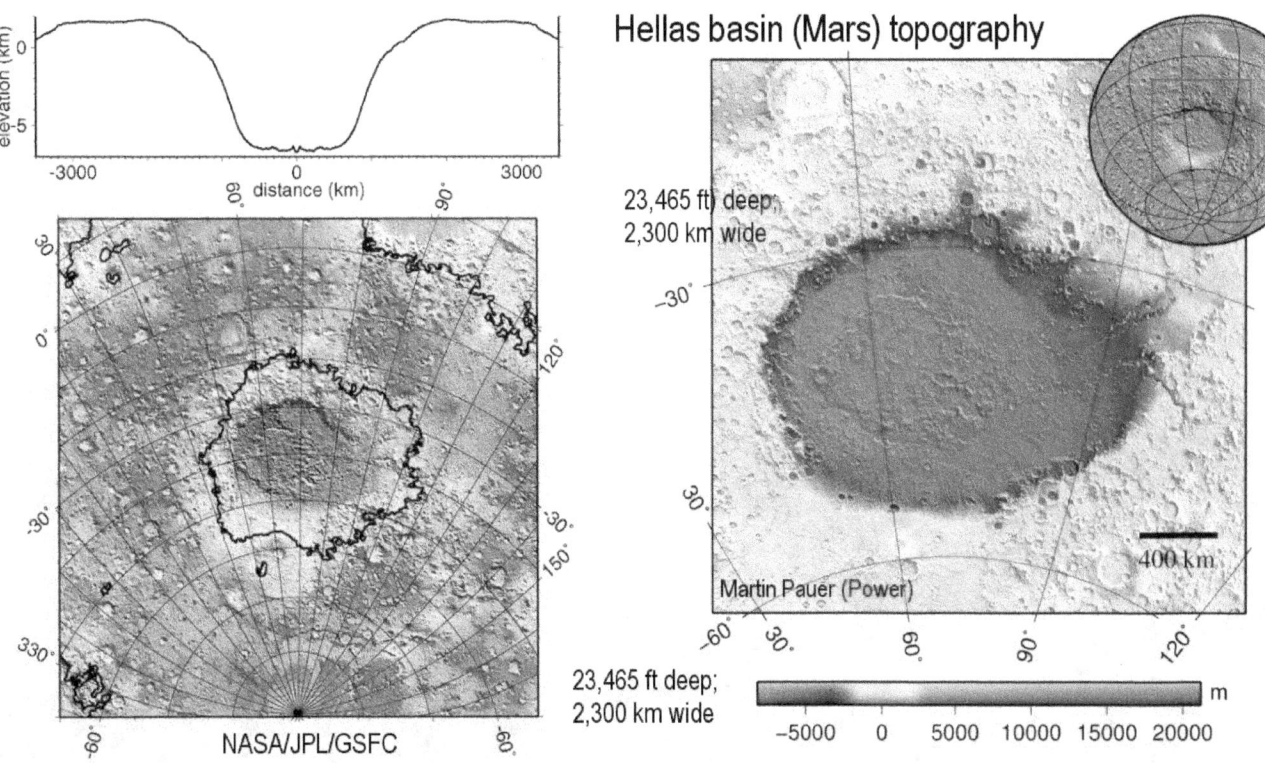

The big discharge-impact crater on Mars, named Hellas Basin, was carved 23,000 feet deep, and 2,300 kilometers across. Events like these would have brought the enormous volumes of the heavy CO_2 to Mars, which, over time became its atmosphere, which is fast fading again.

The forming of the Deccan Traps in India

Many ripples are apparent in the measurements for the historic CO2 on Earth. One is linked in time with the forming of the Deccan Traps in India, which may have resulted from another solar lightning impact, a comparatively small one that caused a flood basalt event extending across a mere half a million square kilometers, with an estimated volume of 512,000 cubic kilometers of magma. The event is closely timed with the extinction of the dinosaurs. The resulting volcanism may have blanketed the sky with ash for extended periods that wiped out all food resources.

Evident in the carving of the Grand Canyon in Arizona

How massively effective a large high-intensity lightening discharge from the Sun can be, is evident to some degree in the carving of the Grand Canyon in Arizona. The entire canyon may have been excavated in less then 15 minutes as the Earth rotated beneath the discharge from a high density plasma stream that cuts through rocks and stress-fractures them into powder.

A similar canyon was cut on Mars in distant time

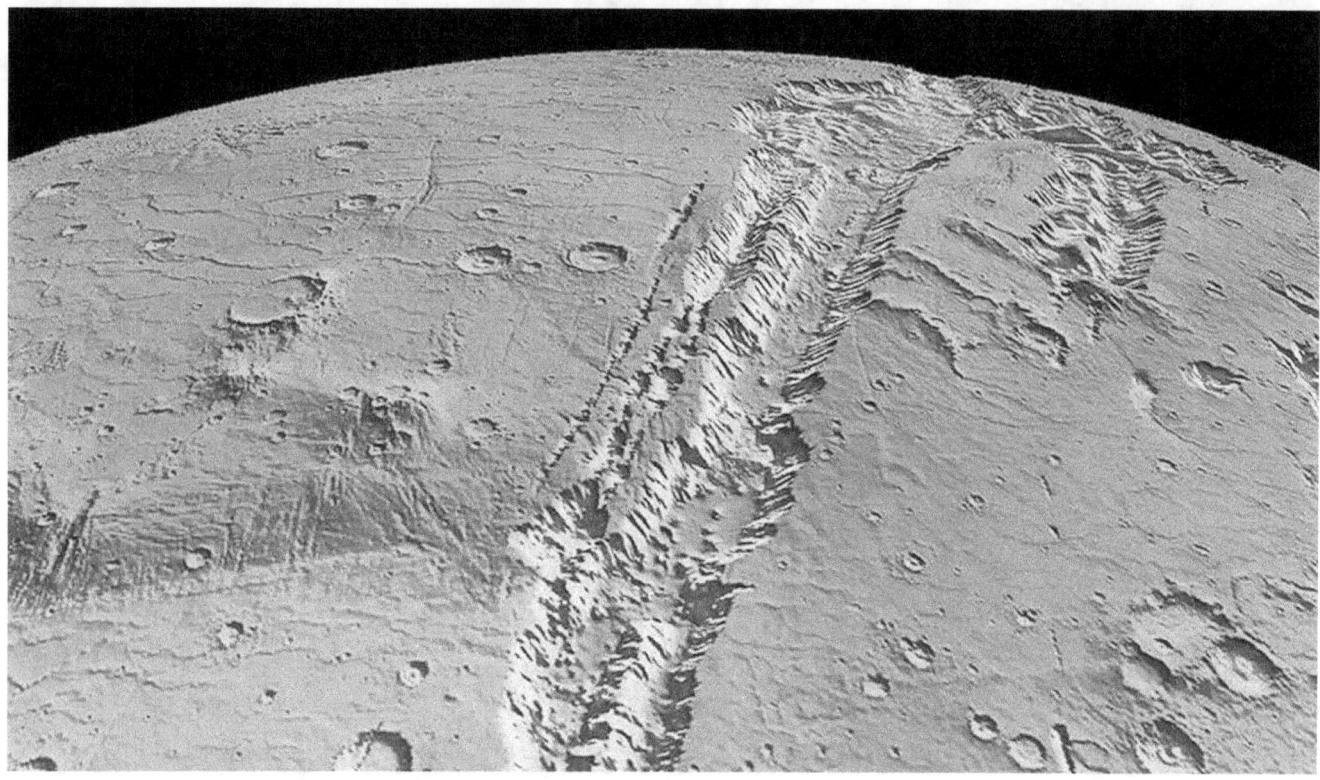

A similar canyon was cut on Mars in distant time, only much bigger in size. Valles Marineris is a system of canyons 4,000 km long, 200 km wide, and up to 23,000 feet deep. It was likely cut in a single large electric discharge event, in which Mars also gained a large portion of its atmosphere, which is made up almost entirely of CO_2.

Life might have vanished on Earth

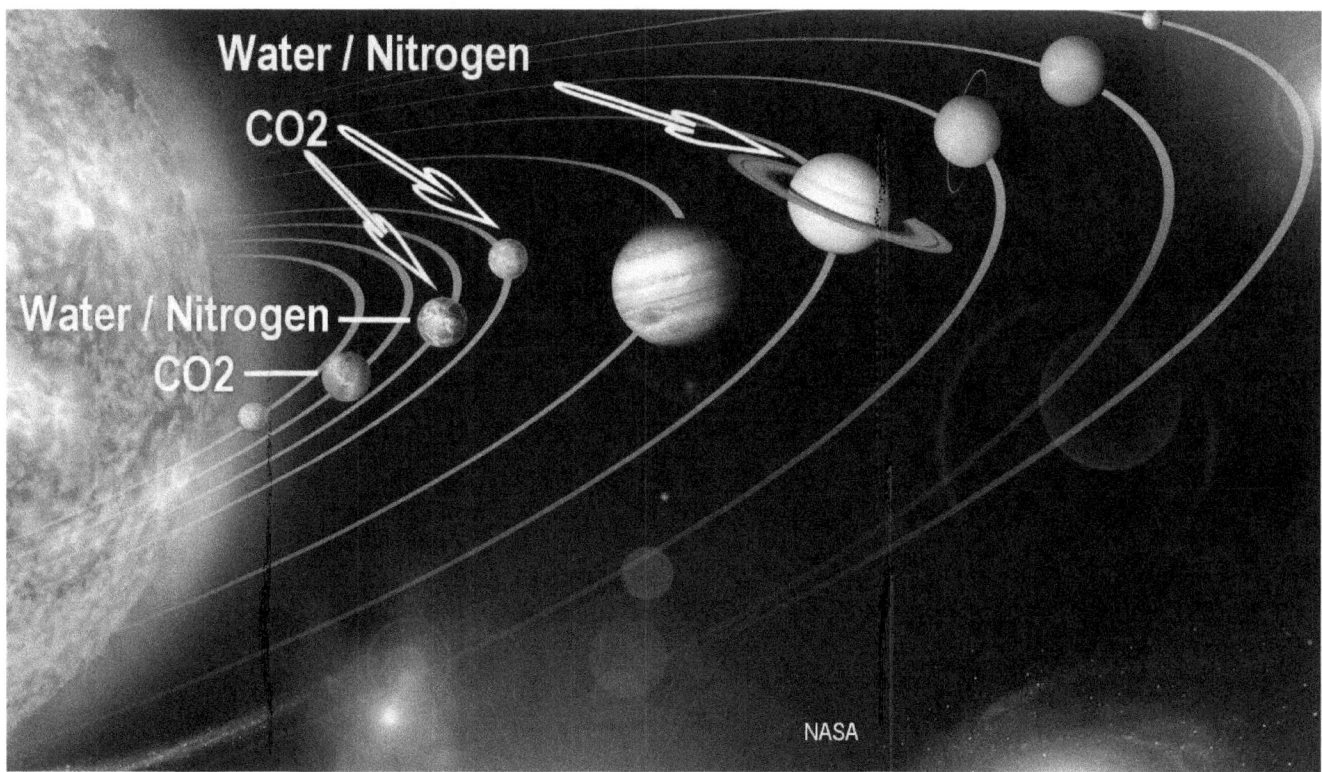

Mars is too far distant from the Sun for the heavy CO2 elements to become carried to it in large volumes by the solar wind. It is save to say this, considering that the Earth, which is closer to the Sun, doesn't get enough of CO2 from the solar winds to maintain its biosphere. The normal distribution channels, evidently carry the CO2, which originated at the Sun, not much further than Venus. The atmosphere of Venus is 96.5% CO2 and 3.5% nitrogen. It appears that the solar lightning is the only efficient transport of heavy gases over long distances in the solar system.

The high gas-transport efficiency of the solar-lightning channel may be the reason why we exist today. Life might have vanished on Earth, except for primitive forms, if it hadn't been for the gigantic lightning events that brought CO2 to our planet, and had uplifted our atmosphere repeatedly with the most basic necessities for carbon-based life to exist, which CO2 is.

Solar lightning is apparently such an efficient long-distance transport channel for heavy gases that it has enables Saturn's rings to form, and its ice moons.

Saturn's rings are made up of water ice

Saturn's rings and some of its moons, are made up of water ice, which Saturn itself, does not have, nor would have, because of its distance from the Sun.

The existence of Saturn's icy moons

The existence of Saturn's icy moons and of the great rings made of ice, has become possible by the great conveyer belt of solar discharge lightning, which Saturn tends to attract.

Solar lightning events are more likely for Saturn

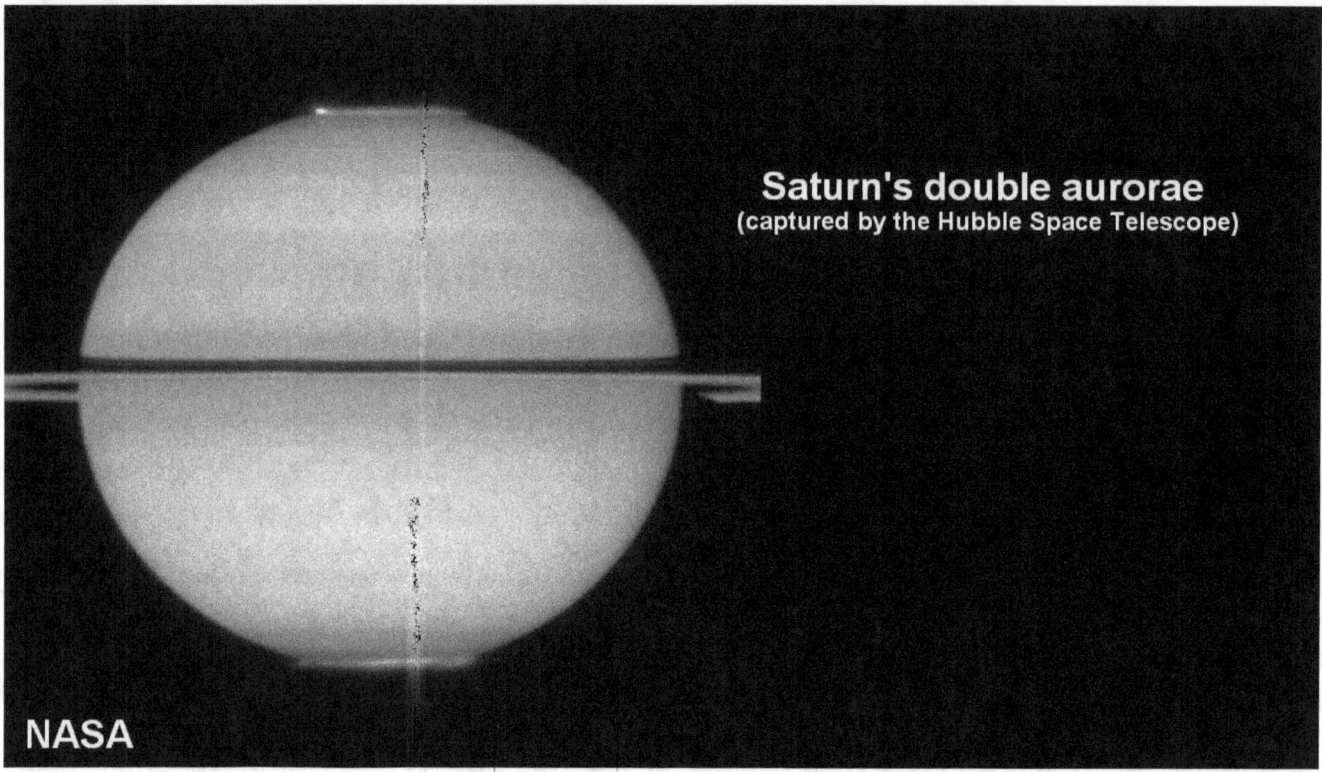

That solar lightning events are more likely for Saturn than for any of the gas planets, reflects the fact that Saturn is the most electrically-active planet in the solar system. The cause for it lays beyond the scope of this video, but the fact itself is evident by the strong aurora structures on both of the poles of the planet, which may be the strongest aurora in the solar system.

Titan, the largest of Saturn's moons

Titan, the largest of Saturn's moons, has in addition to being an ice moon, a nitrogen atmosphere. The heavy nitrogen gas should not be found there. It does not exist on Saturn. It most likely arrived with a solar discharge event, in combination with the water molecules for the ice moon itself.

The great CO2 up-ramping that gave life a new chance

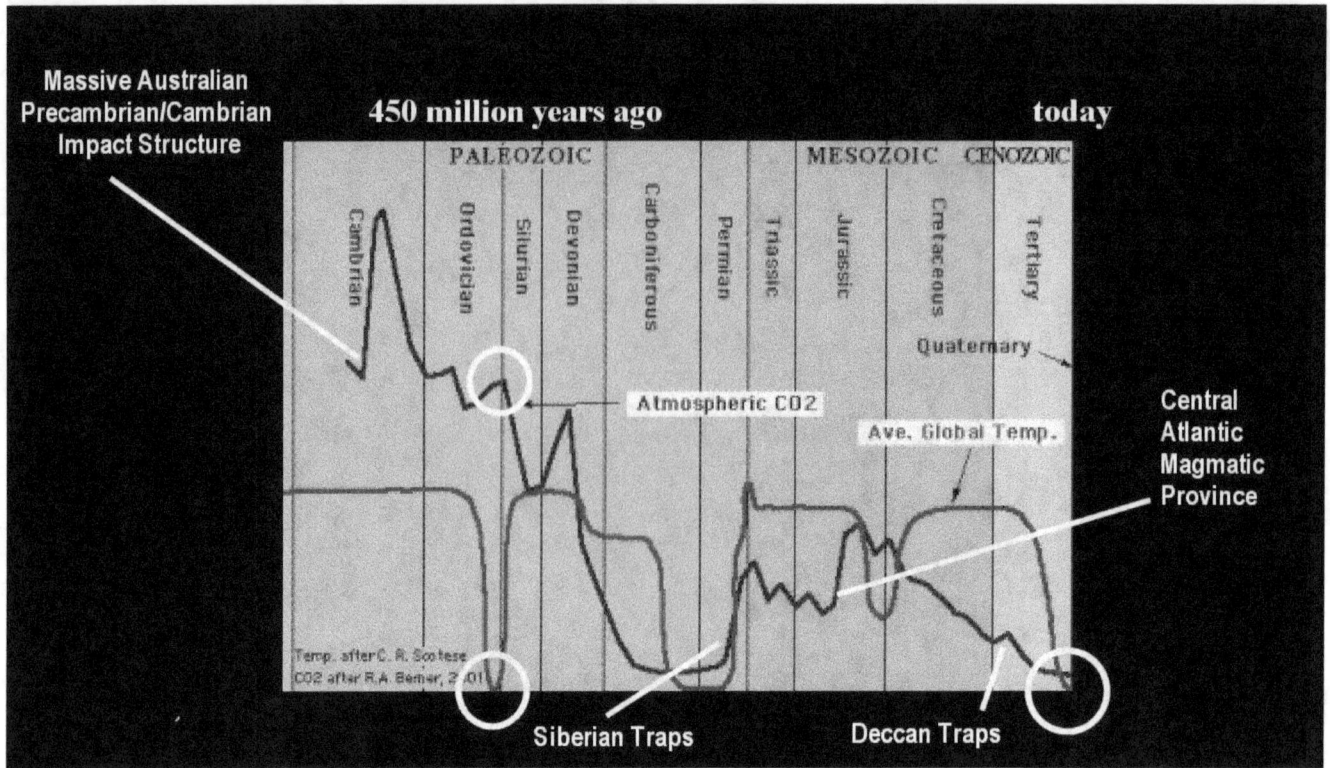

While both of the gigantic Permian and the Jurassic discharge events caused enormous volcanism and some of the largest-ever mass-extinctions of life, they produced the great CO2 up-ramping that gave life a new chance with enough CO2 in the air, by then, to last for the next 250 million years.

CO2 on Earth has diminished back to the CO2 starvation level

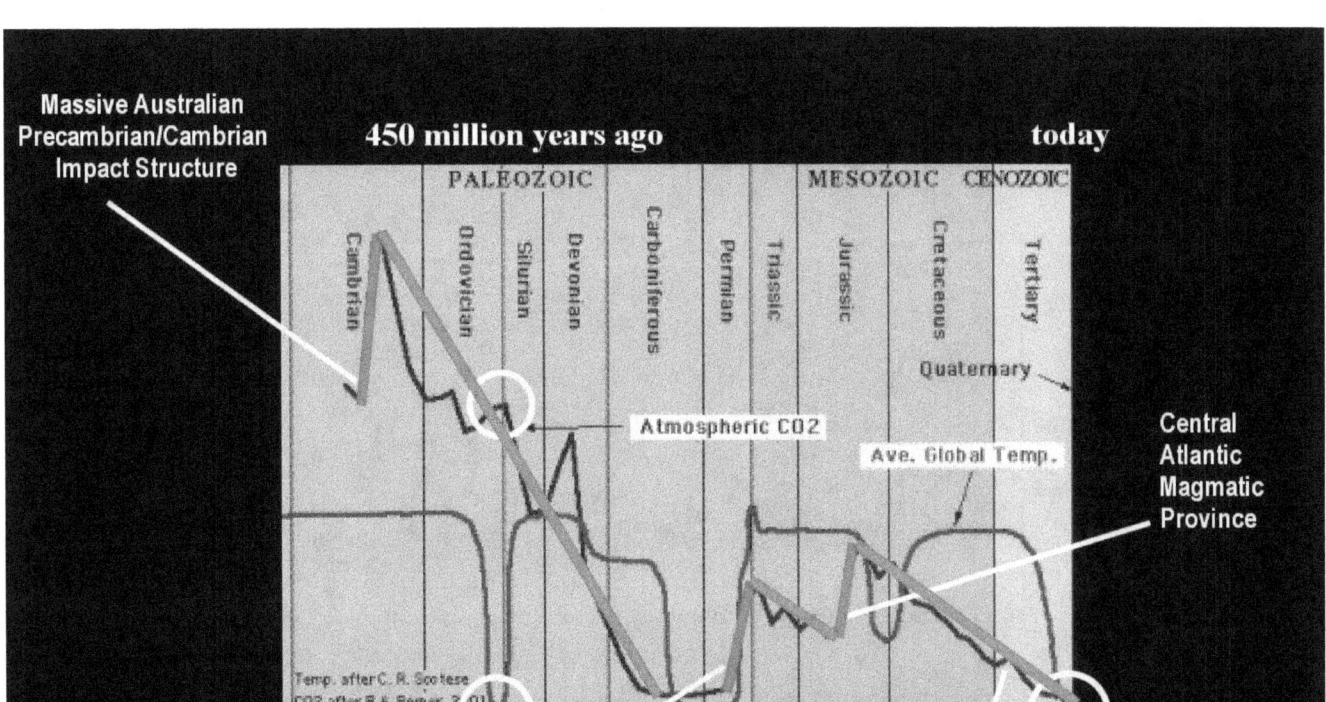

Unfortunately, the CO2 on Earth has diminished back to the CO2 starvation level that may be similar to that of the early Permian time, which was a harsh time for life, before the second round of the great up-ramping events began.

Ironically, it may have been those enormous sharp uplift in global CO2 in Permian and Jurassic time, which together gave life another 250 million years to exist, which had caused as a consequence two huge mass-extinction events.

CO2 is 28 times more soluble in water than is oxygen. The sharp and sudden increase in CO2 concentration in the air, would have massively increased the concentration of CO2 in the oceans. It would have caused ocean acidification beyond what many forms of life could tolerate. The result was, in Permian time, together with other causes, the mass-extinction of 95% of all existing species. The observed extinction pattern is consistent with the preferential extinction of heavily calcified species. While many species became extinct thereby, the extinction itself opened the landscape to the development of new species that could utilize the new and more powerful environment for life, which otherwise might not have existed at all, such as the dinosaurs, and the mammals afterwards that we became a part of.

The normal solar induction of CO_2 is insufficient to maintain life

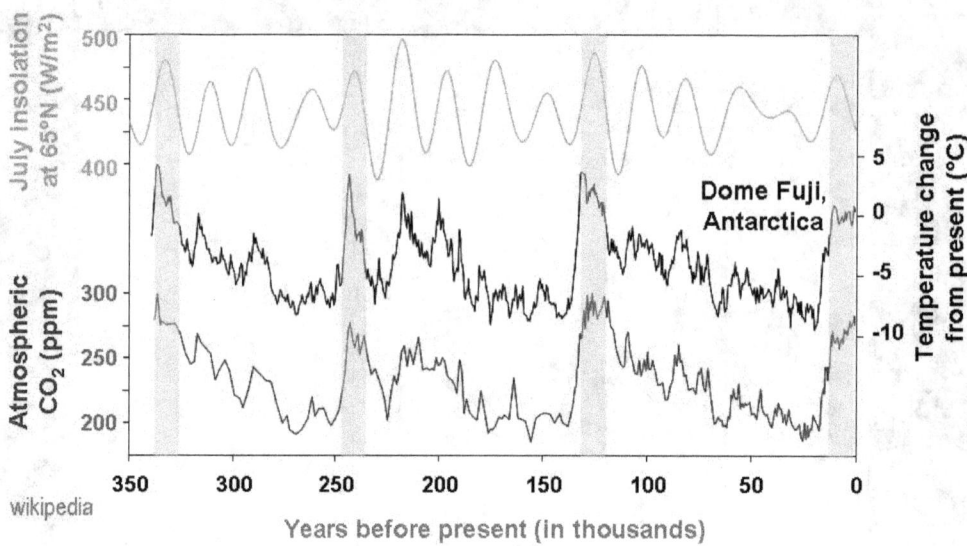

It is critical for us to realize that the normal solar induction of CO2 into the Earth's atmosphere, by the solar wind, is insufficient to maintain life on our planet, especially during Ice Age conditions when no influx from the inactive Sun occurs, and when greater amounts of it are absorbed by the oceans in the colder climate.

Ice core samples tell us that during Ice Age conditions the CO2 concentration in the air may drop to such low levels that plant-growth potentially stops. Plants simply die when the concentration falls below the 150 parts per million level. We may have come close to that during the last Ice Age.

We need to up-ramp the CO2 concentration ourselves, this time

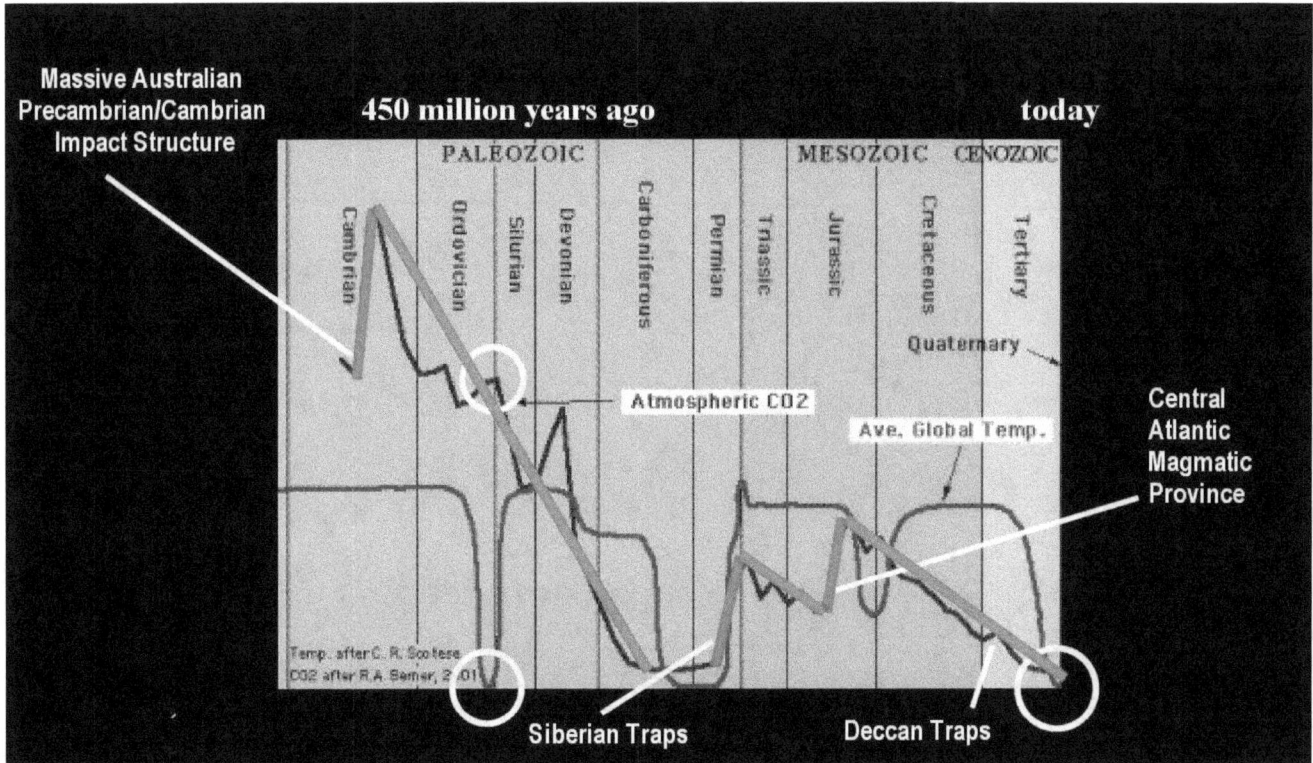

The three big discharge events that ramped up the global CO2 won't be repeated any time soon, probably not for another 100 million years. The electric density no longer exists at the present time, for this to be possible. Our galaxy is presently at the weakest state it has been in 440 million years. Because CO2 has diminished after each ramp-up, it presently stands at a biological starvation level. This means that we need to up-ramp the CO2 concentration ourselves, this time, not by getting more CO2 from the Sun, but by getting substantial amounts out of the great storehouse of CO2 on Earth, that the oceans have become.

Pumping some of the stored-up CO2 back out of the oceans

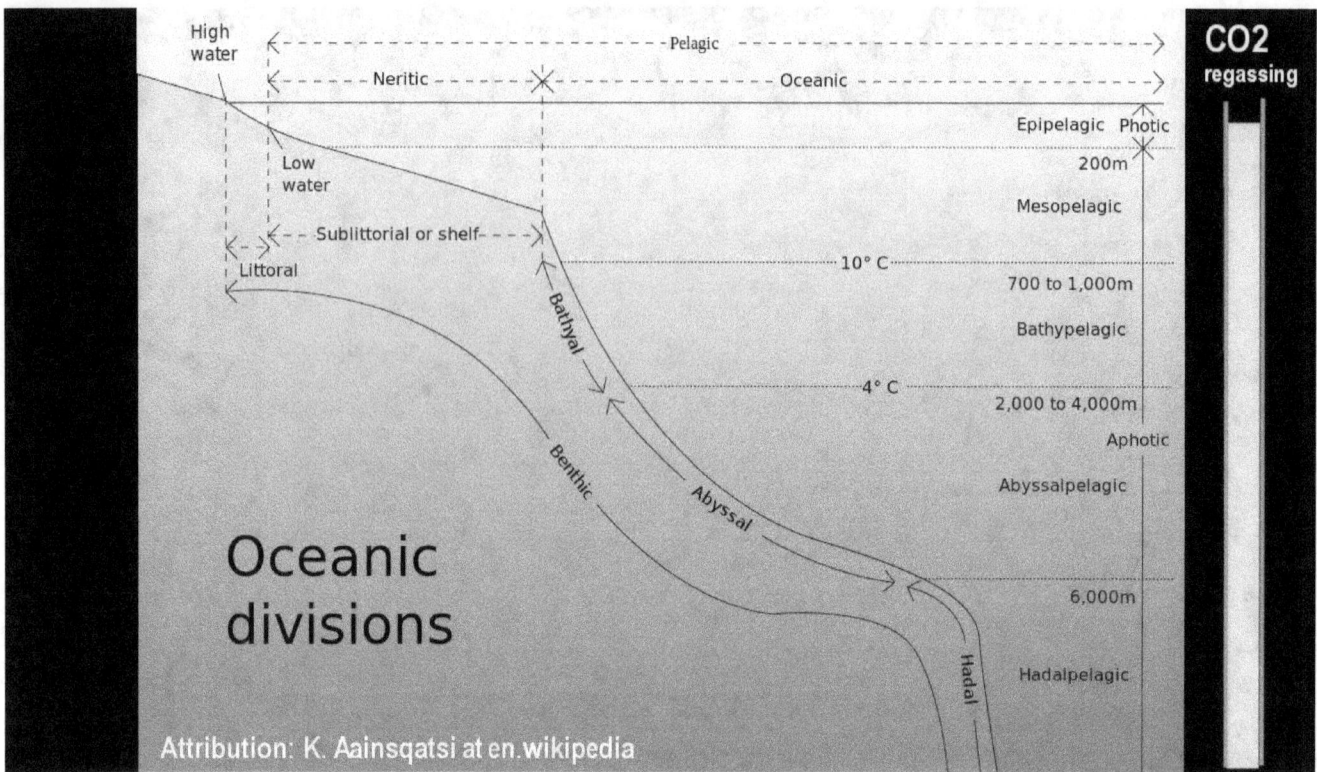

Knowing all this, will hopefully inspire us to uplift the global CO2 concentration, by pumping some of the stored up CO2 back out of the oceans. That's easily done. Close to 40 trillion tonnes of CO2 is stored in deep ocean waters, dissolved under pressure. It is a simple process to liberate 5 trillion tonnes of it from the deep oceans, by simply lifting the water to the surface. With the pressure removed, the CO2 degasifies, like with the opening of a soft-drink can. When the degassing forms bubbles, the resulting lighter weight of the water column, should drive the entire degassing process all by itself, once it is started.

So it is, that an intelligent understanding of the Sun and its dynamics can inspire us to build the required infrastructures that are needed to support life on our planet, which becomes extremely critical during the next Ice Age.

Part 6: The diminishing solar wind towards solar collapse

Our Electric Cold Fusion Sun (Part 6) Solar-wind 'steam' diminishing

170 ** ~sun_topics_15

Solving the impossible paradox before us

The solar winds are diminishing, while the Sun remains constant. That's a paradox.

The weakening has dramatically increased

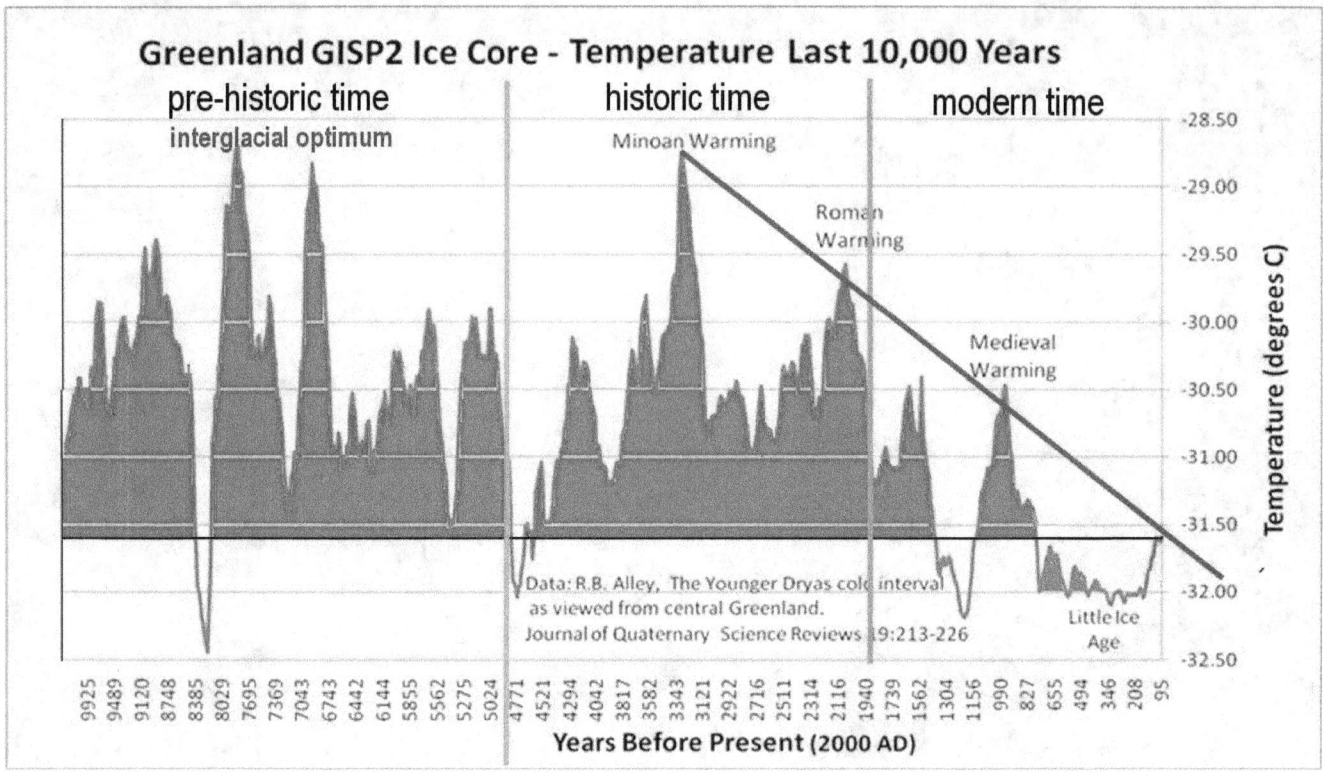

By looking deeper into the past, one can notice that the systemic weakening that Ulysses had measured between its 1st and 3rd orbit, is actually not a new phenomenon. A long-term weakening had began already more than 3,000 years ago, with ups and downs along the way, according to what we can gleam from ice-core samples drilled from the Greenland ice sheet. The Ulysses satellite tells us, however, that the weakening has dramatically increased in modern time.

The measured 30% reduction of the solar wind pressure

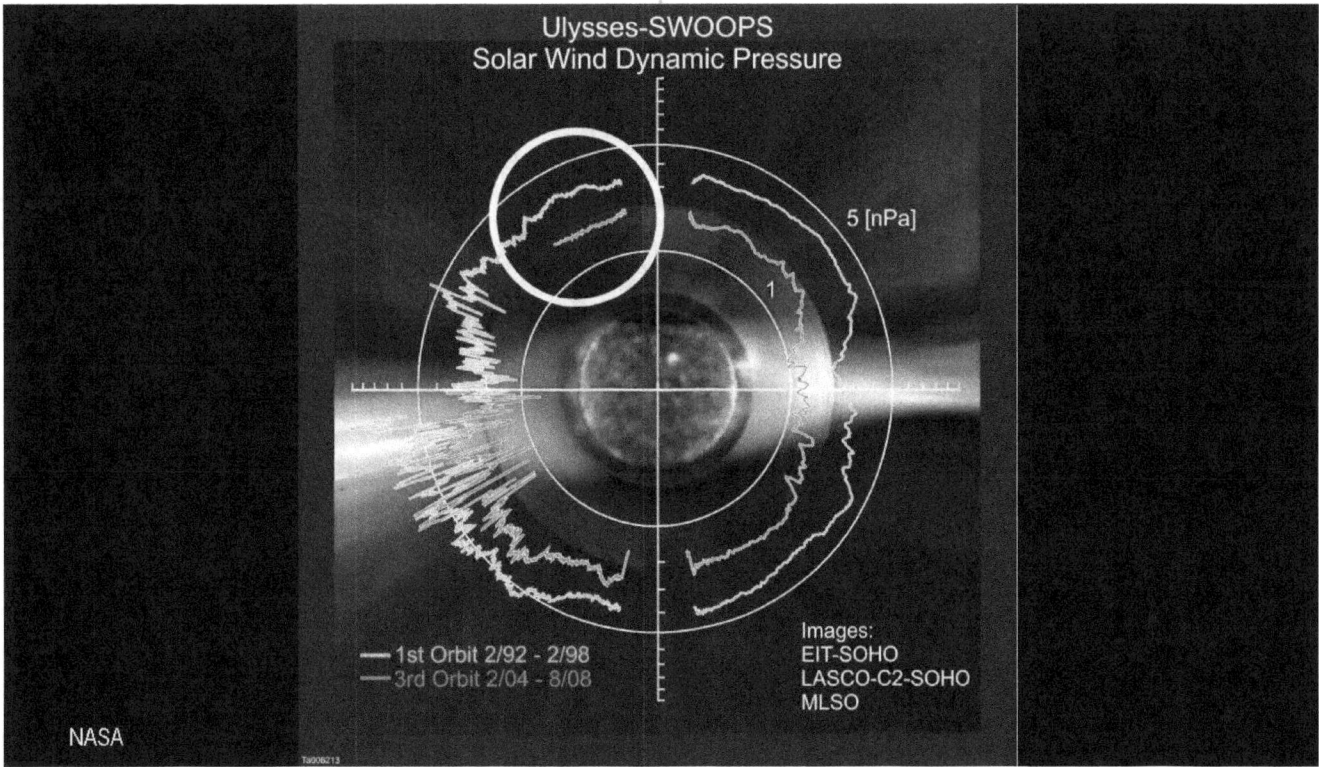

The measured 30% reduction of the solar wind pressure over a span of 10 years, is huge. The entire heliosphere that surrounds the solar system at a distance equal to 100 times the distance of the Earth to the Sun, is getting measurably weaker as the result of the diminished solar wind.

The heliosphere attenuates galactic cosmic ray flux

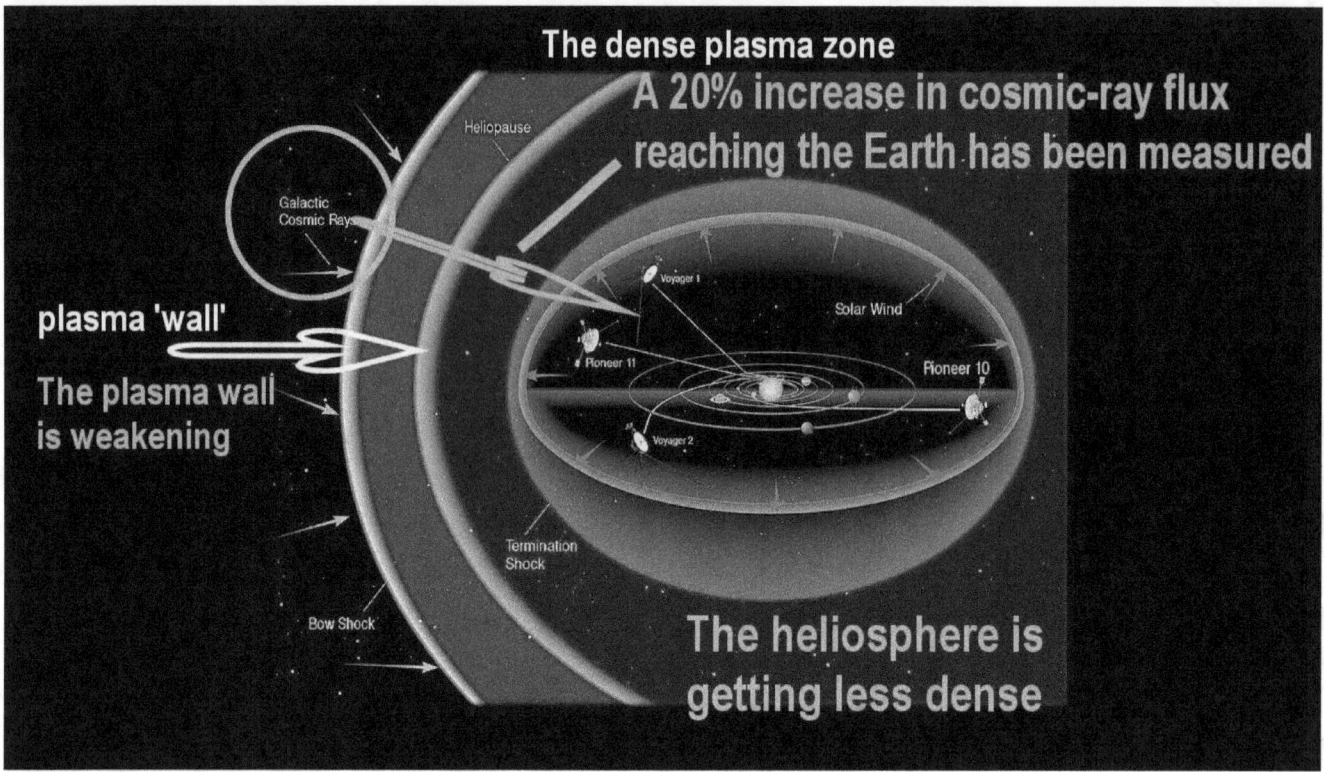

Because the heliosphere that forms around the Sun at this great distance, where the solar wind is coming to a halt, creates a shell of plasma at the wind's termination zone, which shields the solar system against galactic cosmic radiation, we should see an increase of galactic cosmic radiation getting through, as the result of the shielding getting weaker.

The heliosphere attenuates galactic cosmic ray flux. When the heliosphere gets weaker, more of the galactic radiation flux gets through.

A corresponding 20% increase in galactic radiation

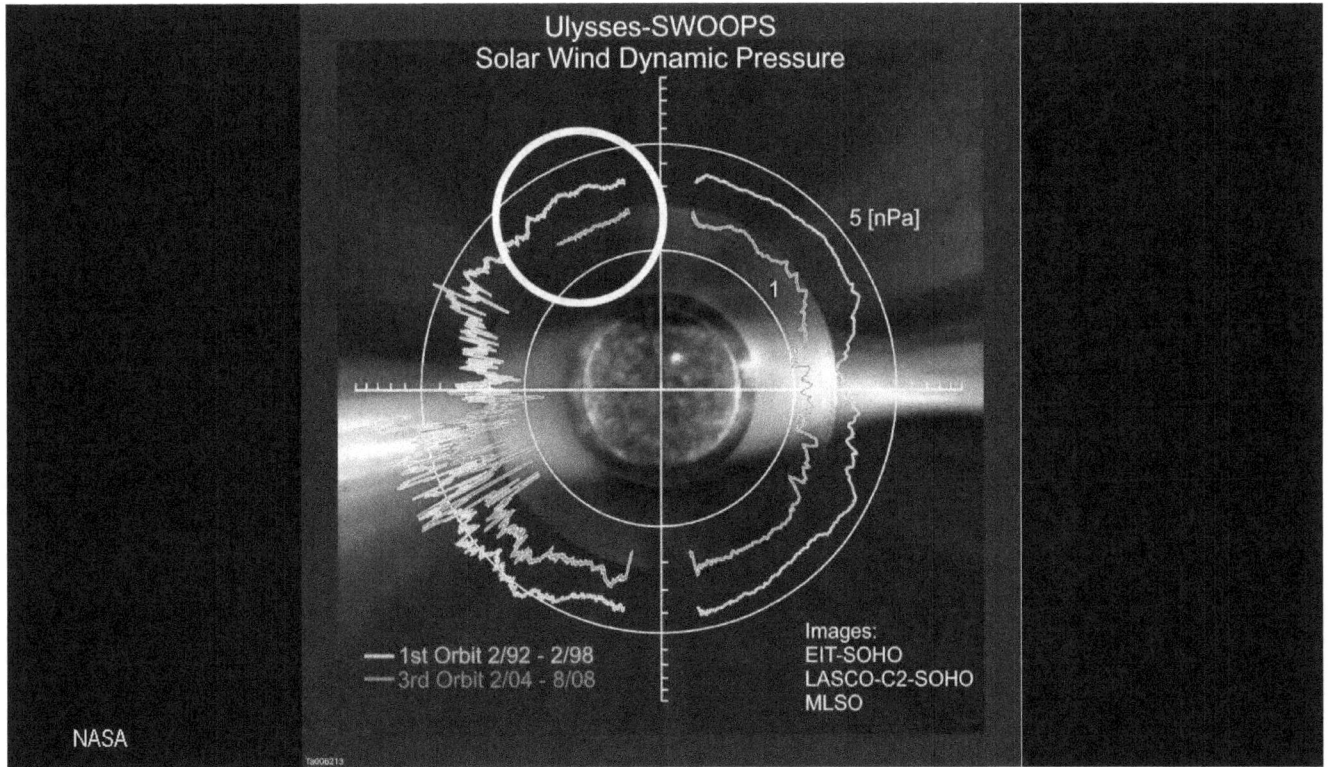

NASA's Ulysses satellite had measured a corresponding 20% increase in galactic radiation. This is precisely what one would expect as the result of the weakening of the solar winds by 30%.

Thus, with the two measured items of evidence supporting one another, the measurements tell us that the electrodynamic weakening of the solar system is real and is fast progressing. But why isn't this electric weakening reflected in a weaker Sun, when the Sun is electrically powered? Does the paradox disprove the electric-Sun theory?

No it doesn't disprove it. It proves it. It adds one more item of proof. The proof lies in what the solar wind represents.

The solar wind a feature of a regulating system

- an example of the amazing solar eclipse photography of Milloslav Druckmueller

In electro-astrophysics, the solar wind comes to light as a feature of a regulating system that maintains the operation of the Sun at a tightly controlled level of intensity.

The KEY to the solar-wind paradox

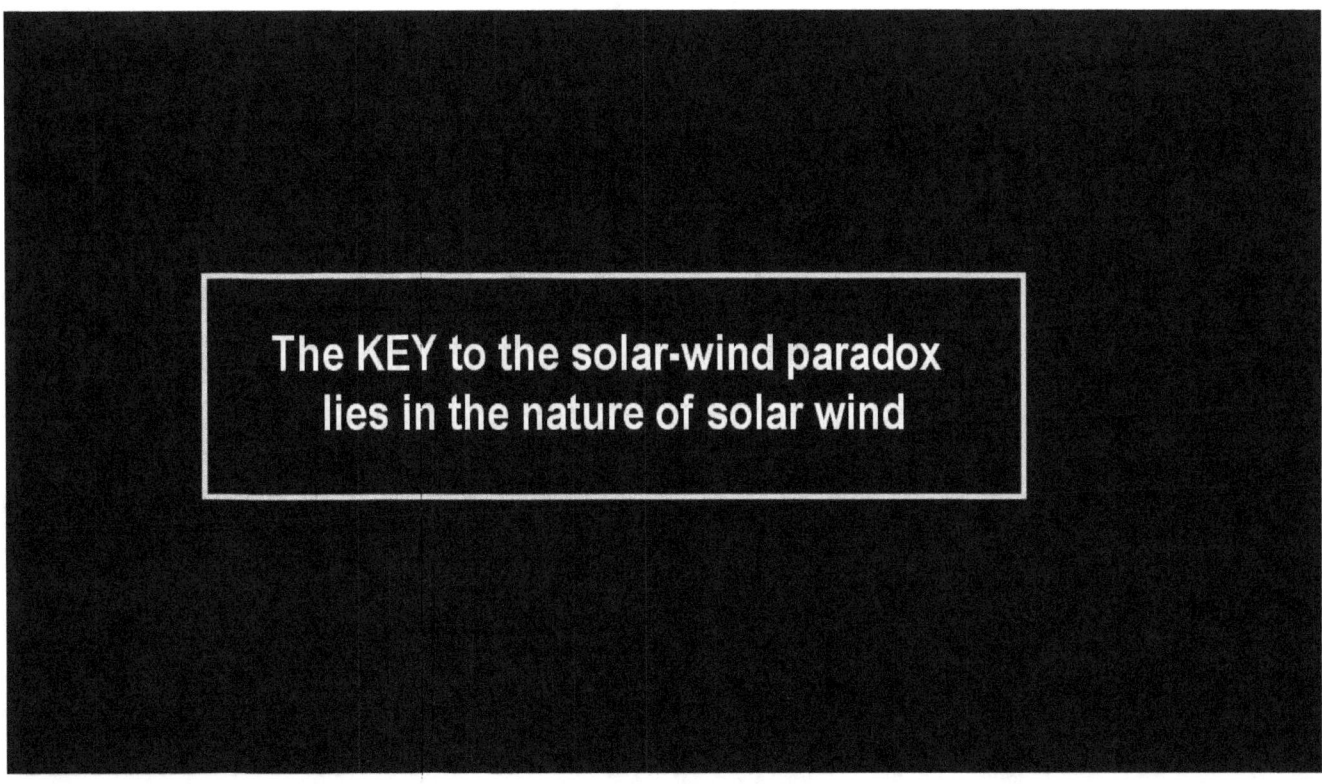

The KEY to the solar-wind paradox lies in the nature of solar wind.

The solar wind may be seen in terms of a tea kettle

The solar wind phenomenon may be seen in terms of a tea kettle. The kettle is placed on a stove. The heat is on high. The kettle is venting off steam as it boils.

It is elementary in chemistry that it is not possible to heat the water in the kettle to a temperature hotter than the boiling point. If one turns up the flame, the water in the kettle doesn't get hotter. The extra energy is devoted to producing more steam.

The solar wind can be likened to steam being boiled off

http://www.zam.fme.vutbr.cz/~druck/Eclipse/ - an example of the amazing solar eclipse photography of Milloslav Druckmueller

In the electrodynamics of the cosmic system that powers our Sun, the solar wind can be likened to steam being boiled off from a kettle. The solar wind is the excess plasma that the regulating system lets go. The regulating feature keeps the Sun at a relatively steady temperature. This also means, at the present time, that when the input plasma density around the Sun diminishes, only the solar winds diminish, while the Sun itself, remains constant in its operation.

The solar wind is the 'steam' that flows from the 'kettle'

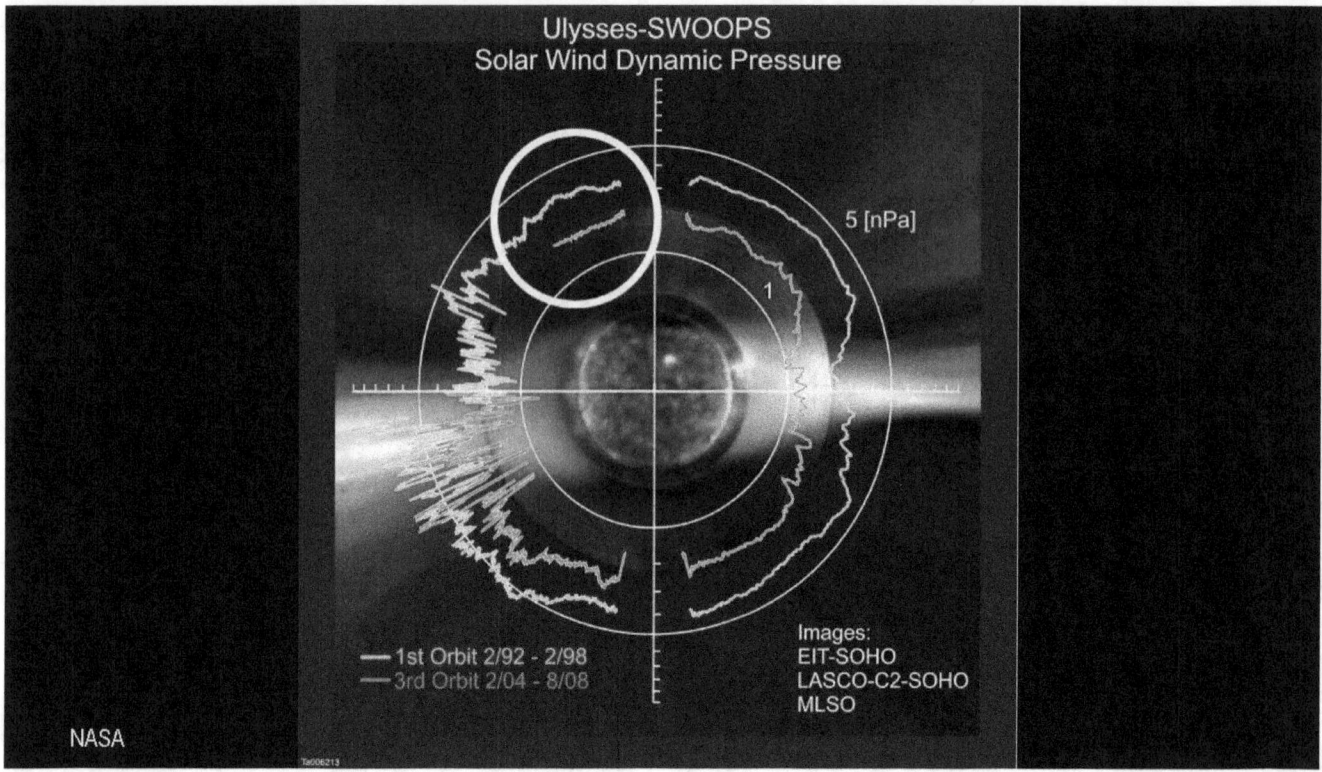

The comparison with a kettle tells us that the solar wind pressure can be used to some degree as a measure of the effective power density in the supply line that is feeding into the the Sun at its reactive surface.

In terms of the comparison, the solar wind is the 'steam' that flows from the 'kettle'

Below the boiling point

Maximum temperature of liquid water at ambient pressure is 100 degrees Celsius: The Boiling Point

If one turns the flame up, the kettle boils faster, and a stronger stream of steam comes out of it. If one turns the flame down, a weaker stream of steam results. By turning the flame down still further, one gets to the point when no steam at all is being produced. Should one reduce the flame even more, the water would become colder in the kettle. Below the boiling point, the temperature of the water diminishes with the energy input. This comparison is significant.

Water can be heated many times hotter under high pressure

However, the comparison isn't as simple. Although the water in the kettle can never exceed 100 degrees at ambient pressure, the water can be heated many times hotter under high pressure. This is utilized in electric-power plants. High pressure steam is needed to drive the electricity generating turbines. To make this possible, water is heated to hundredth of degrees, under great pressure. The same, in principle, also applies to the Sun. When the plasma around the Sun becomes intensely concentrated, the fusion reaction on the surface of the Sun intensifies and creates more fusion products, and more light and heat, especially in the high-energy band, and faster and stronger solar winds. As a consequence the Sun gets hotter, and the climate gets hotter.

The Earth's climate is the coldest in 440 million years

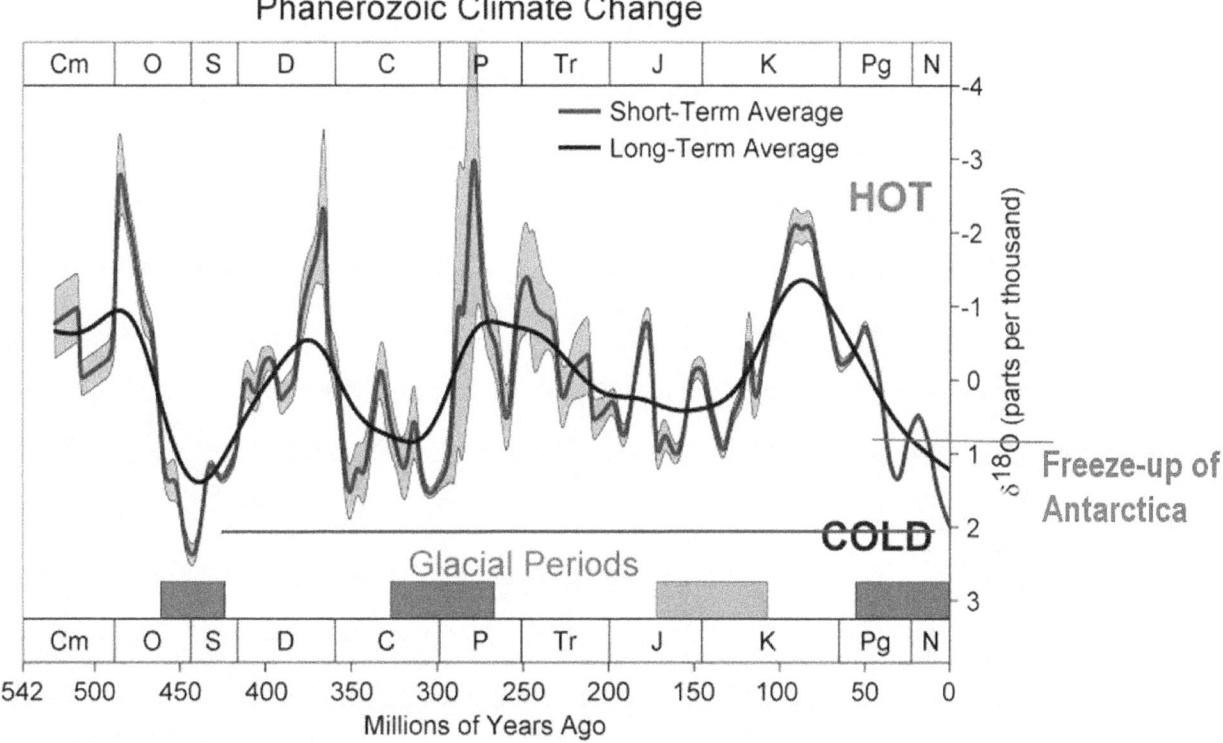

On the big scene, of the Earth's climate history spanning the last 500 million years, the Earth's climate is presently the coldest it has been in 440 million years. During the big spikes 100 and 300 million years ago, the Sun was likely tremendously hotter. Its present temperature of 5505 degrees corresponds with the present plasma density in the solar system, and its current state as the coldest in 440 million years. During the peak hot times, the Sun may have been operating at 9,000 degrees or above that, for which it would need an environment of much greater plasma density. Correspondingly, it would have generated a stronger solar wind, probably exceeding 1,500 Km per second.

The big climate oscillations that we see here result from two long resonance cycles overlaid on each other. The dominant one is a 150-million-year cycle. The lesser one is a 31-million-years cycle. These two long cycles can be seen as resonance cycles in the plasma streams that flow in and out of our galaxy. Their combined effect determines the over-all plasma density in the galaxy. When both cycles add up, we have an extremely high plasma density everywhere, including in our solar system. Inversely, when both cycles come together at their low point, the galaxy experiences the lowest plasma density. We are presently close to that. We are far below the freeze-up point for Antarctica, and way below the high point of a 100 million years ago. This means that we are in a marginal environment for the Sun, where the slightest reduction in plasma pressure can cause the Sun to go inactive.

The threshold line becomes important

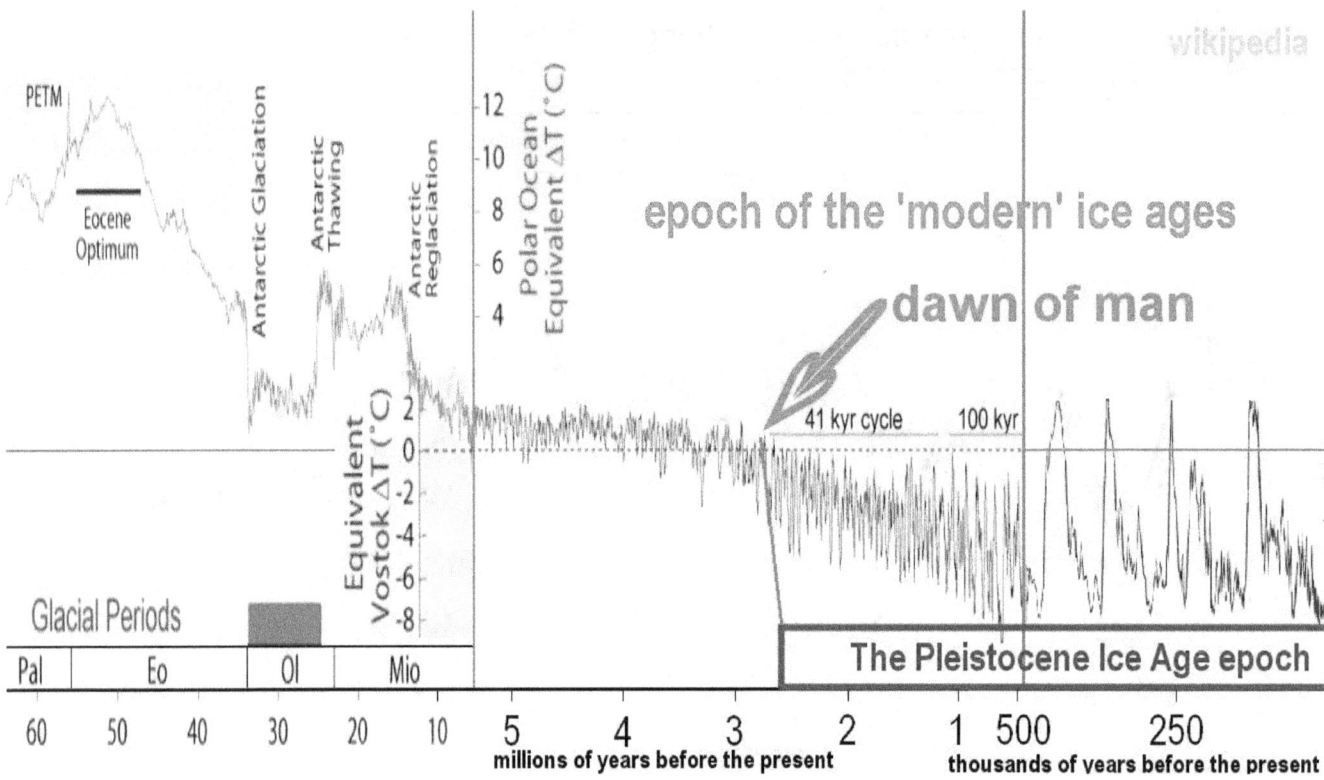

That's where the threshold line becomes important. If we drop below this line the Sun becomes inactive periodically, and the Ice Age begins. At first, the cycles of solar inactivity were shorter, with a 41,000-year beat. Then a million years ago, a different resonance became dominant that gave us the longer Ice Age cycles spanning roughly 100,000 to 120,000 years each. During the entire Ice Age Epoch so far, of the last two million years, the Sun has been inactive for roughly 85% of the time. And this may be the reason why we exist as human beings.

When the Sun is inactive

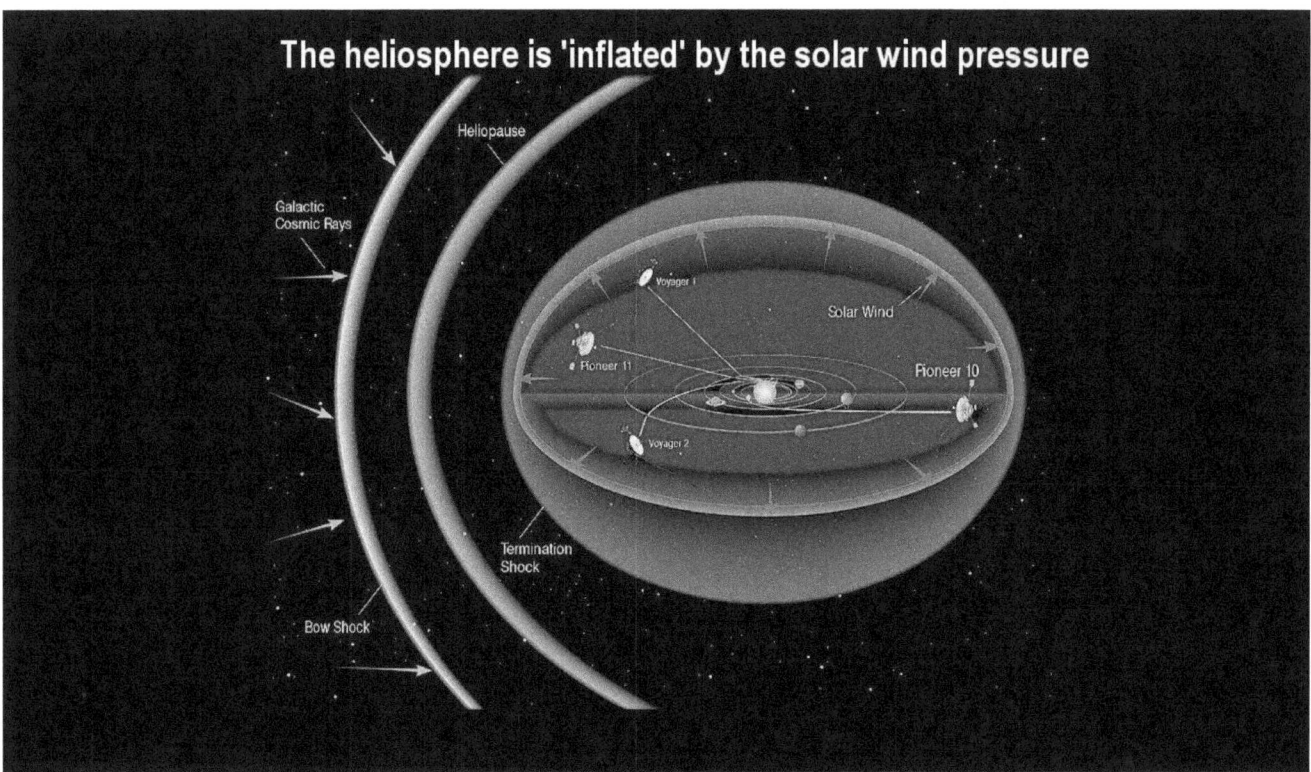

When the Sun is inactive, the solar heliosphere that is created by the solar wind and surrounds the solar system like a shield against incoming galactic cosmic ray flux, no longer exists. This means that for the last 2 million years, the Earth was exposed to the full dose of galactic rays for 85% of the time.

Cosmic rays are fast moving protons or electrons

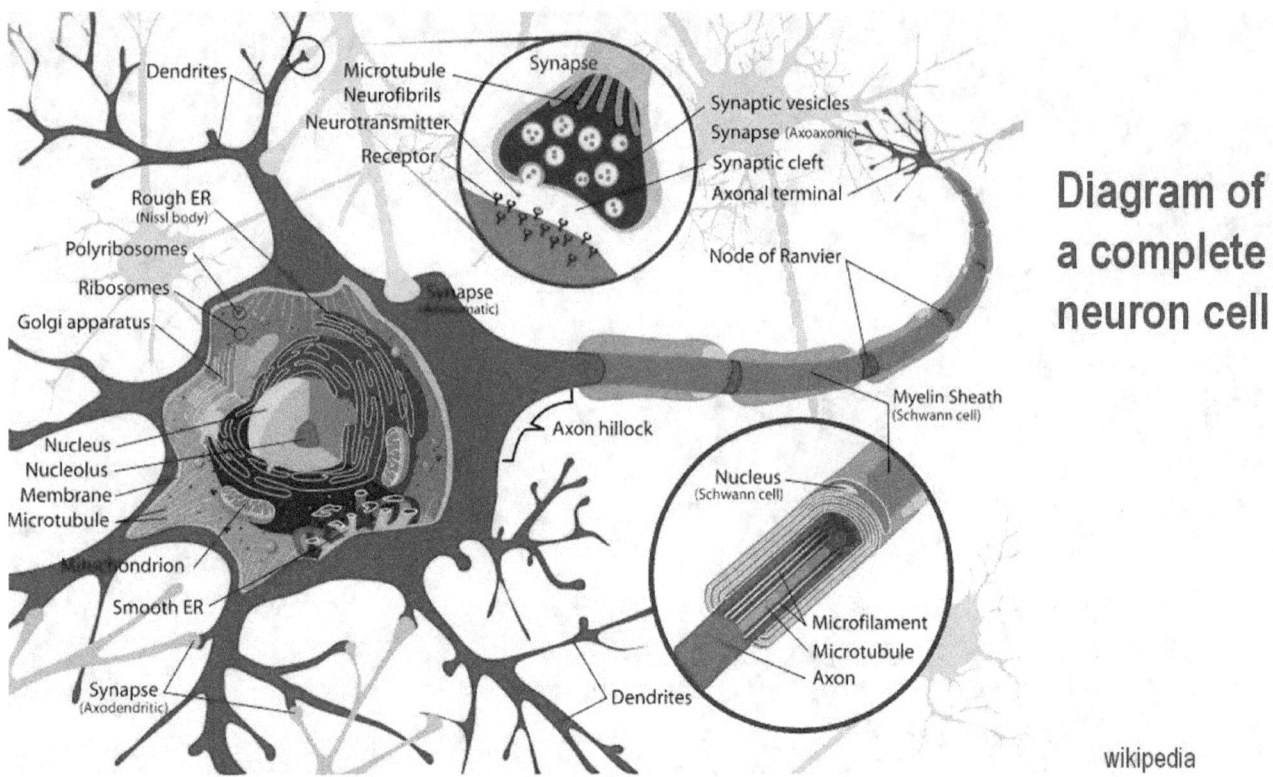

Diagram of a complete neuron cell

wikipedia

Cosmic rays are fast moving protons or electrons that pass right through the human body. While they don't collide with anything, which their electric charge prevents, they create a magnetic field as they flow, which induces a secondary electric current in the human tissue. This electric current appears to be highly beneficial for the neurological development, since the neurological system operates electrically.

The dawn of humanity didn't begin until the Ice Ages began

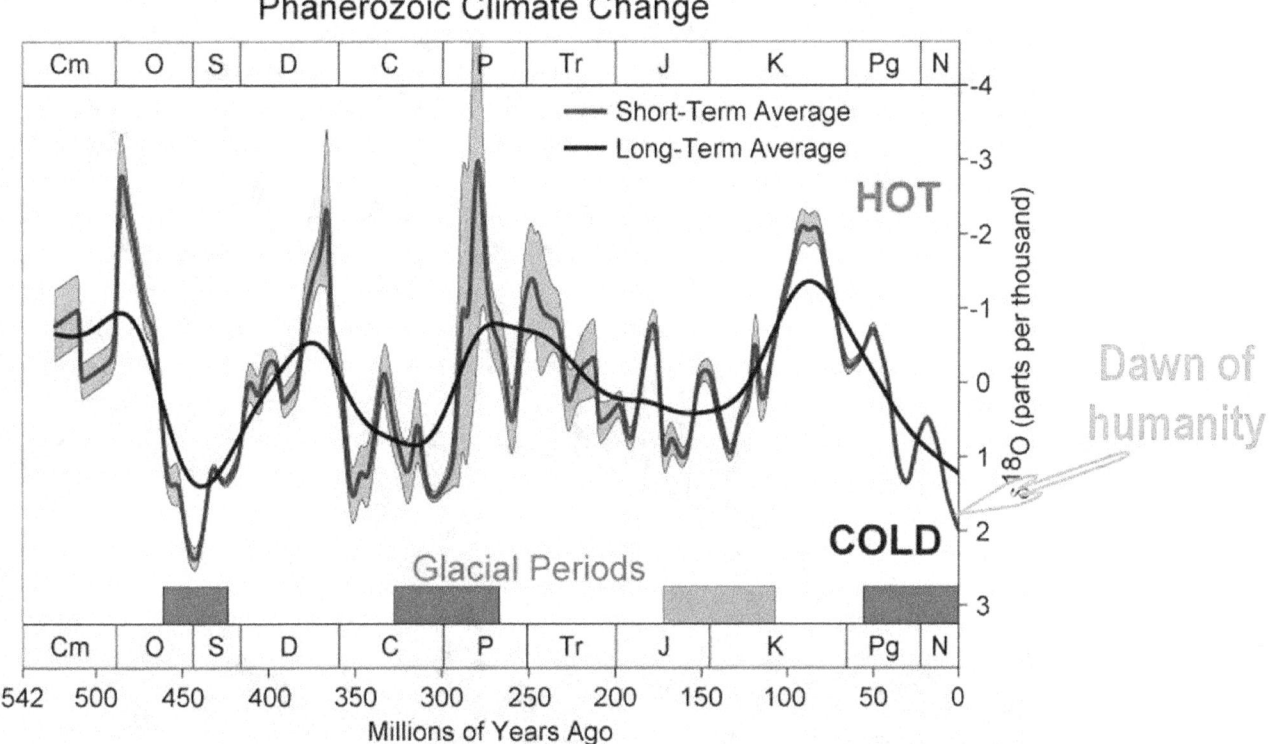

The fact that the dawn of humanity didn't begin until the Ice Ages began, seems to suggest that the development of humanity had not been possible prior to the times in which the Sun could go inactive for long periods, which had enabled strong galactic, cosmic-ray flux to reach the Earth.

This means that our very existence might be proof that we live in fragile times in which the Sun can, and will, go inactive for long periods. The present plasma density in the solar system appears to be the rock-bottom minimal density, with the solar wind fast becoming weaker towards its cut-off point.

The story of the human development

And still, the story of the human development isn't as simple as that. It may be directly linked with the forming of stars. The artist rendering of Cygnus X-1, contrary to prevailing theories, illustrates the case of immensely concentrated plasma, focused by primer fields onto a hyperactive sun. It appears that the hyperactive sun is spinning off what is said to be a companion sun via a continuous stream of dense plasma that may be perceived as a type of interstellar lightning.

The Cygnus X-1 system is roughly 6000 light years distant. Its hyperactive sun has been 'measured' to have a surface-fusion temperature of 31,000 degrees. It is said to be a large sun, with a 20 to 40 times greater mass than our Sun. The entire system has been measured to be 300,000 to 400,000 times more luminous than our Sun. The resulting, immense fusion-intensity makes the Cygnus X-1 system the strongest x-ray emitter in the galaxy.

A black hole is theorized to be at the center of this system, as gravitational anchor for the orbiting star. Since black holes cannot exist in the real world, what is deemed to be a high-gravity black hole, is simply a large mass of magnetically concentrated plasma furnished by its Primer Fields. Plasma doesn't emit light, only its interaction with atomic matter can be visually detected, which creates electromagnetic radiation. The particles of plasma, themselves, are too small to be visible. What is theorized in astronomy to be "black matter," detected by its gravitational effects, is in reality nothing more than concentrations of plasma that does have a substantial, invisible, mass and gravity.

The interaction of concentrated plasma in the Cygnus X-1 system is so intense, that the X-1 system emits strongly in the x-ray band. It would be surprising if it didn't. It is in this context where we find another possible connection of astrophysics with human development.

Another star-system named Cygnus X-3

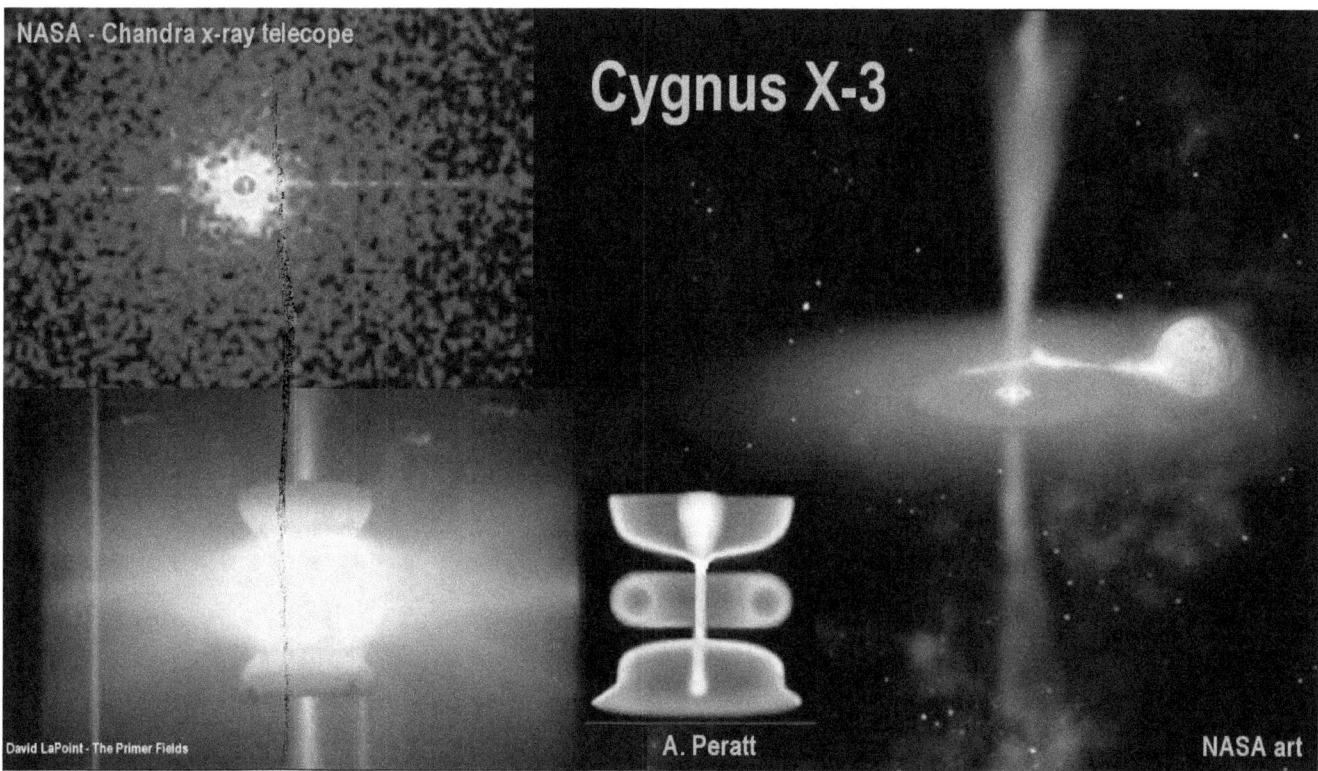

There exists another star-system of a similar type, 30,000 light years distant, named Cygnus X-3. This system features large jets flowing from its confinement domes. This system appears to be even more-intensely active than the X-1 system, which the strong jets indicate. The interstellar lightning that extends to the still developing companion star appears to be of such a great intensity that a shower of a high-energy cosmic-ray particles is being created, of a type that has not been encountered before. The unknown type of ration has been detected to occur at intervals of 4.79 hours, which could reflect the orbital period of this binary star system in a manner in which the interstellar lighting stream is pointing towards us.

It has been discovered in October 1985, in underground experiments as a part of the SOUDAN experiment series, that the X3 star-system causes anomalous muon events that can penetrate 2000 feet of rock.

The Cygnus X-3 pulses

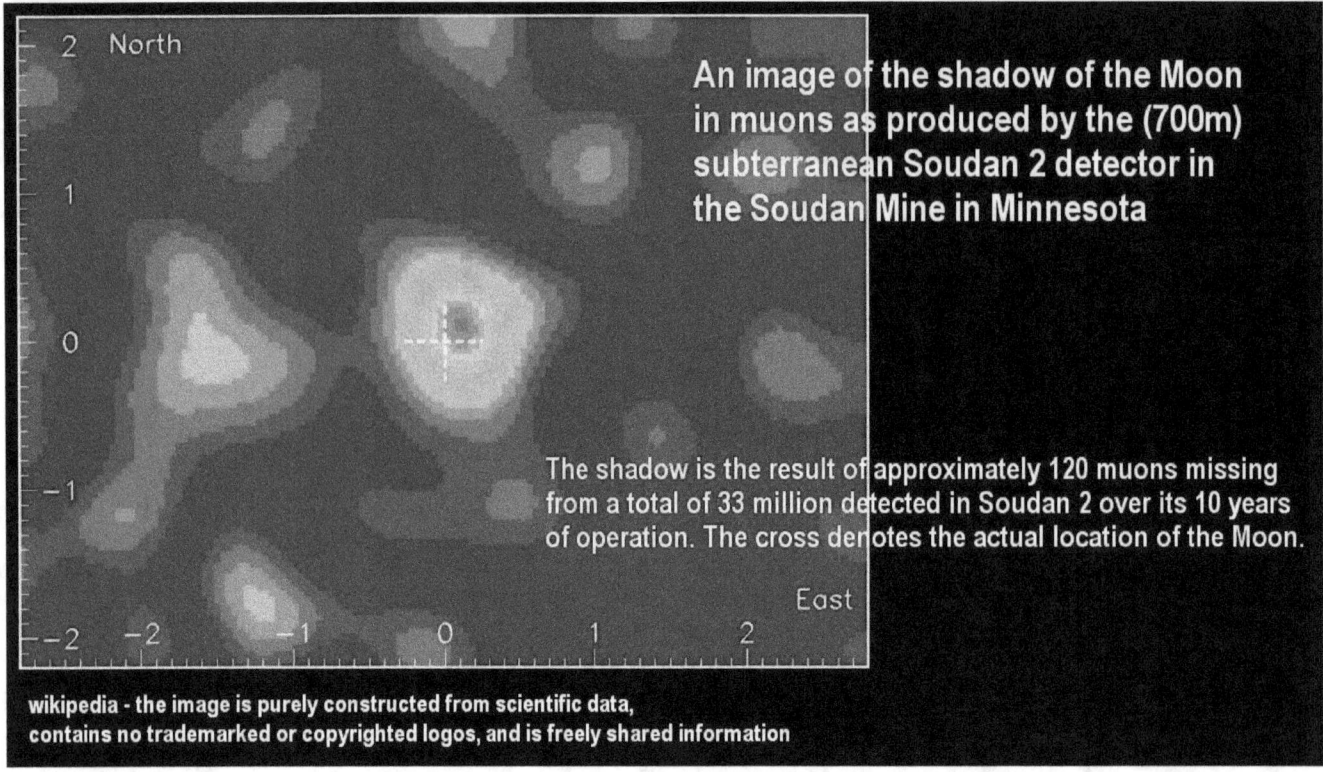

An image of the shadow of the Moon in muons as produced by the (700m) subterranean Soudan 2 detector in the Soudan Mine in Minnesota

The shadow is the result of approximately 120 muons missing from a total of 33 million detected in Soudan 2 over its 10 years of operation. The cross denotes the actual location of the Moon.

wikipedia - the image is purely constructed from scientific data, contains no trademarked or copyrighted logos, and is freely shared information

The muon is a short-lived unstable subatomic particle with a mean lifetime of only 2.2 μs, It typically results from high-energy cosmic-ray interaction with atomic matter, typically in the atmosphere. The muon is a type of electron with a 200 times greater mass than the electron, which due to is greater mass. This allows muons to penetrate more deeply into matter with lesser deceleration due to energy loss, than electrons. They were detected 2000 feet below the ground in the Soudan deep mine laboratory.

The interesting part is that the Cygnus X-3 pulses detected in this deep-mine facility, were not detected as normal cosmic-ray pulses. Their origin, thus remains unknown, a puzzle that is yet to be solved.

The Cygnus X-3 source became active around 700,000 years ago

The interesting part is, that it is believed that the Cygnus X-3 source became active around 700,000 years ago. This means that it may have had a decisive impact on human development.

The 700,000-year timeframe

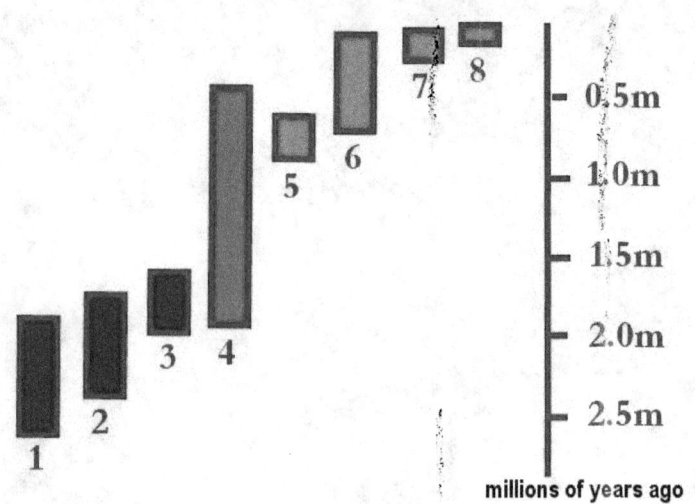

Main human species

australopithecus rudolfensis (1),
australopithecus habilus (2),
homo ergaster (3),
homo erectus (4),
homo antecessor (5),
homo heidelbergensis (6),
homo neandertalensis (7),
homo sapiens (8)

We, the homo sapiens (8), are the only surviving, and the shortest lived of all the the human species, at barely 200,000 years of age.

The 700,000-year timeframe coincides roughly with the qualitative dividing line between the early sequence of human development, and the start of the modern sequence beginning with homo heidelbergensis, number 6, that took us past homo erectus towards modern man, who we became as a species of immense intellectual capacity that is precious beyond compare on the carpet of life, that is worthy of the greatest care and protection.

It is reasonable therefore, to speculate that the highly capable species that we have become, second to none, had been made possible in part by a series of unique astrophysical events in the electric universe that we are just beginning to understand, which take astrophysical exploration ever-further outside the primitive concepts in cosmology where only gravity is deemed to rule.

The phase shift in human development

Of course we will never know what type of gigantic event has created the Cygnus-x3 system 700,000 years ago. Nevertheless, the anomaly that came out of it, that remains a puzzle to the present day, suggests that something extraordinary may have happened back that still affects us in our time.

The phase shift in human development with Cygnus-x3 becoming active, if there was such a coincidence, suggests strongly that human development may be built on vastly more than primitive evolution.

The heliosphere is already becoming noticeably weaker

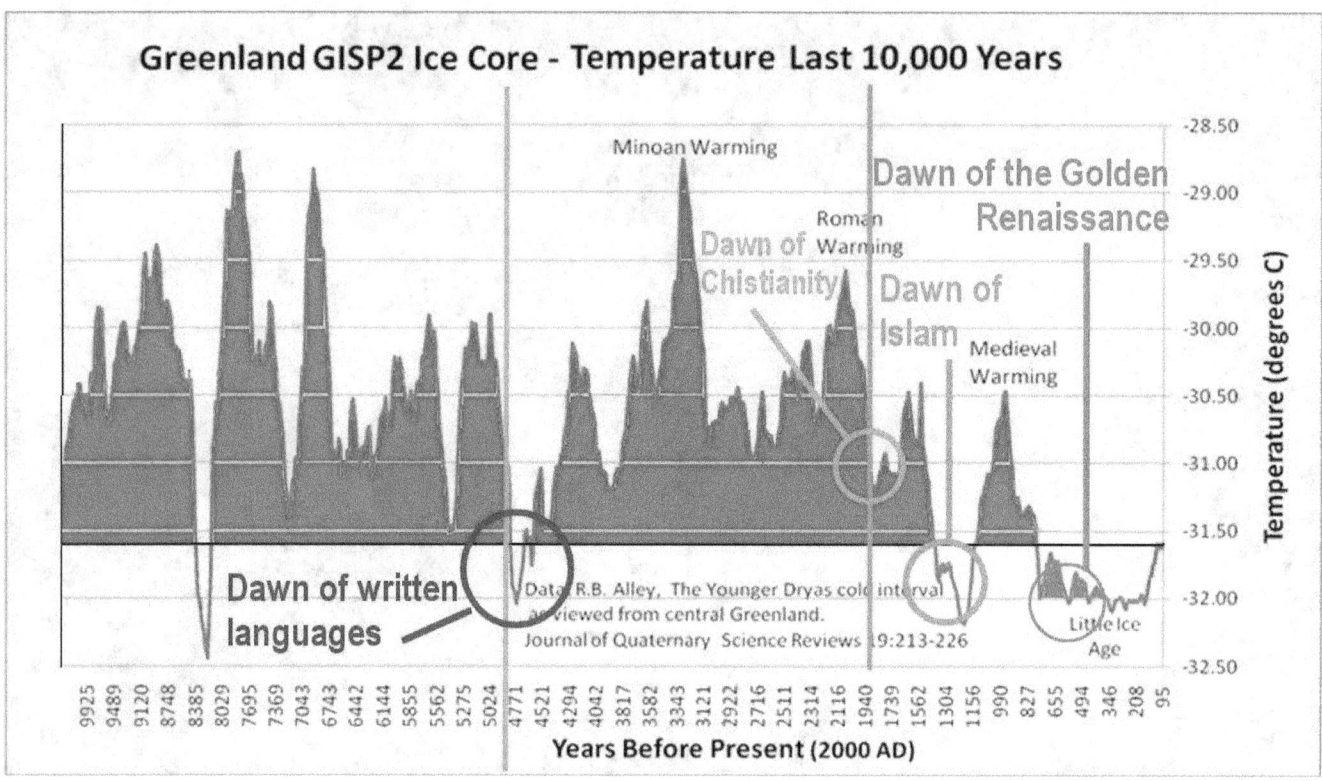

It is extremely likely that humanity may find itself snapping out of its small-minded mode of thinking that has been cultivated in modern tine. Historically, the greatest intellectual and scientific breakthroughs occurred during the cold times that would have been times of weak solar wind and greater cosmic-ray flux reaching the earth.

Since the heliosphere is already becoming noticeably weaker, because of the weaker solar wind pressure, the current increase in cosmic-ray interaction will therefore become stronger over the next 30 years as the result of the dynamic electric weakening in the solar system. Hopefully, this will aid us to end the madness of preparing for nuclear war, and idealizing human depopulation by starvation as the current policy.

The current electric weakening in the solar system

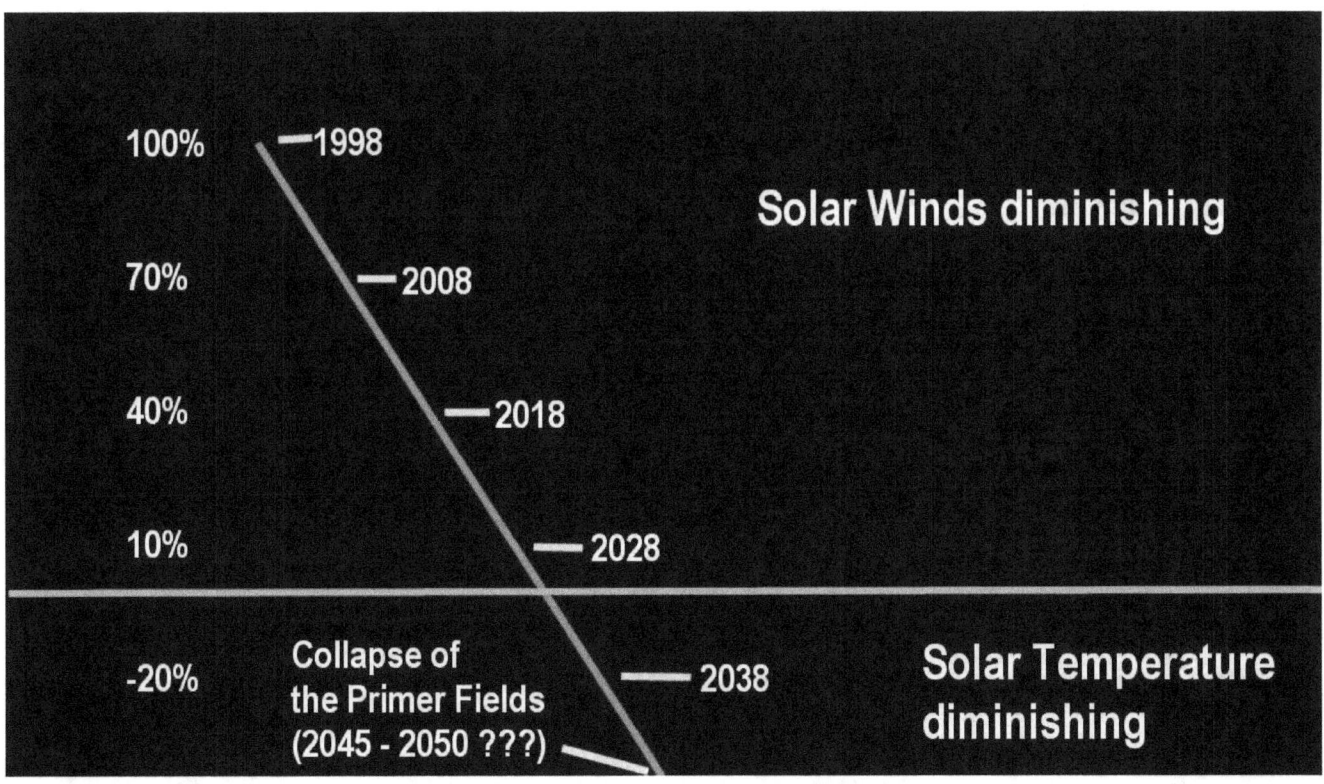

The current electric weakening in the solar system, and its apparently beneficial effect, will most likely continue until the Sun's Primer Fields collapse completely and the heliosphere vanishes, at which point we will receive the full dose of cosmic-ray intrusion with unknown effects. This may happen in the 2050s. The process has already begun. The solar 'kettle' is generating less and less 'steam' as it were.

Ulysses has measured 30% less 'steam' coming from the Sun

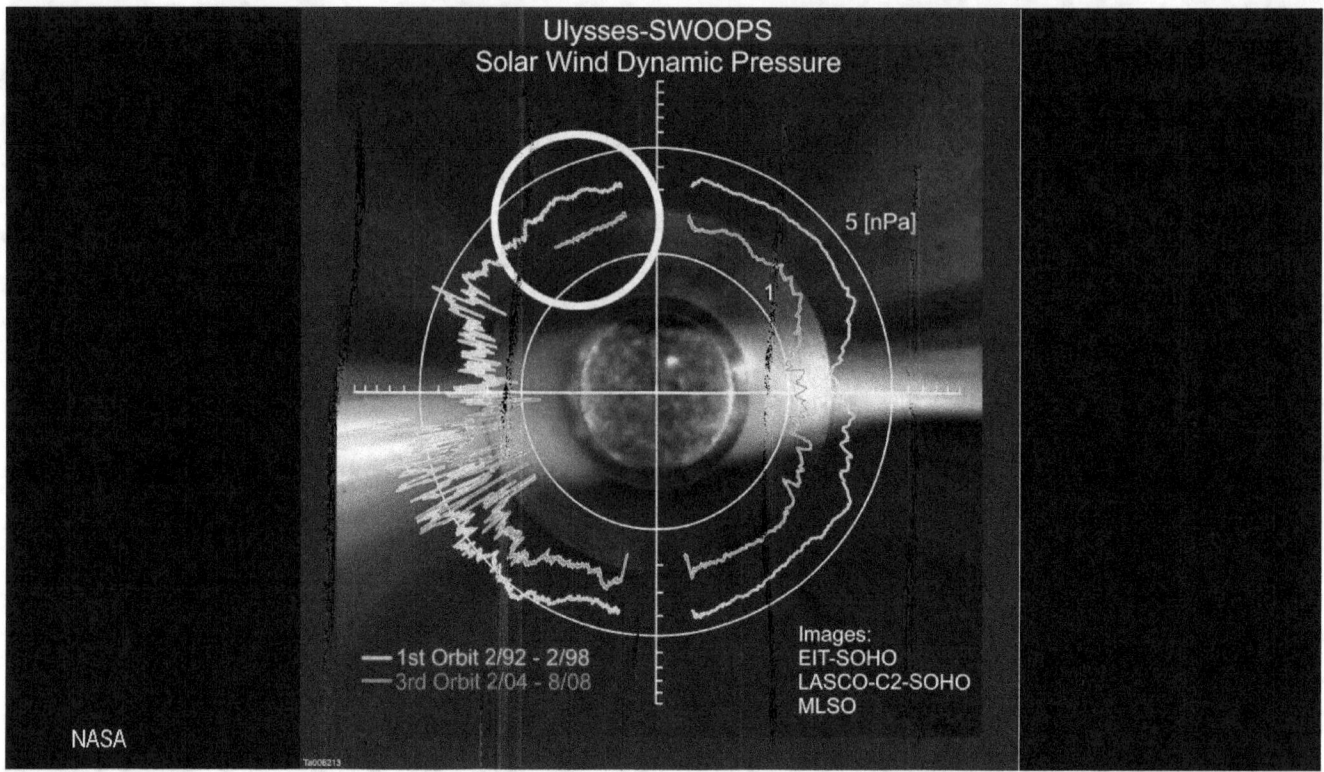

Ulysses has measured 30% less 'steam' coming from the Sun over the span of 10 years. Now, 5 years further down this path, the solar wind pressure is likely reduced by 45% from what it had been in 1992 when the measurements started.

We might not have much time left

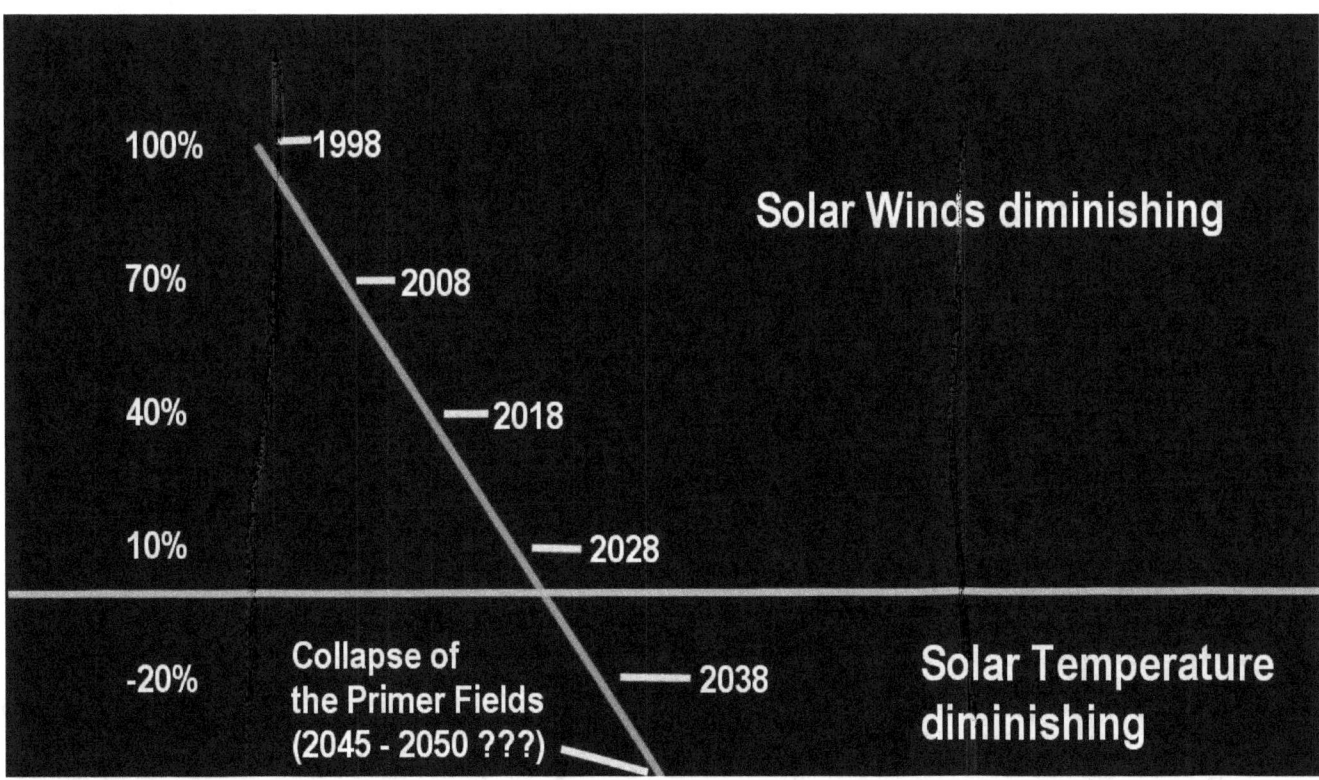

Fifteen years from now, in the 2030 timeframe, the 'steam' flow will likely have diminished by a total of 90%. This means that we will have only 10% left then. And still, even at this highly diminished stage, we won't likely see the slightest change in the Sun's energy-radiation intensity. Only when the input energy drops below the point where solar winds are no longer ejected, will the Sun itself begin to diminish. This 'fading' would happen progressively from the 2030 timeframe onward. The Sun would still function at this point, but with a diminished intensity. When we get to this point, the weakening of the input streams, would of course continue. This means that we might not have much time left at this point, to the Sun becoming inactive completely. When the supply factor gets interrupted, the entire plasma powered system begins to collapse.

The magnetic dynamic-flow geometry of the Primer Fields

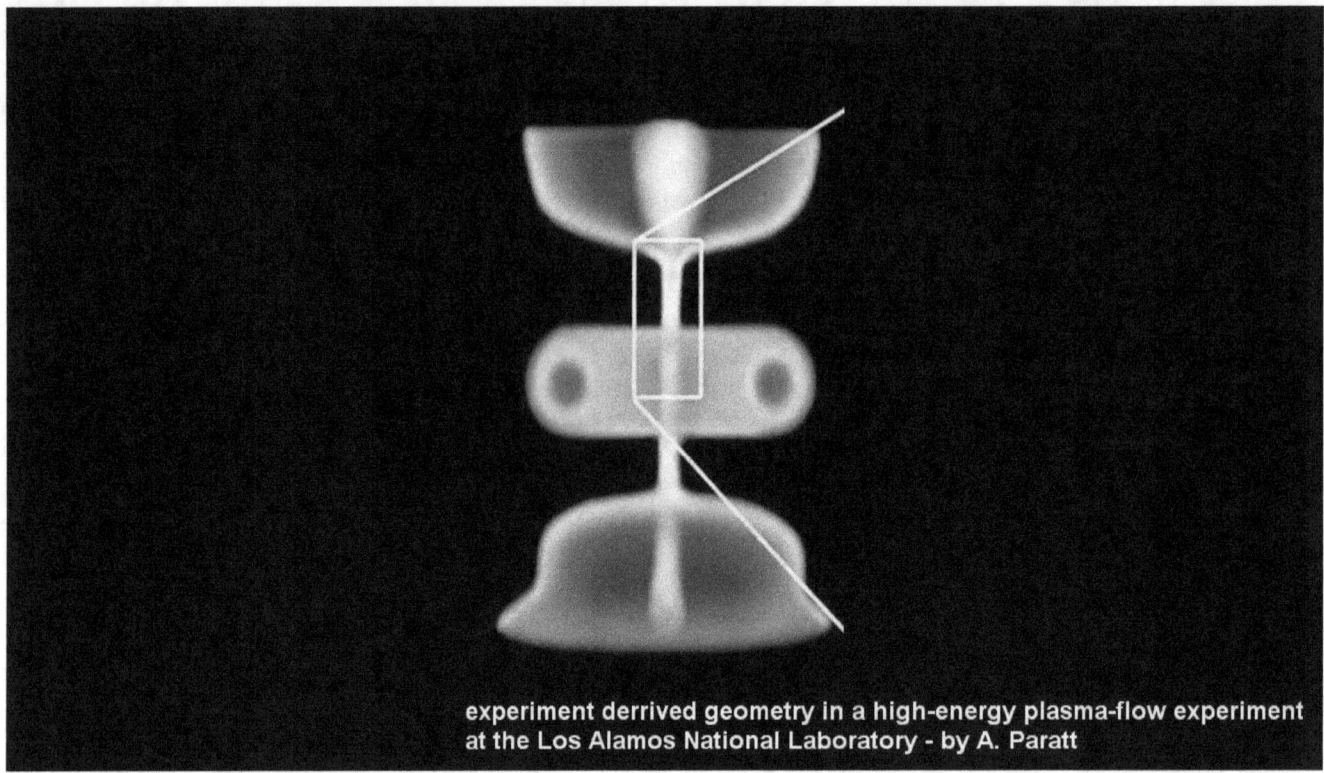

experiment derrived geometry in a high-energy plasma-flow experiment at the Los Alamos National Laboratory - by A. Paratt

The magnetic dynamic-flow geometry of the Primer Fields is a function of the intensity of the flow. When the flow is diminishing, as it already is, the resulting geometry reshapes itself accordingly. When the reshaping affects the dynamic functioning of the system, a threshold point will be reached when the entire interlocked structure can no longer be maintained and vanishes as if it had never existed. When this happens to the supply system of the Sun, which powers its nuclear-fusion process, the Sun will simply go inactive.

Part 6: The diminishing solar wind towards solar collapse

The inactive Sun will in time become a dim star

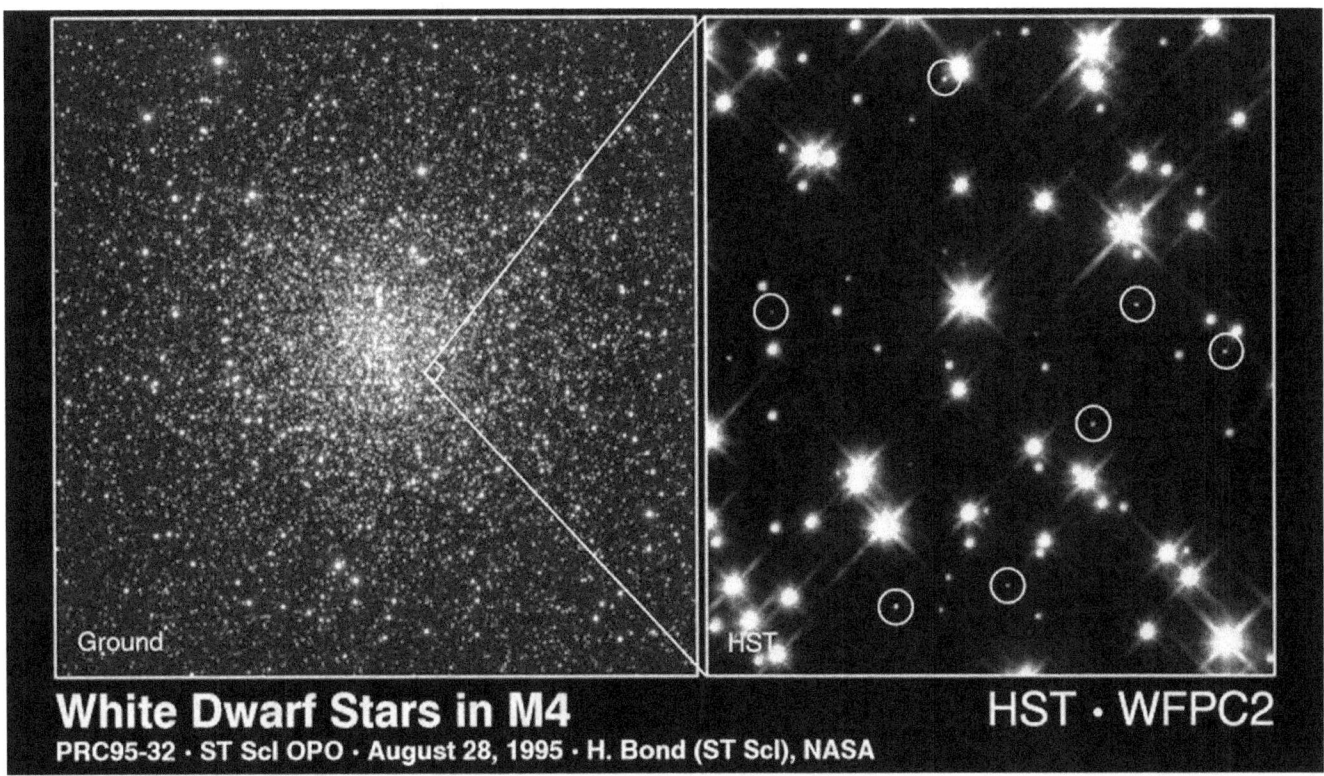

The inactive Sun will in time become a dim star that initially glows by its residual energy and some remaining plasma interaction with its corona, as a red dwarf. The inactive Sun may also become in time a tiny white dwarf, if it comes to that, which then glows only by nuclear decay within it.

The energy output of the inactive Sun

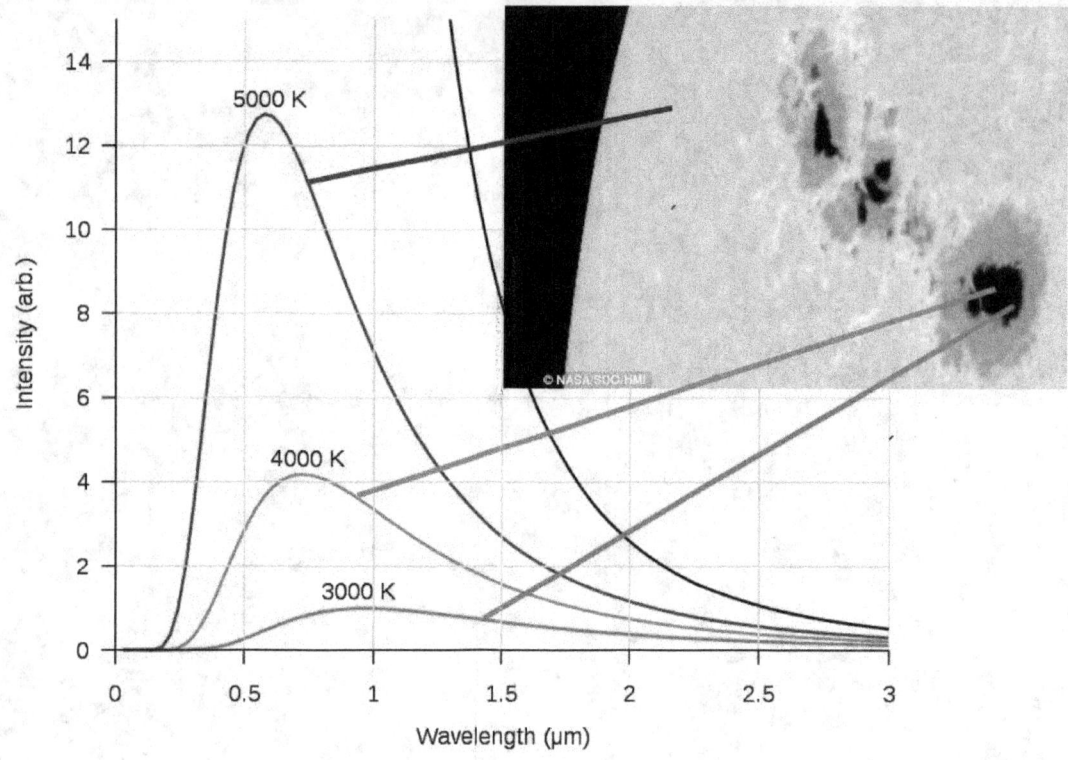

The energy output of the inactive Sun would likely reflect what we presently are able to measure in the umbra of the sunspots. The measurements the we make there tell us that the residual energy level of the Sun will likely be roughly 30% of its active level that we presently enjoy. At the inactive energy level, agriculture will still remain possible in the equatorial region.

The world's agriculture will have to be relocated

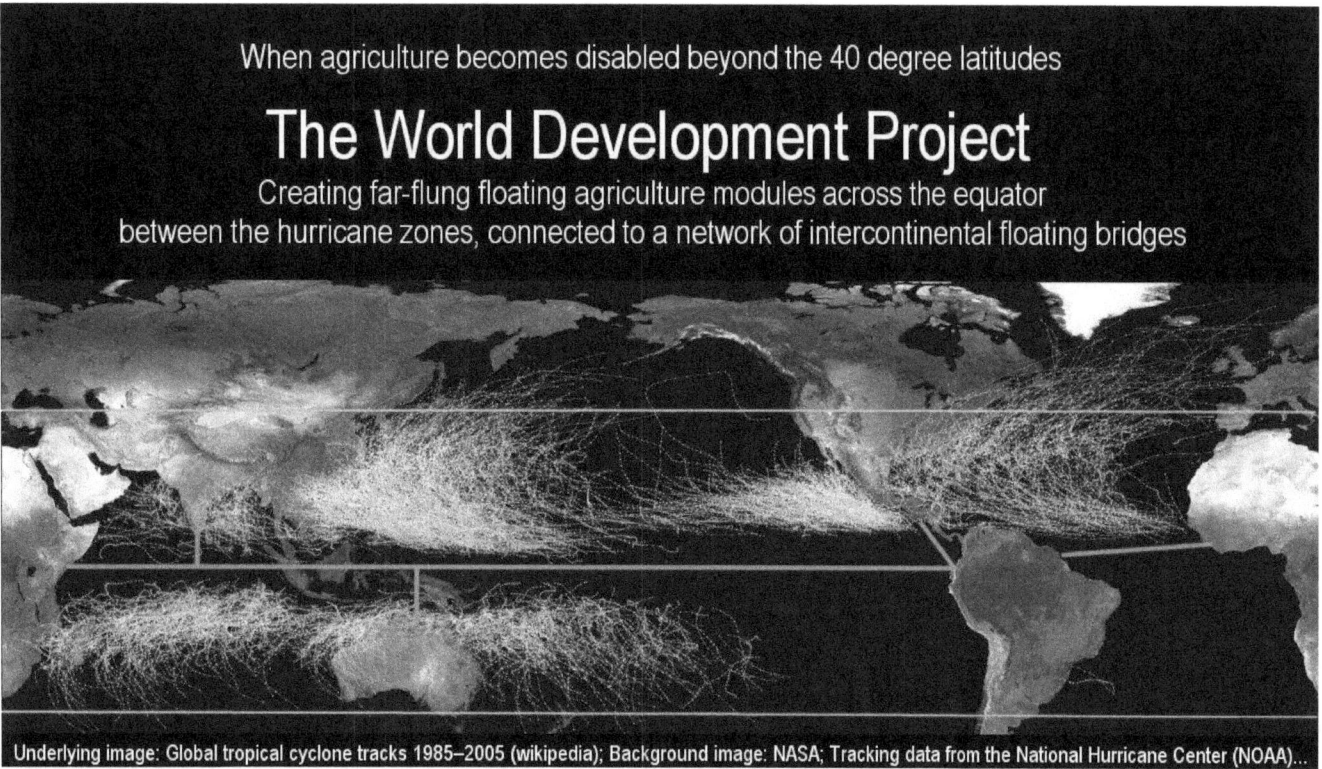

This means that most of the world's agriculture will have to be relocated to there. Since little land exists in the tropics, agriculture will have to be put afloat across the equatorial sea, onto infrastructures that do not yet exist. Two sea-bridges will have to be built, roughly 500 kilometers wide, along the equator, between the continents, serving as a new place for the agriculture that can no longer be maintained in the regions that become uninhabitable by the cold. These extensive infrastructures will have to be built to maintain the existence of humanity.

The building is not optional. The materials exist for it, and so do the technologies and the energy resources. But will humanity create what is necessary for it to survive, and in time? This becomes a spiritual question, which remains yet to be answered.

The observed rate of diminishment

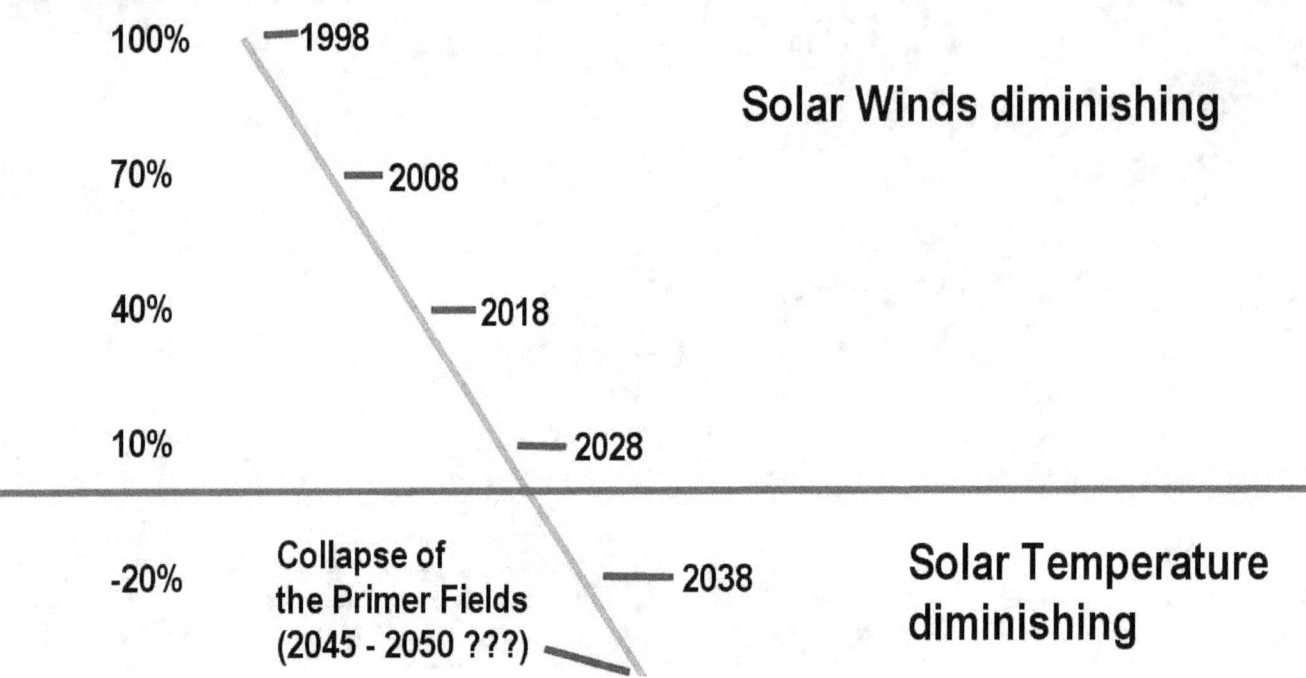

The principles that have been discovered, tell us that the potential exist for the electromagnetic fields to collapse that prime the Sun with its highly concentrated sphere of plasma that surrounds it. The observed rate of diminishment, such as measured by the Ulysses satellite, projected forward in time in a linear manner, suggests that the Sun may go inactive in the 2050 timeframe. With the cause for this collapse being located in the diminishing plasma streams in interstellar space, far outside the solar system itself, the human capability is far too feeble to master the large cosmic events that operate in interstellar space.

We are perfectly able to know with great certainty

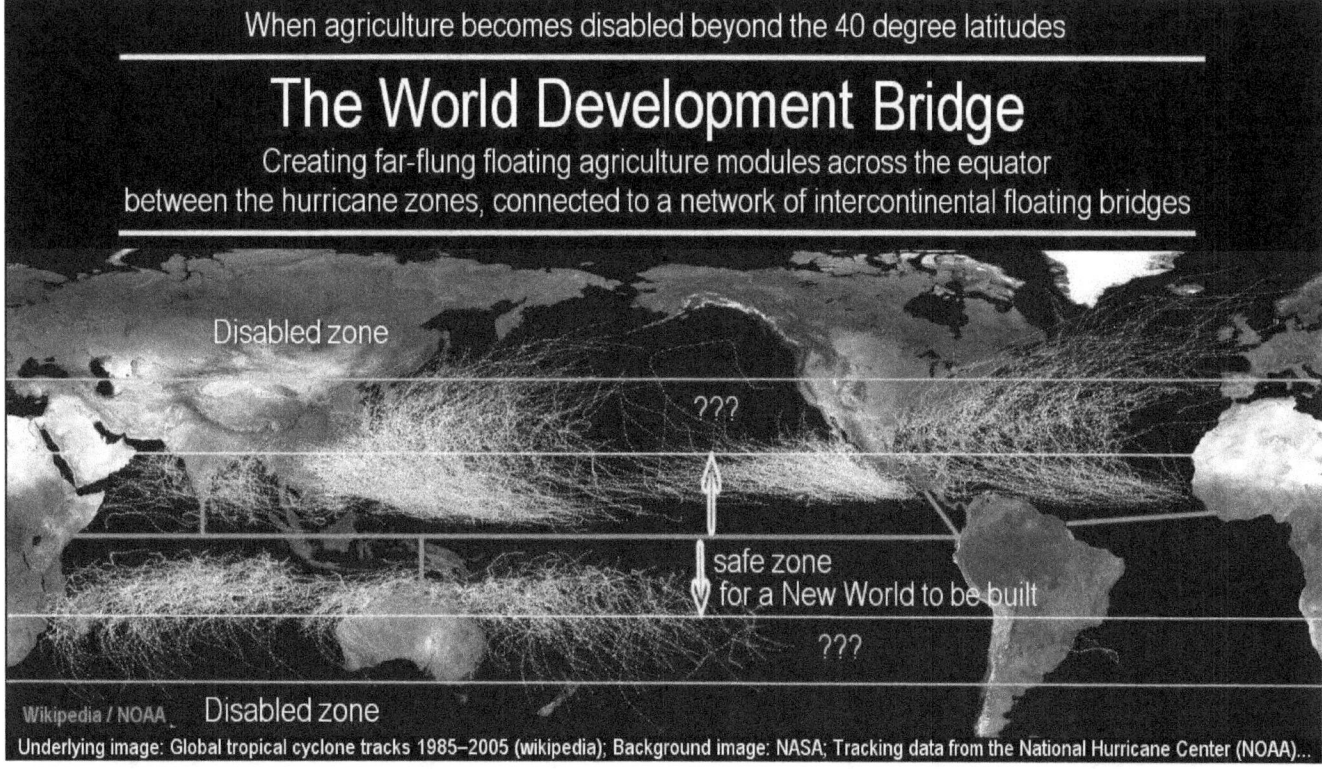

The big question is whether we can muster the capability to master ourselves, and thus to rouse ourselves to upgrade our world for the requirements of the future that we are able to forecast by the dynamics of the causative principles, instead of events.

We are perfectly able to know with great certainty that we will kill our children and ourselves likewise, if the infrastructures will not be created to place most of the world's agriculture into the tropics before the next Ice Age begins, which may happen in 30 years time. Most of the world's agriculture lies presently in the zone that becomes uninhabitable when the Ice Age begins, or at best, becomes useless for agriculture. Without the continued operation of agriculture, where most of the world's food originates, humanity will simply perish by starvation. This needs to be avoided.

Harsh conditions under an inactive Sun

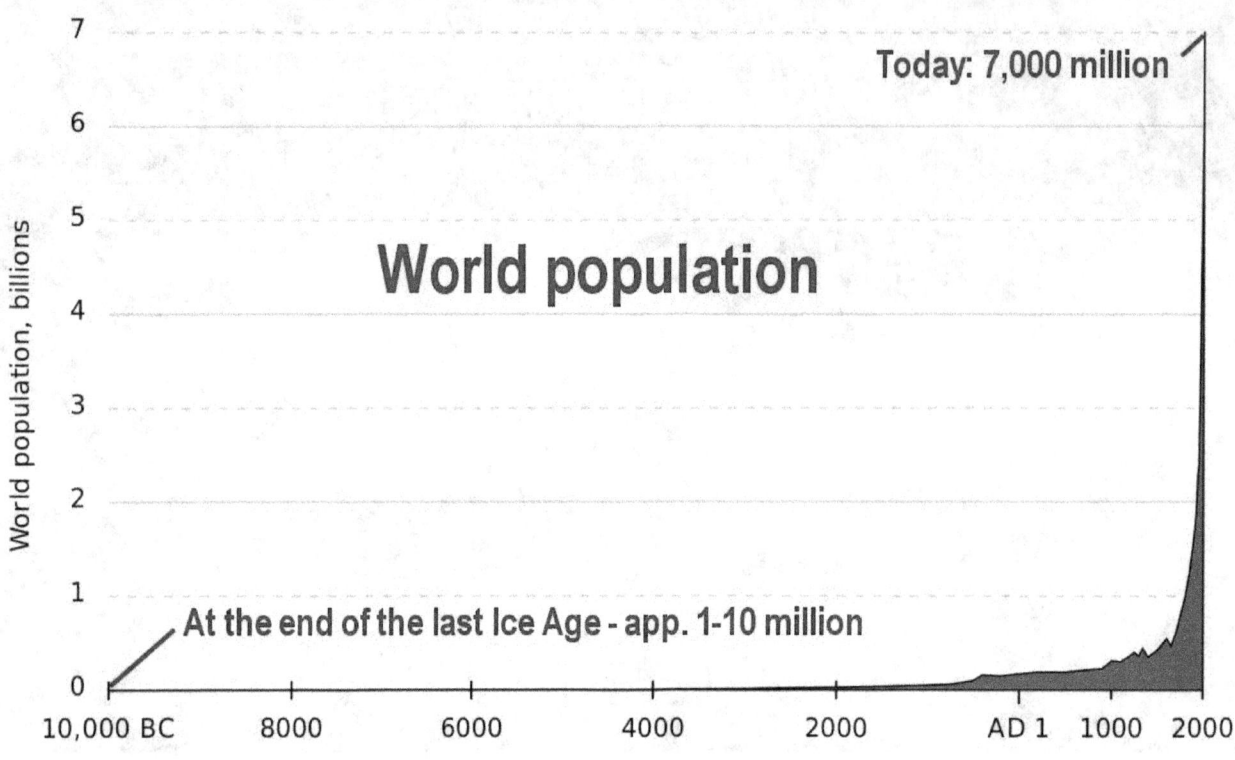

The harsh conditions in the world under an inactive Sun, may have been the reason why only 1 to 10 million people had survived through the last Ice Age, worldwide. That's all we had after more than 2 million years of human development. The climate conditions that have placed this immense chokehold on humanity will be upon us potentially in the 2050s.

We need to decide, therefore, if we want to prevent us from becoming choked to death, globally, this time around.

The Ice Age choke-hold may have been a major contributing reason

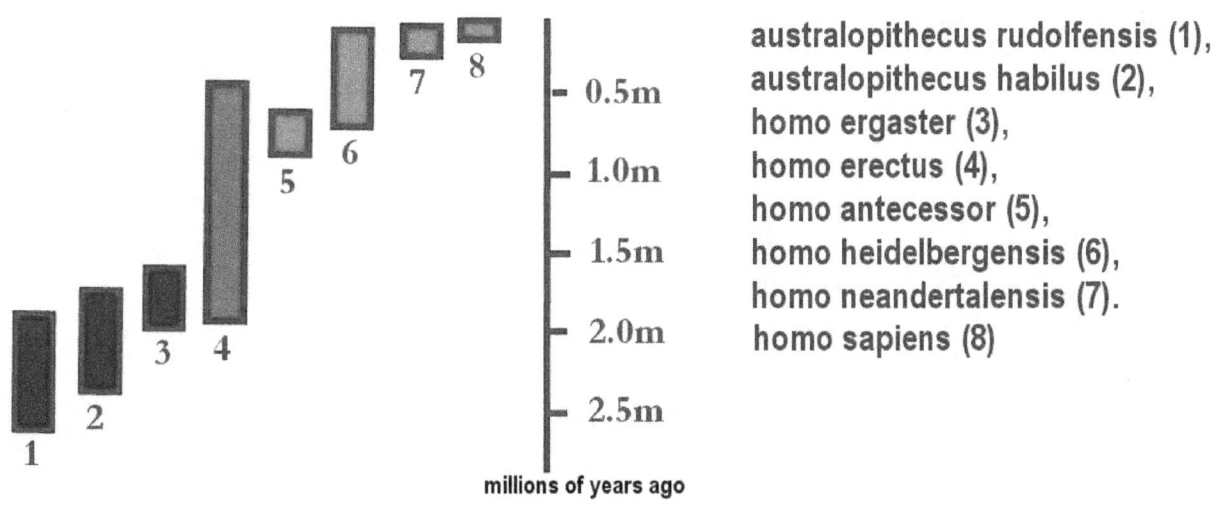

We, the homo sapiens (8), are the only surviving, and the shortest lived of all the the human species, at barely 200,000 years of age.

The Ice Age choke-hold may have been a major contributing reason why the seven previous human species have all become extinct.

Event-driven reactions are dangerous traps

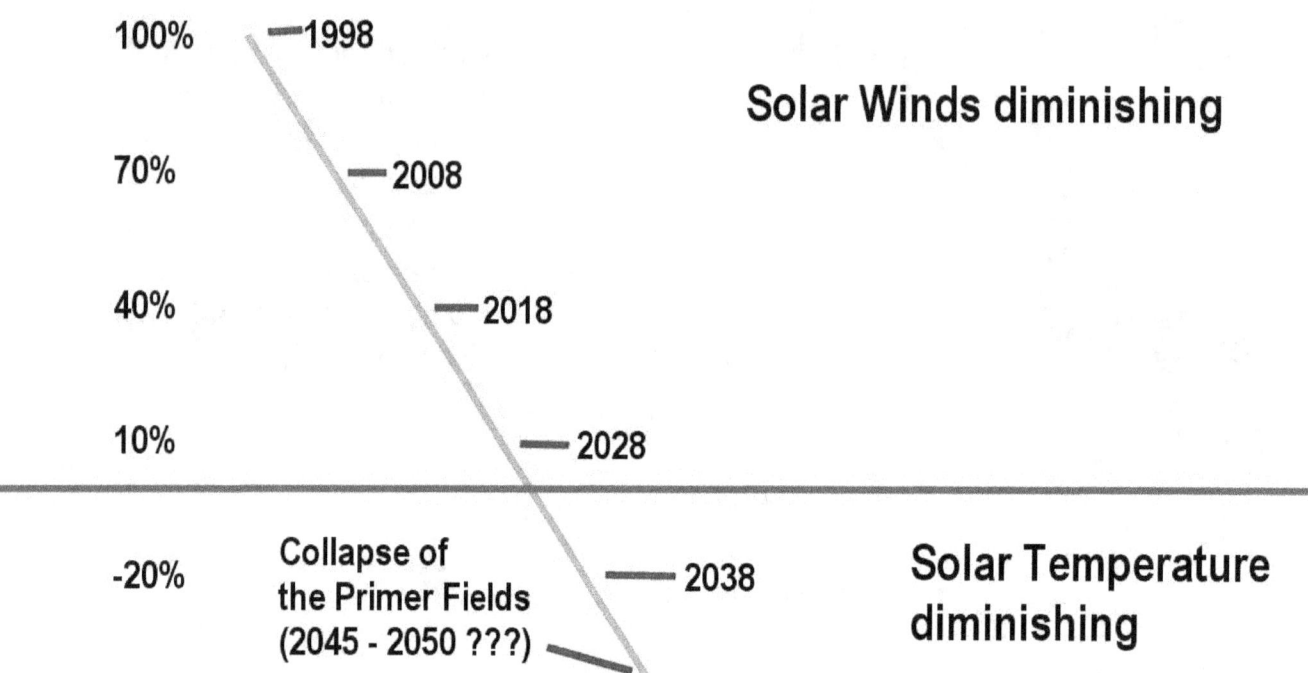

Of course, we cannot predict the future rate of the energy diminishment in the solar system. The current rate of diminishment could slow down. Or it could accelerate. The effects could become manifest faster, or unfold slowly. This means that we cannot orient the existentially critical world development against such potential events.

Event-driven reactions are dangerous traps. Economic policy systems must become principle oriented. We cannot predict events, but we can understand the principles and the dynamics that are in operation, by which events occur.

With being guided by the imperatives of known principles, it becomes possible that preparations for the critical events are made before the events occur, whenever this may be. If the known principles project an existential danger, then by responding to the principles, we eliminate the danger. With any lesser approach, we effectively lay ourselves down to die, possibly even before the Ice Age transition begins.

Part 7: Shaping the future is a spiritual issue

Our Electric Cold Fusion Sun (Part 7) A spiritual issue

The deception of normality

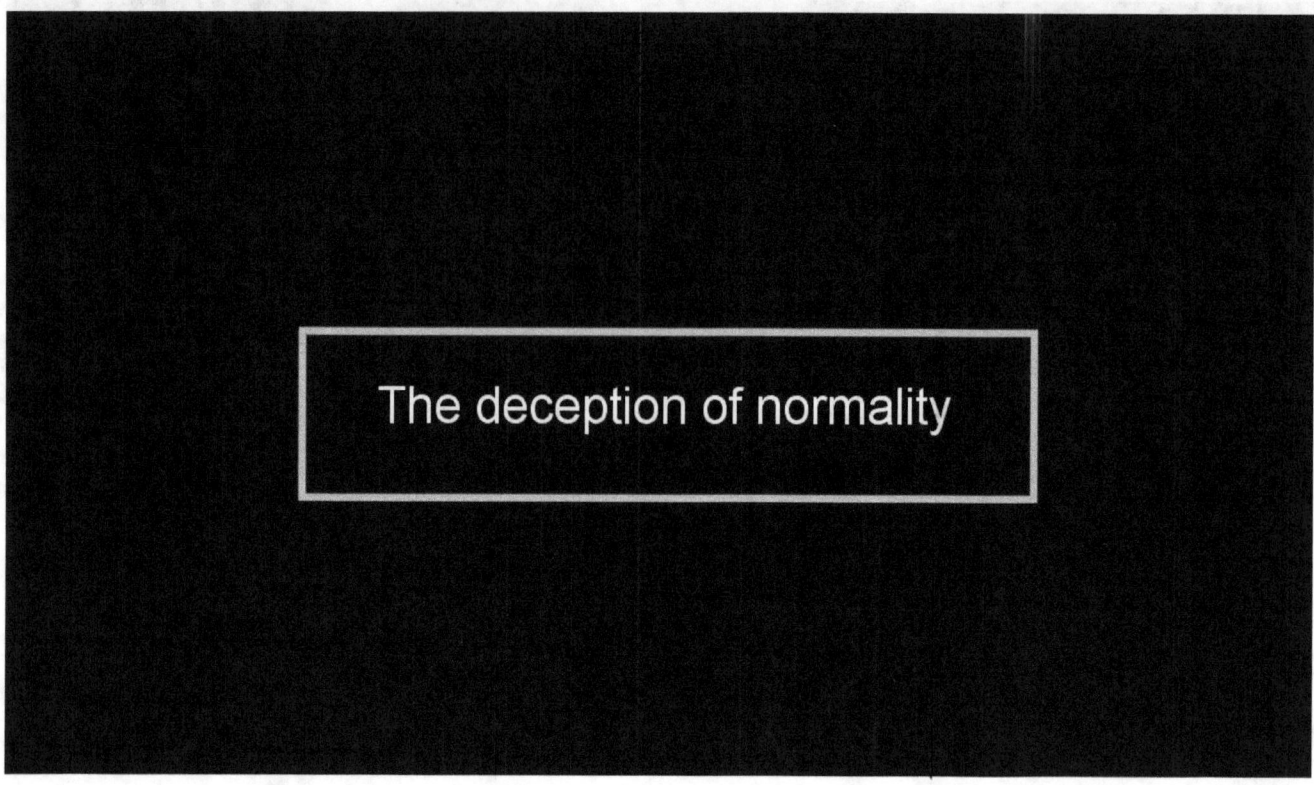

The deception of normality

The perception of normality is deadly in times when a dramatic collapse happens behind the scene that remains largely invisible by cultivated blindness in perception.

Our solar system is collapsing in numerous ways

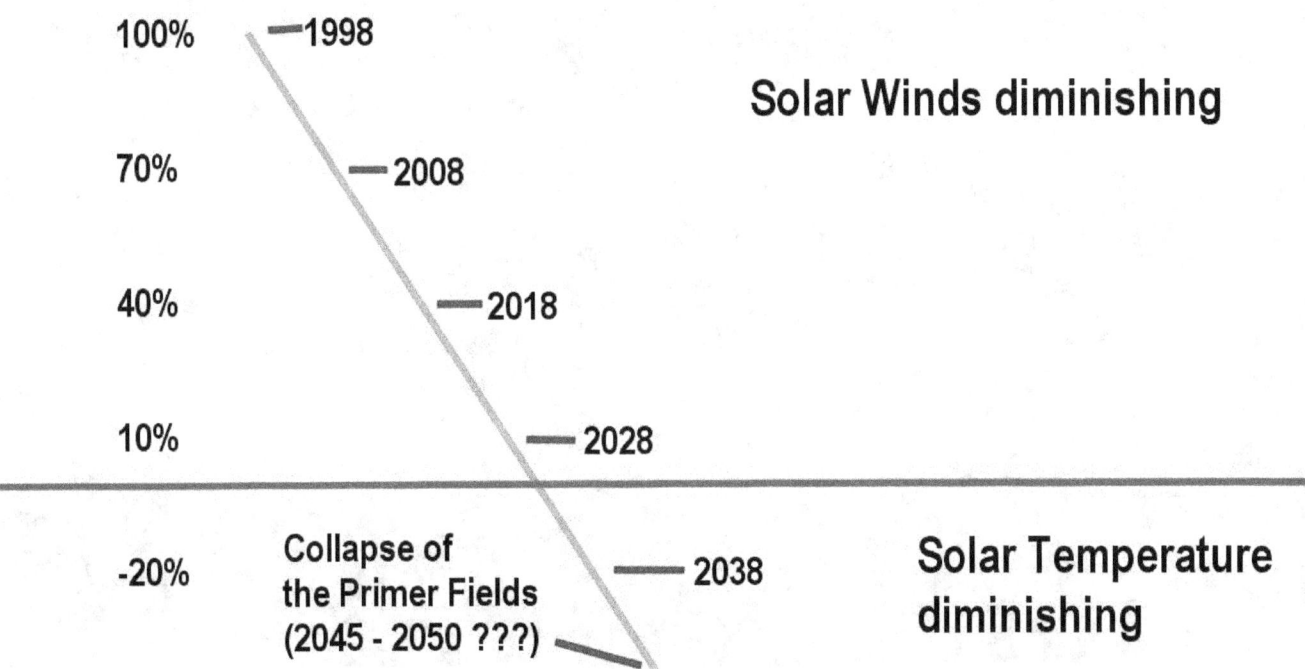

The same happens in astrophysics. Our complex solar system is collapsing in numerous ways while the visible scene remains serene. Here too, the facade of normality is maintained by numerous types of regulating systems that keep the facade going. This hidden collapse continues till one day, the illusions collapse. Before this point is reached, rarely more than a few fringe effects penetrate the facade with dire forebodings, which presently far too few take note of in astrophysics.

Multi-level regulated astrophysical systems

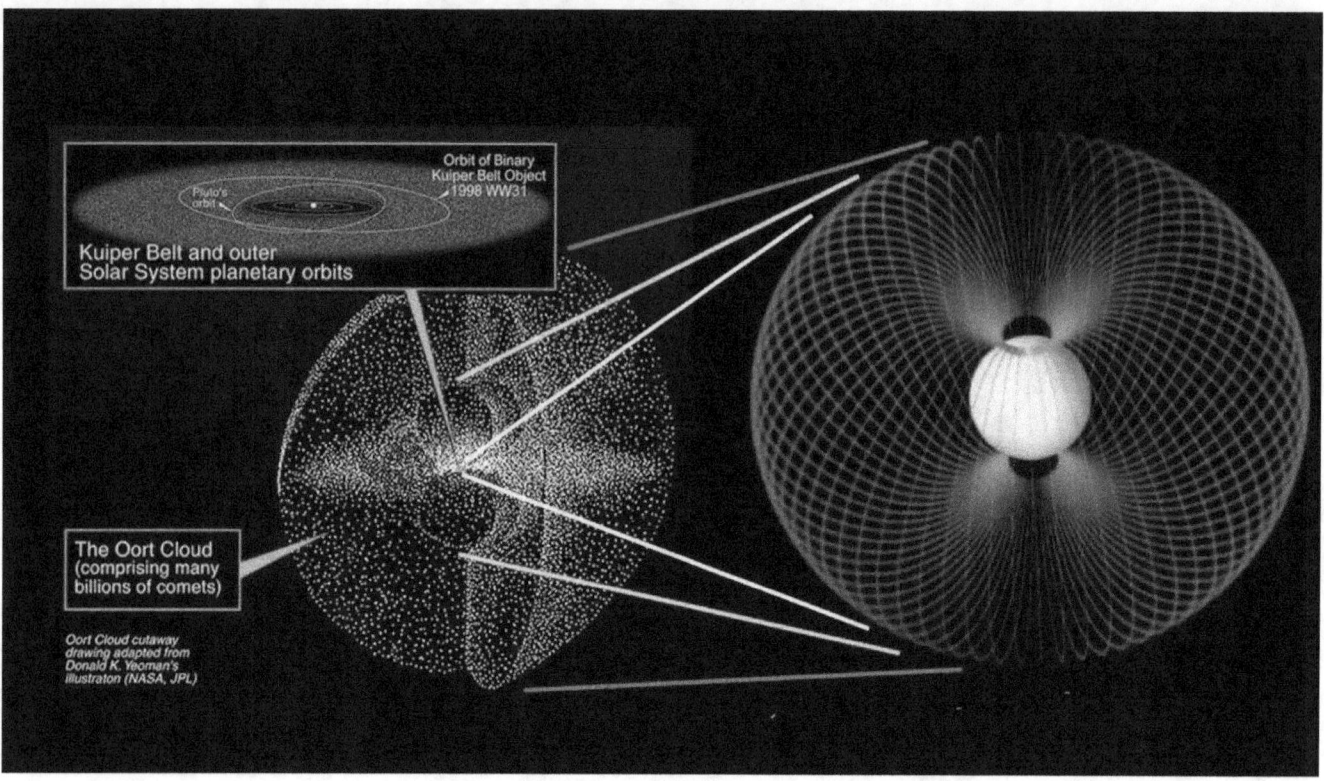

Multi-level regulated astrophysical systems can only be understood in terms of principles, because it is in the nature of multi-level regulated systems to prevent potential events. This means that everything looks normal for the end process, until all the reserves at the various levels are depleted.

The first-level regulating stage

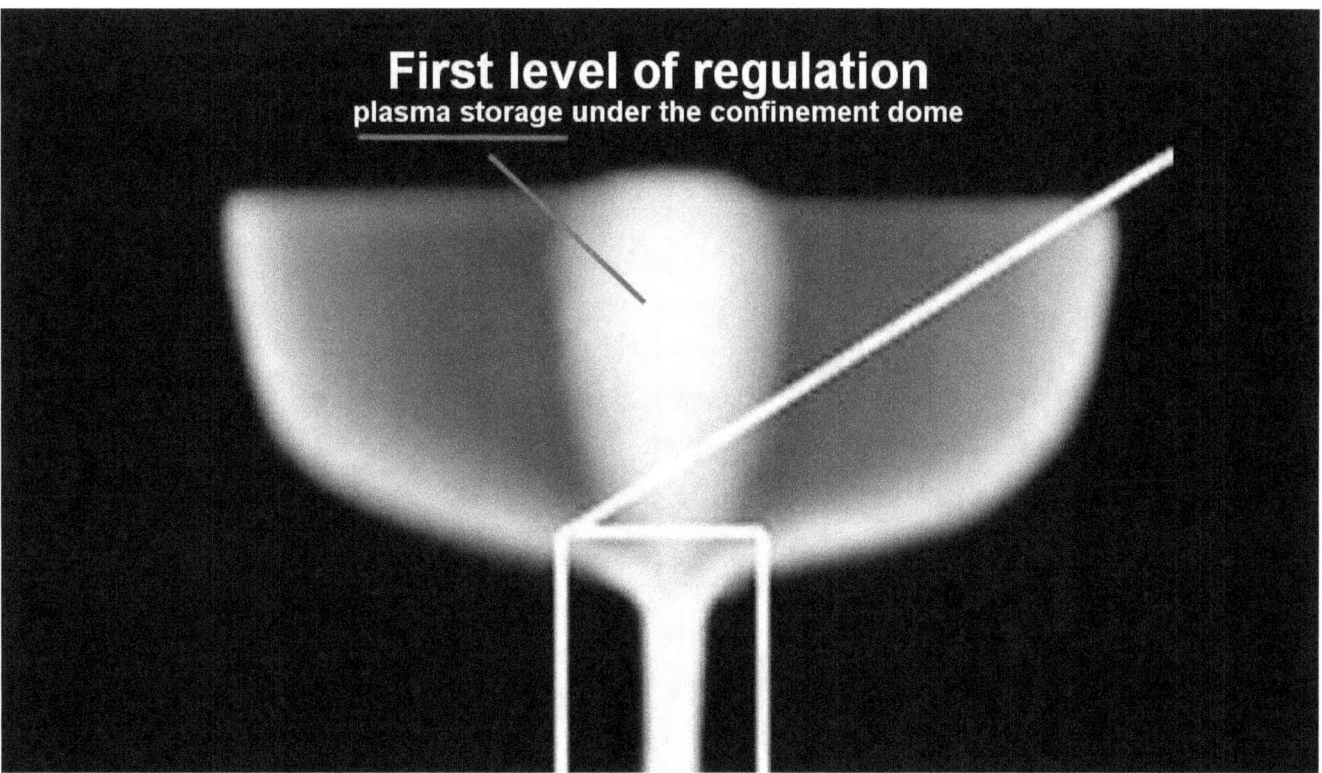

The first-level regulating stage may be as large as the entire inner solar system. It holds large volumes of plasma in reserve to smooth out supply fluctuations. This regulating effect keeps the fluctuations averaged to long trends, to prevent shock events.

The regulating system can't affect the very long trends

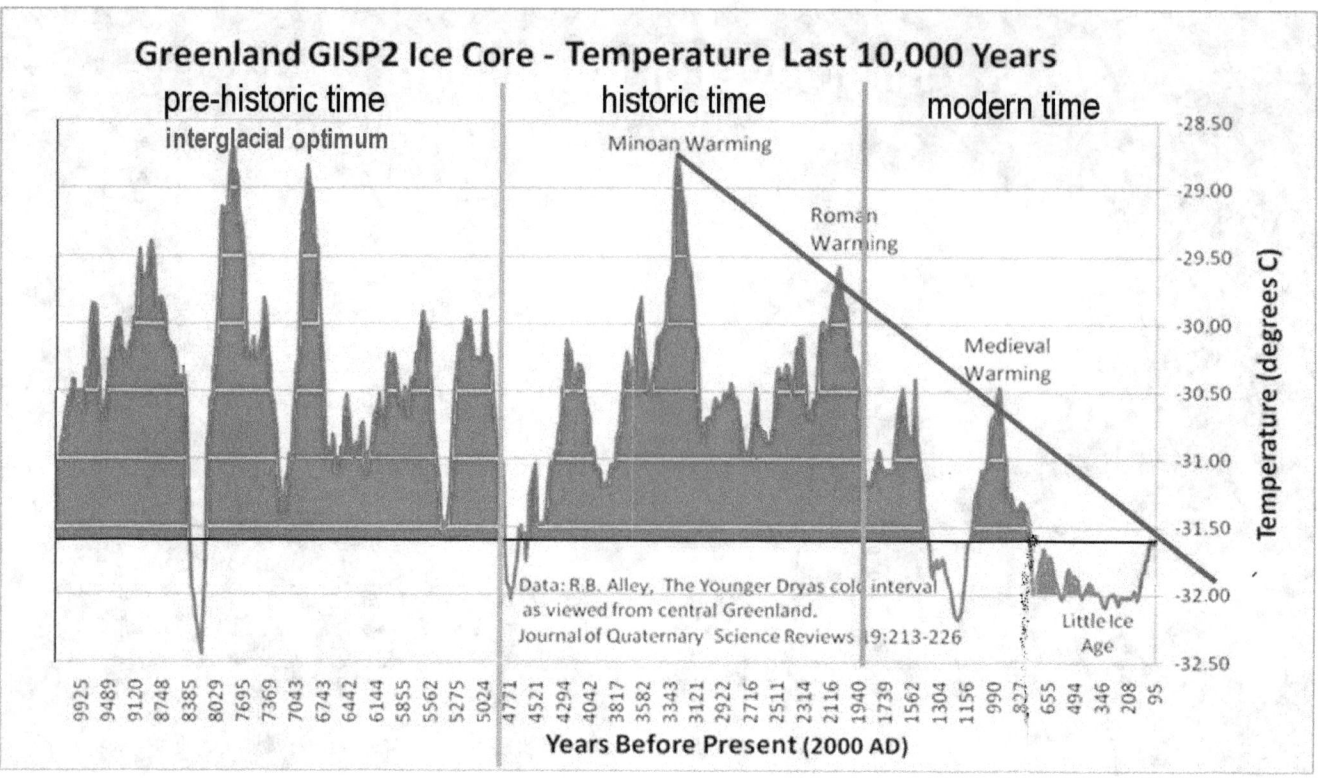

The regulating system may be able to smooth-out fluctuations over hundred-year periods. But it can't affect the very long trends.

The second-level of the regulating system

The second-level of the regulating system is the large plasma reservoir that is the Sun's corona, which is several times larger in volume than the Sun itself. It appears to be gravity held. It likely prevents the solar system's 22-year resonance cycle from affecting the radiation intensity in the visible light band.

The 22-year resonance cycle

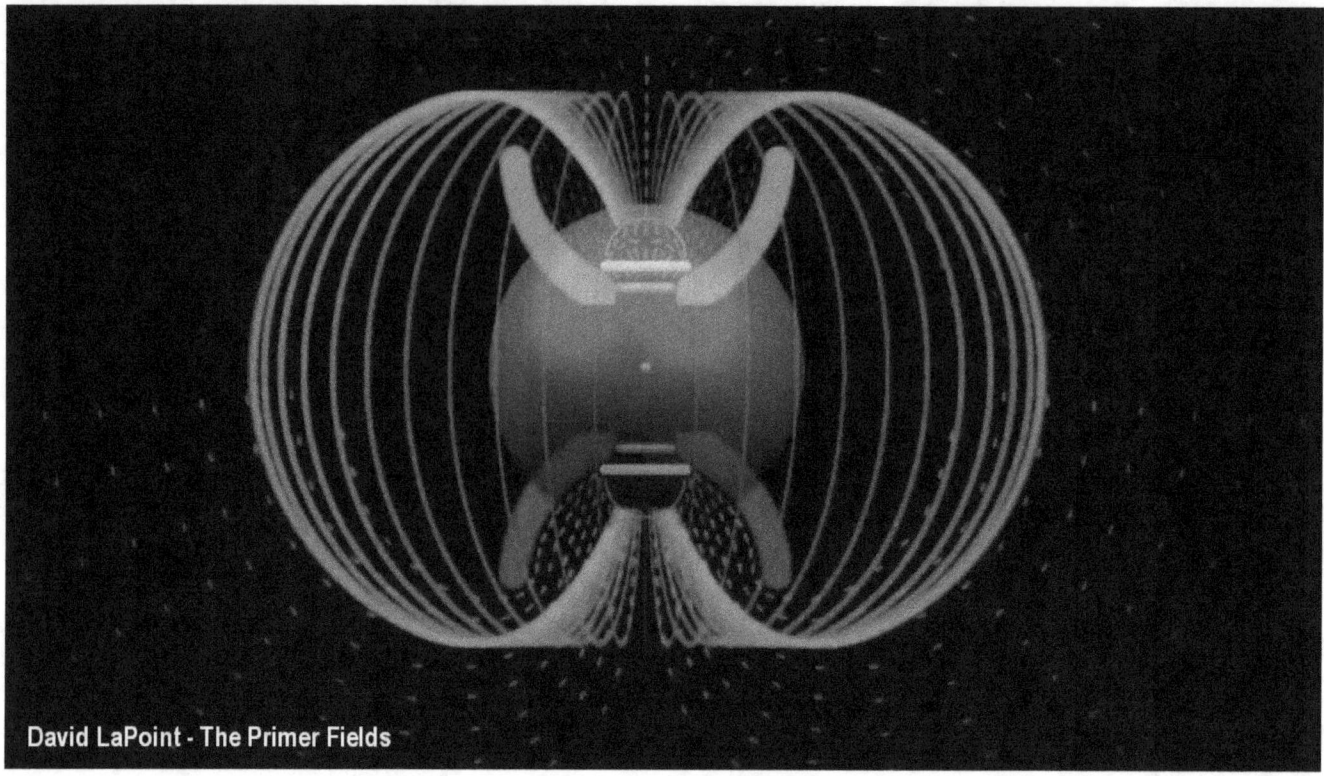
David LaPoint - The Primer Fields

The 22-year resonance cycle, when it is divided into two 11-year solar activity cycles, has been explained by David LaPoint as a cyclical change of magnetic dominance between the upper and lower Primer Fields that determines the effective alternating magnetic polarity of the Sun at the center between the two fields.

This means that that the magnetic field of the Sun is an induced phenomenon that originates with the Primer Fields. When the Ulysses spacecraft measured the reduction of the underlying magnetic field of the Sun, by 30% over the 16 years of its mission, it gave us a measurement of the overall weakening of the Primer Fields during its timeframe. This may be the only direct measurement we have been able to gleam, of the intensity of the system of the Primer Fields.

The third level of the regulating system

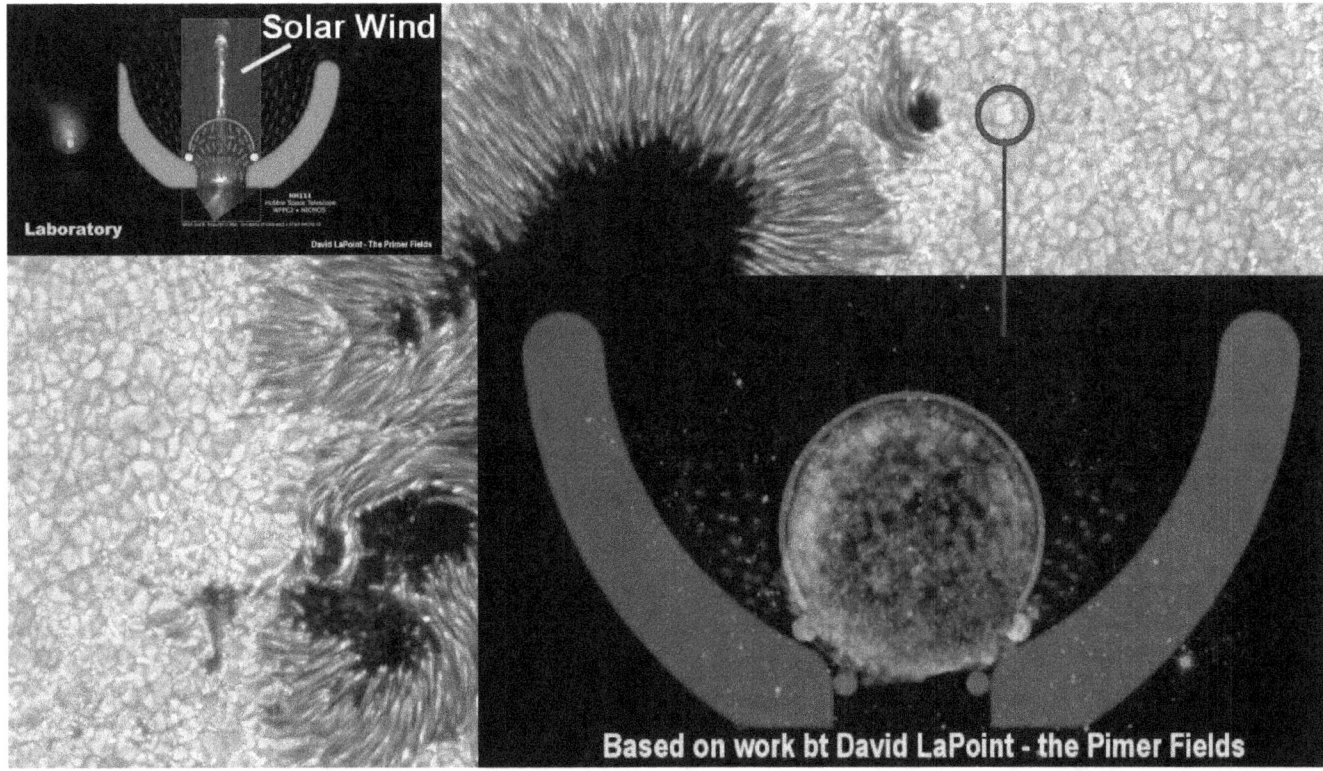

The third level of the regulating system maintains the plasma pressure in the magnetic confinement dome of the reaction cells themselves, where the fusion reactions take place and the solar wind originates. The solar wind can therefore be seen as the final thermometer of the health of the system. When the wind diminishes to zero, big events will likely follow.

The Little Ice Age between the 15th to the 18th Century, was still a fringe event.

The next event on the diminishing slope

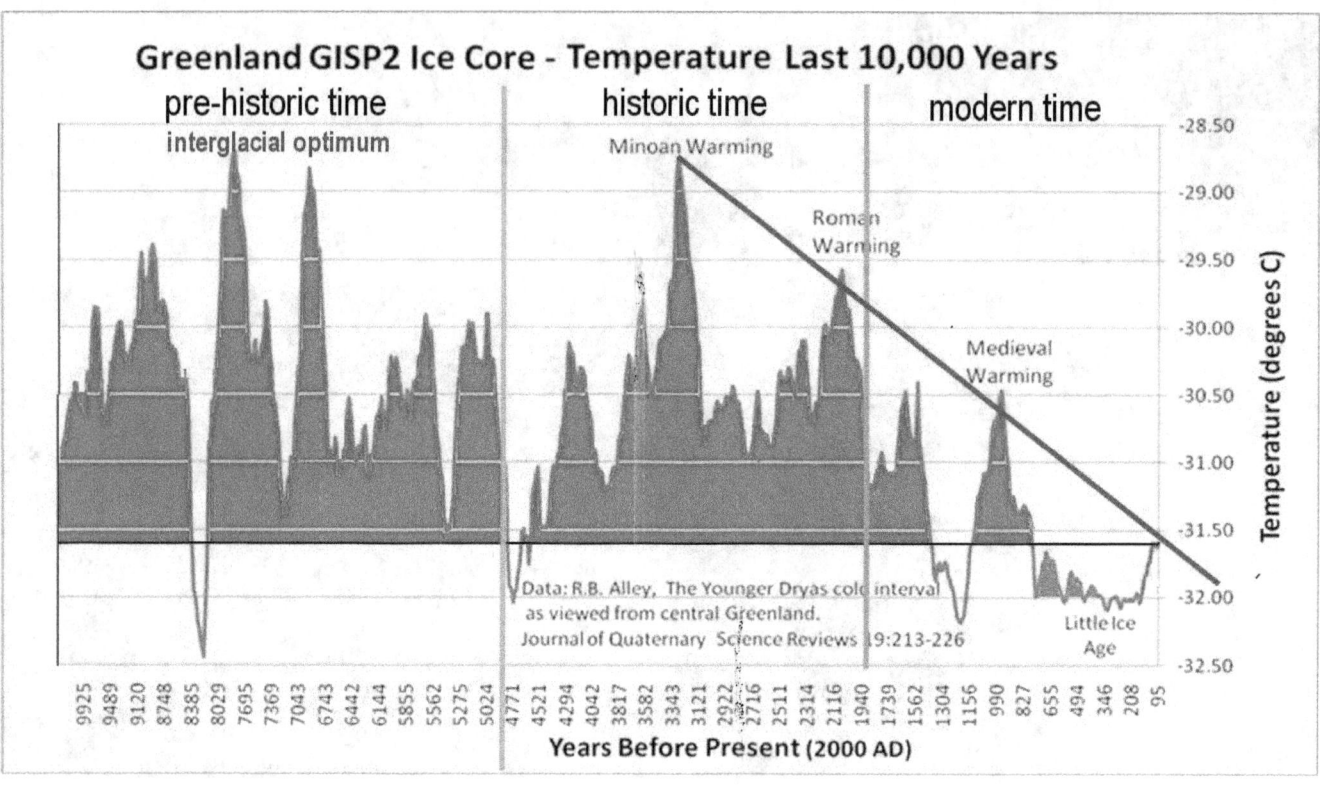

The next event on the diminishing slope of the solar system may be the terminal event that ushers in the big Ice Age when the entire interlocked system stops functioning and the Son goes inactive.

If the astrophysical principles become understood, preparations will likely be made before the potential big event happens. This IF, remains uncertain.

When the astrophysical principles become understood

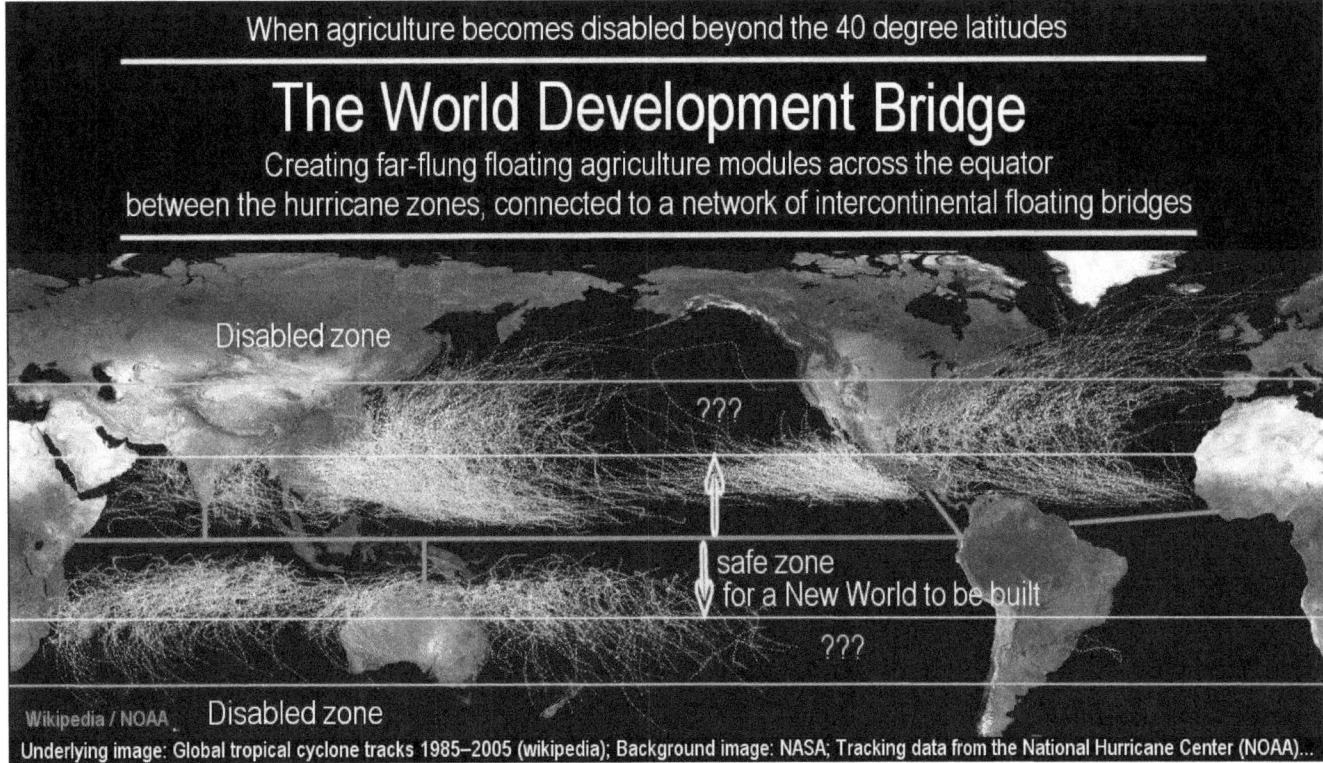

When the astrophysical principles become understood, most of the world's agriculture will be relocated into the tropics along the equator onto networks of floating modules a thousand kilometers wide, spanning the seas between the continents, connected with floating bridges, serviced from floating industries and floating cities with free housing as an investment by society into itself. Then, when the potential event happens, as it may by its own timing, it becomes a non-event.

On the platform of understood principles

On the platform of understood principles, humanity as a whole becomes uplifted by its own creative power. The humanist creative power already exists, as a latent resource for a New Age.

Ironically, the physical construction that is required to save our existence appears to be the smallest part of the challenge that we need to master in the near term in order to have a future in the changing physical environment.

The larger part is spiritual in nature

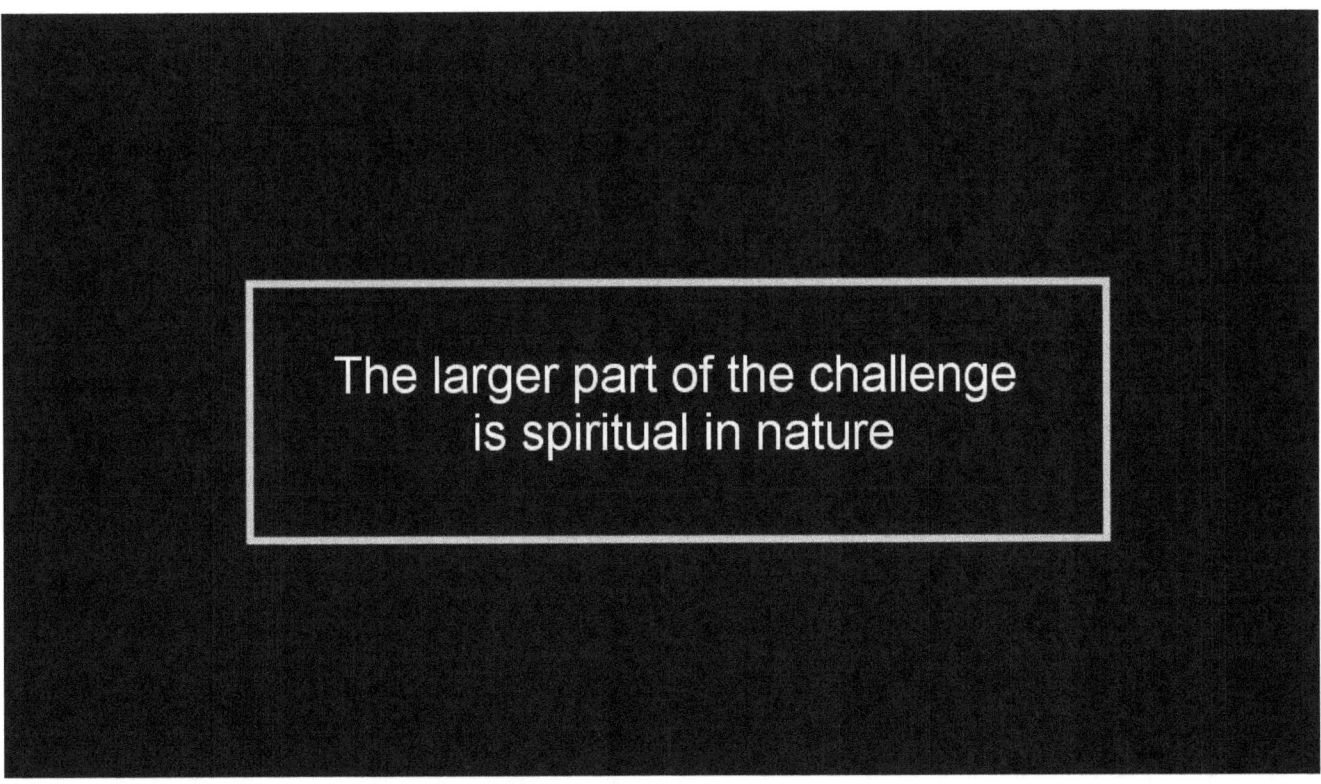

The larger part of the challenge is spiritual in nature.

Here the question comes to the surface: will we do what needs to be done?

Will we become sufficiently human

Will we become sufficiently human to support one another in universal love, for the general welfare of all humanity, with a commitment to meeting the common aims of mankind, including the aims for continuous development?

Or will we remain stuck in the current smallness of thinking

Or will we remain stuck in the current smallness of thinking and encumbered relationships in universal isolation, and privatization into poverty and worse, especially at the grassroots level where the living of humanity unfolds?

In order to explore these types of questions

In order to explore these types of questions, I have created a series of 12 novels, 'The Lodging for the Rose', where I explore the challenges involved towards a wider sense of loving that is reflecting the historic Principle of Universal Love as a counter-pole against the long ages of inhibiting conventions, enabling an escape to greater freedom and creative power with the dynamics of the human heart and soul.

We need this exploration

We need this exploration in order to develop the dynamics in loving universally, which an advancing world requires. And this is not a small thing. It soon became apparent when the work was taken up that such a project couldn't be completed with a single novel. Consequently the story was continued and expanded into a long series.

I had thought at the time

Harvest is seedtime, thoughts ripening
Carried as by a great wind
Carriers of secrets to unfold

The Ice Age Challenge

I had thought at the time that the principle of universal love could be explored sufficiently with the addition of one more volume. Ah, but instead it enabling a closure, the additional volume opened numerous doors for moving forward.

As the challenge widened

As the challenge widened, new principles brought greater freedoms into view, but also larger challenges with them, which no one had dared to tackle before, much less to win.

The very notion of winning

The very notion of winning, suddenly came to light as a complex of concepts that has countless dimensions, many remaining unclear, with the future still shrouded.

As the kaleidoscope keeps turning, this time in Russia

As the kaleidoscope keeps turning, this time in Russia, with new questions are emerging. What are peace, grace, love, against the horrors of war, rape, and the pains of impotence?

Is the Earth really flat?

Is the Earth really flat? Some say it is. Some prove that it is. But for those whose 'bus' got stuck in mud, there is still hope. Hope is rooted in a Christmas present of science that sets up a higher stage.

A miracle unfolds in India

A miracle unfolds in India; a universal marriage in the land of the deepest cast-division, deeper than sexual isolation and religious isolation, but flowing unseen with daring commitments, without playing games.

Why should small games continue to dominate

Landscapes of brilliance, ideas of power
Substance for enriching one-another
Substance of the forever maturing

Coffee, Sex, and Biscuits

Why should small games continue to dominate the giant that the human being is? Just as the heart needs healing, and science does so likewise, the continent of Africa needs healing, and humanity with it.

The truly endless horizons are the horizons of our own creating

The truly endless horizons are the horizons of our own creating. In Caracas a daring attempt is made to deny servitude, and to master the future with ideas sown into the winds of time, so that the winds may change.

Still, the world grinds on

Still, the world grinds on. Are there lessons to be learned? Are there lessons to be drawn from the masters of spiritual science, who have in their season, caused the winds of war to be still?

The cup bearer to the king is a dangerous pawn

The cup bearer to the king is a dangerous pawn. He has many names. From this shadow he serves a cup of poison, seen hypothetically in Siberia, located in space, a killer of millions, while not a soul cares to raise a finger to stop it, save a nameless patriot.

When people must flee the land of the free

When people must flee the land of the free, China at last opens a home for healing, a portal for a justified hope, while the healing continues as it must.

Hope is justified

Hope is justified in all this, even on the Ice Age front. The more we discover of the truth, the stronger we will become in mastering the power to meet the challenge, both in the spiritual and in the physical realm.

We live in a universe of vast electric power

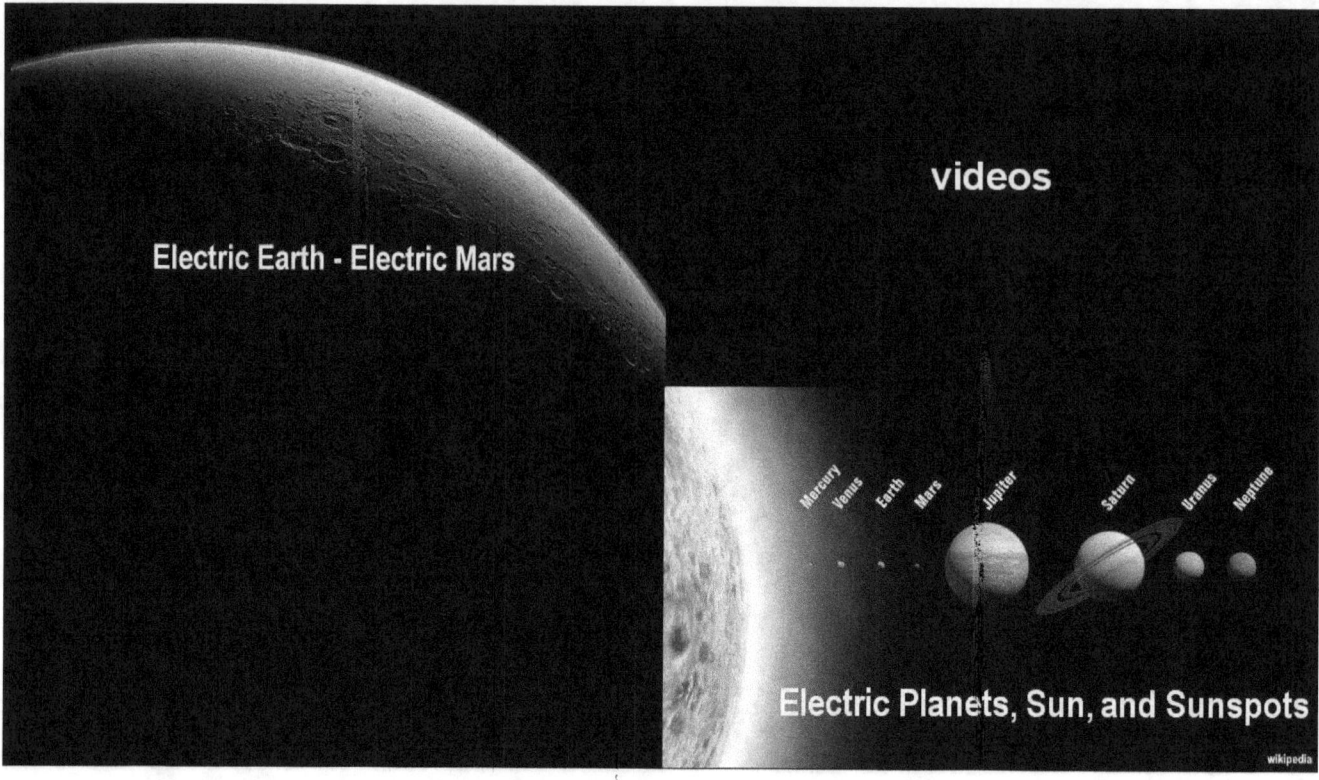

We live in a universe of vast electric power, but before we can use it, we need to acknowledge that this anti-entropic power actually exists. Some day we will do this.

When we stop to keep our mental horizon blocked

When we stop to keep our mental horizon blocked with paradoxical illusions, especially in the sciences, we will find that we have truly infinite horizons before us, both in terms of the critical freshwater resources that we need, and in energy resources more than we can yet imagine, which are all very real.

You are invited to come and visit the great libraries

You are invited here as the video closes, to come and visit the great libraries created for many more explorations of numerous topics, in the form of e-books, videos, and stories from the novels.

Oh, yes, great challenges do indeed lay before us, as we have the next Ice Age on the near horizon in possibly 30 years, and we do have the foreground threatened at the present with nuclear war, depopulation policies, choked science, and economic collapse, and so on. But we also have the greatest force on our side with which to overcome these threats and to move forward towards meeting the Ice Age challenge.

The final seven chapters of my series The Lodging for the Rose

It is appropriate therefore, in this context, that this video ends with the multifaceted celebration of our humanity that unfolds in the final seven chapters of my series of 12 novels, The Lodging for the Rose, focused onto the Principle of Universal Love. While the Ice Age challenge isn't even mentioned in these final chapters of the series, the grand human dimension is strongly focused on and is summarized in these chapters extensively. Hopefully, this type of focus will enable us to meet the greatest challenges that the present and the future may bring.

More Illustrated Science Books by Rolf A. F. Witzsche

www.ingramcontent.com/pod-product-compliance
Lightning Source LLC
Chambersburg PA
CBHW080953170526
45158CB00010B/2792